Global Issues

Series Editors

Paolo D. Farah
West Virginia University
John D. Rockefeller IV School of Policy and
Politics West Virginia University
Morgantown, WV, USA

Jim Whitman
Wakefield, West Yorkshire, UK

The rapid pace of globalizing dynamics has brought in its wake a proliferation of crises, governance challenges and large-scale stresses, unprecedented in their range, seriousness and urgency. These emerging conditions and relations are variously global is their range, inclusiveness and impacts; and many of them are outpacing our legal, political and ethical systems of deliberation and control. Climate change, the ravages of contagious diseases and financial turmoil are all cases in point. Similarly, the transformative implications of new scientific and technological developments, from Big Data to gene editing, are already upon us. This book series is dedicated to providing insights into the complex interaction of human and natural systems; modes of cooperation and conflict; and the ways and degrees to which human values can be reconciled and more effectively enacted. The concentration throughout is on an integration of existing disciplines toward the clarification of political possibility as well as impending crises.

The series is published in cooperation with gLAWcal - Global Law Initiatives for Sustainable Development.

Daniel Iglesias Márquez
Clara Esteve-Jordà • Beatriz Felipe Pérez
Editors

Legal Challenges at the End of the Fossil Fuel Era

Shaping a Just and Clean Energy Transition

palgrave
macmillan

Editors
Daniel Iglesias Márquez
Rovira i Virgili University
Tarragona, Spain

Beatriz Felipe Pérez
CICrA Justicia Ambiental
Barcelona, Spain

Tarragona Centre for Environmental Law Studies
Tarragona, Spain

Clara Esteve-Jordà
Institute of Environmental Science and Technology
Autonomous University of Barcelona
Barcelona, Spain

ISSN 2947-8863 ISSN 2947-8871 (electronic)
Global Issues
ISBN 978-3-031-61765-2 ISBN 978-3-031-61766-9 (eBook)
https://doi.org/10.1007/978-3-031-61766-9

Cover credit: Bernt Ove Moss / EyeEm / Getty Images

This Palgrave Macmillan imprint is published by the registered company Springer Nature Switzerland AG.
The registered company address is: Gewerbestrasse 11, 6330 Cham, Switzerland

Foreword

In the ongoing narrative of our collective history, the epochal trope of the Anthropocene unfolds against a backdrop of unprecedented challenges and nascent opportunities. Central to this narrative is the exigent imperative to confront the environmental, societal and economic entanglements attendant to our entrenched reliance on fossil fuels. Climate change—one of the multiple planetary boundaries trespassed because of anthropogenic pressure from the more industrialised countries—necessitates a concerted and expeditious transition towards a just and sustainable energy paradigm. This transition is fraught with implications from a legal perspective. In this juridical exploration, the present volume, titled *Legal Challenges at the End of the Fossil Fuel Era: Shaping a Just and Clean Energy Transition*, embarks upon a journey to dissect several intricate legal issues that attend to the 'transition away' from fossil fuels.

Instead of an unequivocal 'phase-out' mandate, "*transition away* from fossil fuels in energy systems in a just, orderly and equitable manner" is the compromise and nuanced formula used in the agreement achieved on 13 December 2023 during the United Nations Climate Change Conference (COP28) held in Dubai. COP28 is now lauded as the seminal harbinger of the denouement of the fossil fuel era. UN Climate Change Executive Secretary Simon Stiell, in his valedictory address, articulated this sentiment, stating, "Whilst we didn't turn the page on the fossil fuel era in Dubai, this outcome is the beginning of the end". He urged nations and enterprises to translate these commitments into tangible real-world outcomes expeditiously. The paramount outcome of COP28 is the global stocktake, understood as an inclusive compilation of negotiated elements empowering nations to formulate more robust climate action plans by 2025.

The term 'phase-down' was applied to 'unabated coal power', whereas 'inefficient fossil fuel subsidies' would be phased out, thereby ostensibly leaving open the interpretation of what qualifies as an efficacious subsidy. It also seems doubtful that the agreement calls on the parties to accelerate indiscriminately different technologies, which, far from removing the uncertainty of certain actions, could lead to their proliferation: States are called upon to accelerate "zero- and low-emission technologies, including, inter alia, renewables, nuclear, abatement and removal technologies such as carbon capture and utilisation and storage, particularly in hard-to-abate sectors, and low-carbon hydrogen production".

Beyond the official speeches affirming the value of the results of the climate summit, COP28 has been harshly criticised for the powerful influence exercised by petro-states and fossil energy lobbies. The road to transition will be neither easy, nor straightforward, nor peaceful. The scale and pace of the change of the required energy transition should be conceptually framed as a genuine energy and societal transformation sustained by a new paradigm of Law. However, this dynamic of legal and socio-energy transformation is neither self-evident nor self-executing.

Indeed, as Timothy Mitchell has already convincingly argued, carbon energy, modern law and democratic politics are intensely intertwined. So, states and democracies have developed as a form of politics characterised by an organisation of power and market underpinned by a certain paradigm of Law, geared to producing and using carbon energy, and based on the principle of unlimited economic growth.

The emergence of a new episteme linked to the current geological epoch is also conducive to the deployment of new conceptualisations of the collective and of a new paradigm of transformative law. This new form of law aims to integrate the complexities of the Earth System and the demands of justice.

This book delves into the complexities of this transition, recognising that pursuing a sustainable and just energy future necessitates not only technological and economic innovations but also a robust reflection on the political and legal frameworks that are attuned to the needs of the present and the planetary integrity in the future.

Readers will find an exploration of the legal pathways and barriers that can divert us from or lead us towards a future where justice and sustainability meet. Hopefully, the book can inspire conversations, inform policy and catalyse action to propel us towards a more resilient, equitable and harmonious world.

Tarragona, Spain
22 December 2023

Endrius Cocciolo

Contents

1 **Introduction: Challenges at the End of the Fossil Fuel Era** 1
Daniel Iglesias Márquez, Clara Esteve-Jordà, and Beatriz Felipe Pérez

Part I Environmental and Social Energy Justice 9

2 **Transitioning *All* Together? Participation Rights, Foreign Investment, and the Pursuit of a Just Energy Future** 11
Thalia Viveros-Uehara and Adrian Duhalt

3 **The Right to a Decent Standard of Living: A Practical Tool to Guarantee Basic Needs at the End of the Fossil Fuel Era** 49
Ana García-Juanatey, Bettina Steible, and Luca Volpi

4 **Energy Communities: Why (Sometimes) the Commons Need the State** 73
Clara Esteve-Jordà and Marcos de Armenteras Cabot

5 **Linking Energy Poverty, Environmental Justice and Forced Displacement: Controversies and Ways Forward** 99
Francesca Rosignoli

6 **Theorising a Human Rights-Based Approach to Energy Transition and Its Justiciability in International and Domestic Jurisprudence** 119
Simona Fanni

7 **Just Transition Litigation in Latin America: Striking a Balance Between Economic Development and Environmental and Social Energy Justice** 151
Natalia Urzola, Maria Antonia Tigre, Riccardo Luporini, Lorena Zenteno, Marlies Hesselman, and Pedro Cisterna-Gaete

Part II Legal Challenges of a Clean and Just Energy Transition 187

8 **The Challenge of 'Phasing Out' Fossil Fuels: A Climate-Changing Transition** 189
Guilherme Pratti

9 **The Role of Hydrogen for a Sustainable Energy Transition in the European Union: Between a Fossil-Based and a Renewable Path** 217
Aloia López Ferro

10 **The Legal Protection of the Territory of Catalonia Against Specific Renewable Energy Projects** 249
Ivan Hortigüela Bolsa

11 **Critical Mineral Supply in the European Energy Transition: Towards Just Institutional Approaches** 275
Asmaa Khadim, Ron Janjua, and Chengjian Xu

12 **'Sustainable Energy' under Reformed International Investment Agreements** 311
Monika Feigerlová

13 **Energy Charter Treaty and Investor-State Dispute Settlement: Legal Challenges in Energy Transition** 341
Magdalena Bas-Vilizzio

Notes on Contributors

Magdalena Bas-Vilizzio is Professor of Public International Law and International Relations at the University of the Republic, Uruguay, and at the University of Monterrey, México. She holds a PhD in International Relations from the National University of La Plata, Argentina. Her current research focuses on investor-State dispute settlement, Mercosur, multilateralism, International Economic Law, and didactics of International Relations.

Pedro Cisterna-Gaete is a qualified Chilean lawyer. He obtained an LLM in Global Environment and Climate Change Law with distinction from the University of Edinburgh, where he is currently undertaking his PhD in Law. He worked in 'TECHO', an NGO that works with informal settlements. There, he was the National Deputy Director, managing the organisation's advocacy. In addition, he has been a research assistant at the Edinburgh Law School and an observer of the University of Edinburgh delegation at COP26. He is also a member of the GNHRE and is collaborating with "*Nuestra America Verde*" NGO, which supports a Latin American response to climate negotiations.

Marcos de Armenteras Cabot is Assistant Professor of Public International Law and European Union Law at the Department of Public Law of the University of the Balearic Islands. Previously, he was a postdoctoral researcher at the Scuola Superiore Sant'Anna di Pisa, Italy, and a predoctoral researcher at the Rovira i Virgili University. He holds a degree in Law from the University of Barcelona, an MA in Sociology of Law from the Oñati Institute of Sociology of Law, a Master in Environmental Law from the University of the Basque Country and a PhD in Law from the Rovira i Virgili University.

Adrian Duhalt is a research scholar at the Center on Global Energy Policy at Columbia University's School of International and Public Affairs. Prior to joining Columbia, Adrian was a Postdoctoral Fellow in Energy Studies at Rice University's Baker Institute. From 2013 to 2017, he was an associate professor at the School of Business and Economics at Mexico's Universidad de las Américas Puebla (UDLAP). Adrian completed his master's degree and PhD in Economic Geography at the University of Sussex, in the United Kingdom, where he authored a thesis on the development trajectory of Mexico's petrochemical industry and for which he received the Best PhD Thesis Award from the Economic Geography Specialty Group of the Association of American Geographers. His research interests sit at the junction between political economy and development issues concerning the energy sector. Specifically, his research projects have focused on natural gas and petrochemicals; however, he has similarly had the opportunity to publish articles and reports on topics such as the ammonia-fertiliser value chain, the liberalisation of the energy sector in Mexico, social conflict in energy infrastructure projects, National Oil Companies (NOCs), and the energy-food security nexus. He is also a non-resident scholar at the Center for the US and Mexico at Rice University's Baker Institute.

Clara Esteve-Jordà is a postdoctoral researcher at the Institute of Environmental Science and Technology of the Autonomous University of Barcelona (ICTA-UAB), Spain. She is also an associate researcher at the Tarragona Centre for Environmental Law Studies (CEDAT). She received her doctorate in law from the Rovira i Virgili University with a thesis that reviewed the Western conception of property through comparative constitutional law, with emphasis on the importance of the commons and its eco-social function. Her areas of interest are property rights, green grabbing, environmental movements, and ecofeminist theory and practice.

Simona Fanni is a postdoctoral researcher from the University of Seville, Spain, on the Fellowship Program "Margarita Salas". In this framework, Simona is working on a project on biobanks under international law and European Union law, and has been a visiting researcher at the University of Trento, Italy. She has a PhD in Legal Sciences *cum laude* from the

University of Cagliari, Italy, and the University of Seville, Spain, and has obtained the title of *Doctor Europaeus.* Simona Fanni is the author of various publications in the field of environmental law—especially on climate change and extractive activities—international bio law and international human rights law, and has participated in many international conferences on environmental law, bio law, health law, data protection and new technologies, her main but not exclusive fields of research. She is a member of the European Society of International Law since 2016.

Monika Feigerlová is a research associate at the Institute of State and Law of the Czech Academy of Sciences, a member of the Centre for Climate Law and Sustainability Studies, and an attorney-at-law, focusing on foreign investment, transnational corporations, and climate change law. She obtained the PhD at the Charles University in Prague and completed the Master in International Dispute Settlement at the Graduate Institute and University of Geneva. She was admitted to the Czech Bar Association and practices law in Prague. She is also a member of the Committee for Doctoral Theses at the Charles University, the Czech branch of the International Law Association, and the Executive member of the Czech Society of International Law. She is the author of chapters on corporate climate responsibility and international economic law in the recently published first Czech book on climate law.

Beatriz Felipe Pérez is a researcher and co-founding partner of CICrA Justicia Ambiental. She currently works as the technical coordinator for the Expert Committee on Climate Change in Catalonia too. She is a research associate at the Tarragona Centre for Environmental Law Studies (CEDAT) at Rovira i Virgili University (URV, Tarragona). Specialised in the legal and multidisciplinary analysis of human mobility induced by disasters and climate change, she has also focused on climate change science, education for global justice, and international cooperation. With a degree in Environmental Sciences from the University of Granada, She is a Doctor in Law from Rovira i Virgili University, with a thesis titled "*Migraciones climáticas: retos y propuestas desde el Derecho Internacional*". She was awarded the Josep Miquel Prats Prize for the best doctoral thesis in Environmental Law in 2017. Among other publications, she is the author of the monograph *Las migraciones climáticas ante el ordenamiento jurídico internacional* (Aranzadi, 2019) and *Migraciones Climáticas. Sobre desigualdades, mitos y desprotección* (MRA Ediciones, 2022).

Ana García-Juanatey is a lecturer at the CEI International Affairs, centre affiliated to the *Universitat de Barcelona*. She is a political scientist and holds a PhD in Law by the *Universitat Pompeu Fabra* (2016). She has been a postdoc researcher at the *Institut Barcelona d'Estudis Internacionals*, as well as a visiting researcher at the UN Organization of Food and Agriculture and the *Universidade da Coruña*. Her research interests focus on the relationship between human rights and the environment, foods systems from a human rights perspective, the challenges of human rights law in the age of internet, and the transparency and accountability of public entities. She has published her research work in several international journals such as *Peace Review*, *South European Society and Politics*, *Administration and Society*, and the *European Journal of Higher Education*, among others.

Marlies Hesselman is an assistant professor at the Transboundary Legal Studies Department of the Faculty of Law, University of Groningen. She is a Board Member of the Groningen Center for Health Law and represents the faculty in the Scientific Advisory Committee for the Aletta Jacobs School of Public Health. Hesselman has published extensively with leading publishers (OUP, CUP, Elgar, Brill, TMC Asser) on her areas of interests: international (socio-economic) human rights law, climate change, disasters, environmental health, and access to energy services as a new human right. She is part of the Global Peer Network on Climate Litigation for the Climate Case Chart, headed by Columbia Law School's Sabin Center for Climate Change, and was lead-editor for the volume *Socio-Economic Human Rights in Essential Public Services Provision* (Routledge 2017).

Ivan Hortigüela Bolsa is a lawyer specialised in Environmental Law, with experience in Water Law, Urban Planning, Land Use Law, and Environmental Criminal Law. He has provided legal support across various legal jurisdictions, both preceding and throughout different phases of legal proceedings. Currently, he is a PhD student at INSTA and Rovira i Virgili University. His field of research is the illegal traffic of plastic waste linked to the criminal liability of legal entities.

Daniel Iglesias Márquez is a lecturer and '*Juan de la Cierva*' Postdoctoral Research Fellow at the Rovira i Virgili University. He is also a researcher of the Human Rights and Business Institute at the University of Monterrey (UDEM) (Mexico) and an associate researcher at Tarragona Centre for

Environmental Law Studies (CEDAT-URV). He is a member of the Catalan Government recognised Research Group in Geopolitics, Conflict and Human Rights (GEOCONDAH), the Global Business and Human Rights Scholars Association, the Council of the Latin American Academy of Business and Human Rights, the '*Red Empresas y Derechos Humanos. Incidencia especial en el extractivismo y los acaparamientos de tierra y agua*' (REDH-EXATA), and the Global Network for Human Rights and the Environment (GNHRE). He is Second Editor of the Catalan *Journal of Environmental Law*. He has work experience in international organisations and civil society organisations. He has been involved in several research projects and has collaborated actively with numerous international and national associations dedicated to promoting environmental protection and human rights. ORCID: 0000-0003-2759-3064.

Ron Janjua is a multidisciplinary expert in law, international relations and resource extraction. His research focuses on intersections between artisanal food production landscapes, community consultation, resource extraction, mineral security and energy transitions. Admitted as a Barrister and Solicitor in Canada and as a Solicitor in Australia, he has represented clients in corporate, business, international tax, land, mining and energy matters, and has served as a policy advisor for the Queensland Resources Council. In addition, he held academic positions as a law lecturer in Australia. Ron's academic background includes a specialised BA(Hons) in Global Political Economy and International Relations from York University, an MSc in International Relations from the London School of Economics, a JD from Bond University and an LLM specialising in International Mining and Petroleum Law and Community Consultation from the University of Queensland. He is currently affiliated with the University of Groningen as a doctoral researcher.

Asmaa Khadim is a legal expert in environmental conflicts over resource extraction and energy transitions. Her research areas include environmental constitutionalism, comparative constitutional law, environmental and Indigenous rights, and just energy transitions, with a regional focus on socio-environmental conflicts in the Americas (including Canada, Argentina, Chile, and Costa Rica). Asmaa holds a BSc in Psychology and International Development Studies from McGill University; an LLB from Osgoode Hall Law School, York University, specialising in International, Comparative and Transnational Law; and a PhD from the Sustainable

Minerals Institute, the University of Queensland. She has lectured in international environmental law, public international law, the law of international organisations, and constitutional law in Australia. Prior to entering academia, she worked in private practice as a Barrister and Solicitor in Ontario and was admitted as a lawyer in New South Wales. She is currently a postdoctoral researcher in Institutions for Conflict Resolution at Leiden University.

Aloia López Ferro holds a Bachelor's degree in Law from the University of Santiago de Compostela (2014). She later pursued the Master's in Advocacy at the University of Santiago de Compostela (2016) and the Master's in Environmental Law (2019) at the Rovira i Virgili University. She serves as an associate professor in the Public Law Department at the Rovira i Virgili University and is a member of the research group Territory, Citizenship, and Sustainability, recognised as a consolidated research group with the support of the Department of Research and Universities of the Generalitat de Catalunya. Additionally, she works as an environmental lawyer at Terraqui.

Riccardo Luporini is a postdoctoral researcher in International Law at the Sant'Anna School of Advanced Studies in Pisa. He is currently working on the projects FLAWS 'Preventing and managing flood events: A technical-legal analysis of the implementation of legislation against hydrogeological instability in Italy', funded by the Italian Ministry of University and Research and EUMA 'Creating an EUropean Higher Education Network for MAster's Programmes in Disaster Risk Management', funded by the European Union. In his research, Riccardo is focusing on rights-based climate change litigation and he is also a national rapporteur for Italy at the Global Climate Litigation Database of the Sabin Center for Climate Change Law, Columbia Law School. Riccardo obtained his PhD from the Scuola Superiore Sant'Anna with a research project dealing with climate change adaptation and disaster risk reduction from a human rights perspective.

Guilherme Pratti is a Postdoctoral Research Fellow in Philosophy of Law and Applied Legal Theory at the Sant'Anna School of Advanced Studies in Pisa, Italy. His research deals with philosophy of technique and the legal rationality of climate change, from the perspective of inter-legality. In this sense, the European Green Deal, the Brazilian Ecological Transition Plan, and the present energy transition are his main research objects at the moment. He obtained his PhD in Legal Philosophy from Sant'Anna and his Master's degree in Legal Theory at the University of Lisbon, Portugal.

Francesca Rosignoli is 'María Zambrano' Postdoctoral Fellow at the Department of Public Law at the Universitat Rovira i Virgili in Tarragona (Spain), where she is also a member of the Research Group on Environmental Law, Citizenship and Sustainability, and researcher at the URV's Centre for Environmental Law Studies of Tarragona (CEDAT). Her research interests include environmental justice, global environmental governance, climate change, climate-induced migration, and collective capabilities. Her current research focuses on climate change-induced migration from a gender perspective in the EU context. Her latest monograph, *Environmental Justice for Climate Refugees*, was published by Routledge in May 2022 and aimed at investigating the legal impasse surrounding the recognition of climate refugees through the lens of environmental justice. She is the co-editor of the volumes *Climate Change Integration in the Multilevel Governance of Italy and Austria* (Brill 2022) and *Racism, Environment, Health. Environmental Racism and Health Inequalities* (PM 2022).

Bettina Steible is Assistant Professor of Public Law at the *Universitat Oberta de Catalunya* since 2020. She also counts with teaching experience at the *Universitat Autònoma de Barcelona*—UAB, the *Universitat Pompeu Fabra*—UPF, the CEI International Affairs, the University of Szczecin, and the University of Montpellier I. She holds a PhD in Law from the *Universitat Autònoma de Barcelona*, for which she received 2018 Jaime Brunet Prize for the best doctoral thesis on the promotion of human rights. Her main research interests revolve mainly around two main axes: European human rights law and European Constitutional Law.

Maria Antonia Tigre is the Director of Global Climate Change Litigation at the Sabin Center. She manages the Sabin Center's Global Climate Litigation Database and heads the Network of Peer Reviewers on Climate Litigation, currently with over 130 volunteers. She also serves as the Deputy Director of the Global Network for Human Rights and the Environment (GNHRE), where she collaborates with scholars and practitioners to study the relationship between human rights and the environment. She is a sought-after speaker at conferences and symposiums and has delivered a TEDx talk. She completed her Doctorate in Juridical

Studies (S.JD.) at the Elisabeth Haub School of Law at Pace University in 2022. In 2013, she earned a dual LL.M., magna cum laude, specialising in environmental law and comparative legal studies, from the same institution. She obtained her LL.B. from the Pontifícia Universidade Católica of Rio de Janeiro, Brazil.

Natalia Urzola is an SJD student at Elisabeth Haub School of Law at Pace University. In 2020, she obtained her LL.M from the University of California, Berkeley. Natalia is the Chief Operating Officer at the Global Network for Human Rights and the Environment and the Colombian Rapporteur of the Sabin Center's Global Peer Reviewers' Network of Climate Litigation. Her recent publications include: 'Gender in Climate Litigation in Latin America: Epistemic Justice Through a Feminist Lens' (*Journal of Human Rights Practice*); 'Climate Litigation in Latin America: Is the Region Quietly Leading a Revolution?' (with Maria A. Tigre and Alexandra Goodman, *Journal of Human Rights and the Environment*); and 'The Human Right to a Healthy Environment and the Rights of Racialized Groups: Applying Critical Race Theory as a Framework for (Re)constructing Environmental Rights Through Foundational Transformation' (in D. Lupin *A Research Agenda for Human Rights and the Environment*, Edward Elgar Publishing).

Thalia Viveros-Uehara is a postdoctoral researcher at the Department of Public Law and Governance at Tilburg University, and a research fellow at the Max Planck Institute for Comparative Public Law and International Law. She earned her Master's degree in Environmental Policy and Regulation from the London School of Economics and Political Science, and her Doctorate in Global Inclusion and Social Development from the University of Massachusetts Boston. Her work focuses on socio-legal approaches to human rights, environmental and climate change law, and their interactions with social development, emphasising the human face of environmental and climate crises. She has over a decade of professional collaboration with decision-makers, including Mexico's National Human Rights Commission, the Boston Human Rights Commission in the USA, and the Global Organization of Legislators for a Better Environment in the United Kingdom.

Luca Volpi is an architect (TU Darmstadt, 2006), specialised in sustainability. His main interests reside in the relationship between habitability and the resources required to achieve it. He is partner at Societat Orgànica, an environmental consultancy cooperative in sustainable urban development and architecture, where he leads teams in research and consultancy projects. He is the director of the postgraduate course in Sustainable Architecture (PAS) at "Josep Lluís Sert" professional practice school of the Architects' Association of Catalonia. He is a guest lecturer in universities, Master's courses and conferences. He studied architecture in Italy, focusing on architectural theory and heritage, and in Germany, specialising in sustainability and energy efficiency.

Chengjian Xu holds a PhD from Leiden University's Institute of Environmental Sciences, specialising in "Lithium-ion batteries and the transition to electric vehicles: environmental challenges and opportunities from a life cycle perspective". His expertise spans industry ecology, critical materials, and renewable energy systems, with a focus on electric vehicles, batteries, and short-term energy storage. With publications in high-impact journals such as *Nature Communications* and *Nature Communications Materials*, Chengjian demonstrates a commitment to impactful research. Notable achievements include investigating material demand for electric vehicle batteries, concentrating on critical components like cobalt, lithium, and nickel. This research was conducted in collaboration with the US Argonne National Laboratory and is cited in the IPCC's "Climate Change 2022: Mitigation of Climate Change" report and the European Parliament's policy document on electric vehicles.

Lorena Zenteno is a Chilean attorney with experience in human rights and environmental issues. She is a former member of the judiciary in Chile. Since 2015 she has been a Law Clerk for the Constitutional Chamber of the Supreme Court, drafting judgements for the Justices, mainly in environmental cases. Lorena is a member of the Environment and Human Rights Commission of the National Association of the Chilean Judiciary, dedicated to train judges and discussing climate change and environmental impacts on human rights. She was a senior researcher for the former UN Special Rapporteur on Cultural Rights, Karima Bennoune, from September 2018 until September 2021. Supported and assisted the UN Special Rapporteur to fulfil his mandate to the UN General Assembly and UN Human Rights Council. Lorena Zenteno got her LL.M. in Environmental Law from the University of Davis, California, in 2020, where she focused specifically on the intersections between human rights, cultural rights, and the environment. She also holds a Master's in Business

Law from the University Pompeu Fabra, in Barcelona, Spain. Lorena's primary research interests include the human rights dimensions of climate change and environmental impacts, climate change justice, gender, and the judiciary's role in the climate change crisis. Currently, Lorena is undertaking her PhD in Law at the University of Edinburgh. Since January 2022, she will start as a Senior Research Fellow for the Global Initiative for Economic, Social, and Cultural Rights, focusing on climate, energy, and gender justice.

Abbreviations

AFIR	Alternative fuel infrastructure regulation
AFOLU	Agriculture, forestry and other land use
BIT	Bilateral Investment Treaty
CBAM	Carbon border adjustment mechanisms
CEC	Citizen Energy Community
CEDAW	Committee on the Elimination of Discrimination Against Women
CESCR	Committee on Economic Social and Cultural Rights
CET	Clean energy transition
CETA	Comprehensive Economic Trade Agreement
CFE	Federal Electricity Commission
CO_2	Carbon dioxide
COP	Conference of the Parties
CRC	Committee on the Rights of the Child
DR-CAFTA	Dominican Republic-Central America Free Trade Agreement
EC	Energy Community
ECP	European Climate Pact
ECT	Energy Charter Treaty
ECtHR	European Court of Human Rights
EGD	European Green Deal
EIA	Environmental Impact Assessment
ENNOH	European Network of Network Operators for Hydrogen
ENTSOG	European Network of Transmission System Operators for Gas
ETS	Emissions trading system
EU	European Union
EV	Electric vehicle
FDI	Foreign direct investment
FPIC	Free, prior, and informed consent

FTA	Free trade agreement
GATT	General Agreement on Tariffs and Trade
GHG	Greenhouse gas
HRC	Human Right Council
IACmHR	Inter-American Commission of Human Rights
IACtHR	Inter-American Court of Human Rights
ICESCR	International Covenant on Economic, Social and Cultural Rights
ICSID	International Centre for Settlement of Investment Disputes
IDMC	Internal Displacement Monitoring Centre
IDP	Internally Displaced People
IEA	International Energy Association
IEM	Internal Electricity Market Directive
IIAJ	Infrastructure Investment and Jobs Act (US)
IIAs	International investment agreements
ILO	International Labour Organisation
IPCC	Intergovernmental Panel on Climate Change
IRA	Inflation Reduction Act
IRA	Inflation Reduction Act (US)
ISDS	Investor-State dispute settlement
JTC	Just Transition Commissions
JTM	Just Transition Mechanism
JTP	Just Transition Platform
LATAM	Latin America
LFP	Lithium iron phosphate
LGEEPA	General Law of Ecological Balance and Environmental Protection (Mexico)
LNG	Liquefied natural gas
MEM	Ministry of Energy and Mines (Guatemala)
MT	Million tonnes
MW	Megawatts
NAFTA	North American Free Trade Agreement
NDCs	Nationally Determined Contributions
NMC	Nickel cobalt manganese oxide
OAS	Organization of American States
OICE	Oficina d'Impuls a les Comunitats Energètiques (Catalonia)
OPCP	Optional Protocol to the Convention on the Rights of the Child on a communications procedure
PAGE	Partnership for Action on Green Economies
PCI	Projects of common interest
RE	Renewable energy
REC	Renewable Energy Community
RED III	The revised Renewable Energy Directive

RED	Renewable Energy Directive
SDGs	Sustainable Development Goals
TIPs	Treaties with investment provisions
TSOs	Transmission system operators
UDHR	Universal Declaration of Human Rights
UN	United Nations
UNCTAD	United Nations Conference on Trade and Development
UNDP	United Nations Development Programme
UNDRIP	UN Declaration on the Rights of Indigenous Peoples
UNFCCC	United Nations Framework Convention on Climate Change
UNGA	UN General Assembly
USMCA	Canada-Mexico-United States Agreement
WCN	World Charter for Nature
WWII	Second World War

List of Tables

Table 2.1 US FTA with Latin American countries (Office of the United States Trade Representative, n.d.) 18
Table 11.1 Socioenvironmental conflicts in Latin American: lithium (Source: Data drawn from EJAtlas) 283
Table 11.2 Elements of a 'just transition' 289

CHAPTER 1

Introduction: Challenges at the End of the Fossil Fuel Era

Daniel Iglesias Márquez, Clara Esteve-Jordà, and Beatriz Felipe Pérez

The events of the twenty-first century, such as the COVID-19 pandemic, underscore the urgency for radical changes in human lifestyles to safeguard the habitability of our planet. This imperative calls for a fundamental shift from the entrenched energy system, traditionally reliant on the combustion of fossil fuels. The zenith of fossil fuels, encompassing coal, oil, and gas, seems to be on the decline. This decline is not solely due to

D. Iglesias Márquez (✉)
Rovira i Virgili University, Tarragona, Spain
e-mail: daniel.iglesias@urv.cat; danniel.iglesias@gmail.com

C. Esteve-Jordà
Institute of Environmental Science and Technology, Autonomous University of Barcelona, Barcelona, Spain

B. Felipe Pérez
CICrA Justicia Ambiental, Barcelona, Spain

Tarragona Centre for Environmental Law Studies, Tarragona, Spain

D. Iglesias Márquez et al. (eds.), *Legal Challenges at the End of the Fossil Fuel Era*, Global Issues,
https://doi.org/10.1007/978-3-031-61766-9_1

their finite and diminishing availability but also because they are the primary contributors to the climate crisis—a crisis identified by studies from the Stockholm Resilience Centre as one of the six planetary limits already surpassed.

Indeed, the energy sector consumes a significantly larger portion of the energy it extracts. Consequently, the dwindling availability of conventional oil resources, coupled with socio-political and environmental tensions, underscores the imperative to decarbonise production. Although nuclear energy remains a contentious topic, the shift towards renewables appears to be garnering more supporters, albeit with a hesitancy to acknowledge the social and environmental impacts it entails, particularly for local communities in the Global South.

Global North actors, including the European Union (EU), are leading the charge in the energy transition without challenging hegemonic narratives. The EU stands at the forefront of this transition with the European Green Deal, aiming to achieve a 55% reduction in net greenhouse gas emissions by 2030 compared to 1990 and climate neutrality by 2050. In alignment with these ambitious goals, the new Directive on Renewable Energy Sources was adopted in October 2023, mandating a rise in the share of renewable energy in the EU's total consumption to 42.5% by 2030. It is essential, at this juncture, to consider the origin of the materials required for the production of renewable energies and the direct and indirect socio-environmental impacts associated with their extraction, production, transportation, and consumption.

Fundamentally, the energy transition presents two primary challenges. Firstly, there is the question of its feasibility: will it materialise in time to avert the irreversible escalation of the planet's average temperature, thus preventing the climate crisis from becoming irreversible? Secondly, there is the issue of fairness. In this context, the energy transition should not exacerbate the already unequal distribution of benefits and burdens generated by our economic model; rather, it should actively contribute to correcting this imbalance. A transition that merely substitutes fossil fuels with renewables within the same power dynamics is not acceptable.

Indeed, opportunities for citizens to participate in decisions regarding energy production, distribution, and consumption remain limited. The current energy system is characterised by stratification and unequal distribution, placing those with less economic and political power at a disadvantage. This disparity is contributing to water and land grabbing, as well as food insecurity and energy poverty among populations that are already vulnerable.

The energy transition presents an opportunity to construct a new system compatible with a higher degree of self-sufficiency and robust democracy. This involves redefining energy generation, distribution, and decision-making, with a focus on fostering environmental sustainability, socio-economic empowerment, and inclusive participation. Regrettably, the current trajectory is not aligning with these ideals. The energy transition, primarily led by hegemonic powers under the umbrella of so-called green capitalism, is, in reality, exacerbating environmental degradation and social inequalities.

This scenario prompts urgent questions: how should we navigate the complexities of the energy transition to effectively address challenges that involve seemingly conflicting but undoubtedly overlapping interests, such as mitigating climate change, reducing energy poverty and social exclusion, and strengthening energy security? In this unfolding process, the role of law becomes pivotal, just as it did during the extended dominance of carbon capitalism rooted in fossil—both national and international—law. Profound questions emerge at various governance levels, from the local to the global. How can we ensure the provision of affordable, reliable, safe, high-quality, and environmentally benign energy services? Are legal frameworks actively supporting a socially and ecologically just energy transition through their regulations?

This book comprehensively addresses all these aspects, bringing together issues that were discussed at the VII Tarragona International Environmental Law Colloquium (TIEC), which took place on May 26 and 27, 2022, at the Universitat Rovira i Virgili in Tarragona, Spain. The colloquium focused on the theme 'Legal Challenges at the End of the Fossil Fuel Era: Shaping Energy Futures Through Legal Intervention'. Since 2016, TIEC has been a platform for researchers with diverse backgrounds to present and discuss their research and works-in-progress annually. This event is jointly organised by the Tarragona Centre for Environmental Law Studies (CEDAT) and the Tarragona Environmental Law Students Association (AAEDAT).

The urgency of the energy transition and its pathways are explored from various perspectives within the book. It is structured into two parts to comprehensively address the subject. The first part delves into fundamental questions concerning environmental and social energy justice. Meanwhile, the second part outlines specific legal and political challenges associated with a clean and equitable energy transition, using concrete examples for illustration.

In the first part, Thalia Viveros-Uehara (Tilburg University) and Adrian Duhalt (Columbia University) emphasise that socio-environmental conflicts in Latin America are poised to increase due to the rising demand for the region's mineral resources, fuelled by the US energy transition. They anticipate a surge in litigation before national courts and caution that favourable outcomes in this realm may face reversals through international arbitration. The authors underscore the crucial need to bolster national environmental regulations by incorporating participatory rights aligned with human rights standards. They advocate for the integration of participatory rights into free trade agreements. Similarly, the chapter by Ana García-Juanatey (CEI International Affairs, Adjunct Center to the University of Barcelona), Bettina Steible (Open University of Catalonia), and Luca Volpi (Environmental consultant, Societat Orgànica) explores the role of the right to an adequate standard of living in relation to basic needs directly influenced by energy, particularly in the context of fossil fuel depletion. They argue that this right can serve as the foundation for prioritising the satisfaction of basic needs such as access to water, food, and housing. Additionally, the chapter sheds light on the current limitations of the human rights framework in integrating the needs of future generations.

Furthermore, within the context of the energy transition, the concept of energy communities emerges as a potentially pivotal player, yet its implementation is not without challenges. Clara Esteve-Jordà (Institute of Environmental Science and Technology, Autonomous University of Barcelona) and Marcos de Armenteras Cabot (University of the Balearic Islands) scrutinise the advantages offered by energy communities for energy governance, the reduction of greenhouse gas emissions, and the promotion of a decentralised and distributed energy system. They analyse these aspects through the lens of the commons. However, they highlight the dependency of this concept on the political, promotional, and regulatory activities of the state. For energy communities to flourish, the state must configure a supportive framework for their development and empower citizens to actively participate in building and sustaining these energy communities.

Similarly, a significant debate revolves around the repercussions of escalating social inequality on access to energy, particularly among the most vulnerable segments of the population, giving rise to the concept of energy poverty. In this context, Francesca Rosignoli (Rovira i Virgili University) delves in her chapter into the legal gap concerning the protection of individuals displaced due to climate change and environmental degradation. She proposes energy communities as spaces that can facilitate governance capable of preventing or anticipating forced displacement.

Simona Fanni (University of Seville) begins her chapter by acknowledging the gap between the mitigation commitments outlined in the Paris Agreement and the promises made by its States Parties. She then centres her analysis on the litigation that has arisen before both national and international courts. This litigation is aimed at compelling states to make more decisive commitments and is grounded in the respect for the human rights of both present and future generations. Fanni sees this litigation as a driving force behind the justiciability of the energy transition itself.

The chapter by Natalia Urzola (Global Network for Human Rights and the Environment), Maria Antonia Tigre (Sabin Center for Climate Change Law at Columbia Law School), Lorena Zenteno (University of Edinburgh), Marlies Hesselman (University of Groningen), Riccardo Luporini (Sant'Anna School of Advanced Studies) and Pedro Cisterna-Gaete (University of Edinburgh) delves into the topic of justice in the energy transition within the context of Latin America. The authors introduce the concept of 'just transition litigation', where citizens challenge the fairness of decarbonisation policies. By focusing on Latin America's unique context, the chapter explores the tension between environmental conservation and extractive development, contributing to a nuanced understanding of just transition policies and litigation in the region.

In the second part of the book, Guilherme Pratti's contribution (Sant'Anna School of Advanced Studies) sheds light on the deep roots of fossil fuels in international law and their historical dominance over environmental protection. This historical legacy implies that current national policies aimed at transitioning to clean energy matrices face substantial hurdles rooted in international regulatory frameworks.

Aloia López Ferro (Rovira i Virgili University and Terraqui) addresses the role of hydrogen in the energy transition plans of the European Union. While the hydrogen strategy boasts advantages such as security of supply, price stability, flexibility, and resilience of the energy system, as well as reduced dependence on sources outside the EU, the author not only analyses the strategies and regulatory proposals but also assesses their contribution to the overall sustainability of the energy system.

An intriguing debate revolves around the continuity of strategies based on large-scale production and transmission through extensive high-voltage power lines, controlled by major electricity companies. This is in contrast to small-scale production, which is more widely distributed throughout the territory, favouring community solutions, collective management of services, and fostering a closer relationship between producer, distributor, and user. In this context, the work of Iván Hortigüela Bolsa (Rovira i Virgili University and Insta Jurídic) delves into the Autonomous Community of Catalonia (Spain). The proliferation of wind and photovoltaic projects concentrated in specific rural areas has environmental and landscape impacts, leading to social opposition movements that have manifested in the courts. Through an analysis of regulations, judicial decisions, and aspects related to citizen participation, the study aims to extract valuable lessons from the Catalan experience.

From a European perspective, a significant challenge arises in questioning the continuation of exporting an extractivist model to third countries to meet the material needs of both old and new materials required for the energy transition. The chapter authored by Asmaa Khadim (Leiden University), Ron Janjua (University of Groningen), and Chengjian Xu (Leiden University) addresses the potential adverse effects on third countries resulting from the European Green Deal. The European energy transition will necessitate the extraction of increasing quantities of minerals located in third countries, potentially leading to the export of new socio-environmental conflicts. The environmental impacts associated with lithium extraction, particularly in Argentina, Bolivia, and Chile, underscore the importance for the EU to consider the assessment of these consequences and strategies to mitigate them in the implementation of its energy transition policies.

In her chapter, Monika Feigerlova (Czech Academy of Sciences) asserts that while the promotion of renewable energies is a key component in the fight against climate change, various states have faced and continue to confront lawsuits from companies before international arbitration tribunals seeking redress for modifications to the national legal framework affecting their investments. Despite the diversity of arbitration decisions, states have started amending investment agreements to acknowledge the role of renewable energies, allowing for an assessment of the extent of these modifications. Likewise, Magdalena Baz-Vilizzio (University of the Republic) underscores this concern in her chapter. In light of the investor-state dispute settlement regime potentially hindering the energy transition, Magdalena argues that the crisis of the Energy Charter Treaty needs a reconsideration of the tensions between climate change, the energy transition, investment protection, and investor-state arbitration.

In summary, this book consolidates contributions from a distinguished group of young researchers from various countries, covering key aspects of the energy transition. The exploration of energy issues in this book spans international law, EU law, and constitutional law, often incorporating legal-philosophical reasoning. Its significance lies not only in its legal examination of energy law but also in its critical connections with environmental and social justice concerns. The book goes beyond a purely technical perspective, delving into the immediate eco-social implications of the transition away from fossil fuels.

The edited volume, *Legal Challenges at the End of the Fossil Fuel Era: Shaping a Just and Clean Energy Transition*, stands as a collective achievement made possible by the dedication and exceptional contributions of our esteemed authors. We extend our heartfelt appreciation to each and every one of them for entrusting us with their valuable insights and for their unwavering commitment to this collaborative project. Our sincere thanks go to the members of the Tarragona International Environmental Law Colloquium Organising Committee, the AAEDAT, the CEDAT, and the Faculty of Legal Sciences at the Rovira i Virgili University for providing us with the forums to discuss the issues reflected in this volume. Special gratitude is extended to Paolo D. Farah for his assistance in preparing the book proposal and for encouraging its publication. We also extend our special acknowledgement to the dedicated team at Palgrave Macmillan for their patience and expertise in the production of this edited book. Their commitment to scholarly excellence has significantly contributed to the quality of the final result. Lastly, we express our deep gratitude to our families, relatives, and friends for their understanding and encouragement throughout the years.

Acknowledgments This book is part of the research project "Challenges of the European Energy Transition: Human Rights, Responsible Business Conduct, and Socio-environmental Conflicts in the Global South" (ICI02724000017), funded by the International Catalan Institute for Peace (ICIP). It has also been supported by the *Agència de Gestió d'Ajuts Universitaris i de Recerca* (AGAUR) and the EU (NextGenerationEU) under Investigo grant (200076ID12).

PART I

Environmental and Social Energy Justice

CHAPTER 2

Transitioning *All* Together? Participation Rights, Foreign Investment, and the Pursuit of a Just Energy Future

Thalia Viveros-Uehara and Adrian Duhalt

Introduction

Latin America and the Caribbean (referred to in this chapter as Latin America) are at the epicentre of global socio-environmental conflicts, accounting for nearly half of the world's disputes related to mining exploration and extraction[1] (EJAtlas, n.d; Martinez-Alier & Walter, 2016: 74). This places the region at the heart of local community resistance against major

[1] According to data retrieved from the Global Atlas of Environmental Justice in September 2023, there were 564 socio-environmental conflicts related to mining exploration and extraction around the world, 262 of which occurred in Latin America (EJAtlas, n.d.).

T. Viveros-Uehara (✉)
Tilburg Law School, Tilburg University, Tilburg, The Netherlands
e-mail: t.viverosuehara@tilburguniversity.edu

A. Duhalt
Center on Global Energy Policy (CGEP), Columbia University, New York, NY, USA

D. Iglesias Márquez et al. (eds.), *Legal Challenges at the End of the Fossil Fuel Era*, Global Issues,
https://doi.org/10.1007/978-3-031-61766-9_2

extractive corporations. On the one hand, affected groups are increasingly turning to domestic courts, seeing judicial mobilisation as a primary avenue for achieving redress (Anglés Hernández & Rovalo Otero, 2019: 290; Heffron, 2022; Sieder, 2007; Schilling-Vacaflor, 2019; Spalding, 2023). On the other hand, Latin America is among the regions with the highest number of investment claims lodged against its constituent countries, with foreign enterprises contending that their economic interests have been compromised[2] (Bottini & Méndez Bräutigam, 2019: 408; Fernández Masiá, 2020: 115; ICSID, 2023: 12). Given this backdrop, and due to the rising demand spurred by the US energy transition roadmap for Latin America's abundant mineral reserves, confrontations between foreign mining companies and frontline communities are likely to multiply.

With the rise in mining conflicts on its doorstep, Latin America possesses a myriad of legal tools from which it can draw, potentially aiding stakeholders in preventing and countering injustices. Specifically, the participation of local communities in the decision-making process related to extractive projects is intended to balance the asymmetrical power dynamics in the mining sector, where states and businesses often hold dominant positions (de Castro et al., 2016: 7; Seck, 2015: 385; Walter & Urkidi, 2016). With this objective, participation rights—in this chapter collectively referring to public consultation and free, prior, and informed consent (FPIC) for Indigenous peoples—constitute the procedural engine of environmental and climate justice (Gonzalez, 2015, 2021: 112). In this capacity, these rights have provided a legal foundation for affected groups to appeal to courts.

Can the escalating judicialisation of participation rights by affected communities foster a just energy transition? This chapter argues that this will seldom be the case. While domestic courts in Latin America have previously paved the way for significant transformations (Couso, 2006; von Bogdandy, 2017: 33)—albeit related to different societal aspirations[3]—the

[2] As of 2023, 33% of all disputes submitted to the International Centre for Settlement of Investment Disputes (ICSID) involved countries in South and Central America and the Caribbean, followed by Eastern Europe with 26%. Furthermore, 25% of all such cases worldwide were related to oil, gas, and mining, thereby representing the economic sector with the highest share. This sector is closely followed by electric power and other energy industries, which constitute 17% of the cases (ICSID, 2023: 12).

[3] For instance, resolutions issued by Colombia's Constitutional Court related to the right to health have had profound impacts on health policymaking. A notable example is Decision T-760/08, which led to a comprehensive reform of Colombia's national health system (Parra-Vera, 2016: 161; Yamin, 2022: 767).

US-centric political economy of the energy transition poses distinctive challenges. Specifically, it curtails the current legal paradigm's ability to enforce behavioural change in those responsible for socio-environmental impacts at a relevant scale. In other words, the *transnational* nature of the energy transition, which involves a vast array of actors, introduces complex and highly asymmetrical power dynamics that exceed the traditional legal boundaries usually confined to nation-states (Lachapelle et al., 2017; Newell, 2019; Pastukhova & Westphal, 2020). More concretely, the interests of foreign investors, predominantly based in high-income countries like the US, along with national elites in host countries, overshadow the voices of historically marginalised groups such as *campesinos*, Indigenous peoples, individuals of African descent, and other ethnic minorities (UN General Assembly, 2022b). These hierarchies which the current investment system replicates are likely to further undermine the effectiveness of domestic judicial mobilisation (UN General Assembly, 2023: para. 8).

Using Guatemala and Mexico as case studies—both countries' mineral wealth, socio-economic landscape, and proximity to US economic interests create a fertile ground for socio-environmental conflicts (Martinez-Alier & Walter, 2016)—it is illustrated how the hard-fought victories of local communities in national courts, when asserting their participation rights, may increasingly be jeopardised by international investment arbitration processes as more foreign companies venture into these countries to extract critical minerals. In essence, the chapter sheds light on how the current intersection of international human rights with the trade and investment legal frameworks, both fundamental to energy transition governance, operates on a "lose-lose logic": a loss for the communities whose judicial victories are undermined, and a loss for the countries at large as they face mounting economic liabilities from investor mining companies.

The notion that *trans* in energy *trans*ition is for *trans*national underpins this chapter's socio-legal approach to the analysis and the proposed responses. The analysis underscores the importance of strengthening domestic environmental regulations by integrating participation rights in line with human rights standards and, *concurrently* and *complementarily*, promoting global solidarities that urge free trade agreements (FTAs) to adopt participation and FPIC as guiding principles. This approach embodies what Roberto Unger describes as an "ad-hoc corrective response". It stands in the large intermediate space between two limit boundaries (Unger, 1983: 626): one where power asymmetries between local communities, states, and businesses grow unchecked—thereby intensifying dispossession—and the other where these asymmetries are preemptively prevented from emerging by terminating the global exchange that the energy transition requires.

This chapter discusses, in a progressive manner, the potential of participation rights as a means to transition together. First, the following section provides an overview of how the US energy transition impacts Latin America, focusing particularly on the repercussions of the Infrastructure Investment and Jobs Act and the Inflation Reduction Act—signed into law by President Biden in 2021 and 2022, respectively. Section "Mapping Participation Across Energy Transition Governance" then explores how energy transition governance, especially in the realms of international human rights and trade and investment law, currently addresses participation. Section "Participation Rights, Domestic Courts, and Investment Arbitration: Mutual Losses in Guatemala and Mexico" delves into the cases of Guatemala and Mexico, presenting relevant socio-legal context on mining conflicts and participation before examining landmark investment arbitration cases brought mainly by US foreign investors against these countries. This highlights the lose-lose logic that opposes the current reliance on domestic courts to tackle challenges stemming from the transnational nature of the energy transition. Finally, section "The Role of Courts in Transitioning *All* Together" reflects on the implications of the chapter's analysis and presents some concluding thoughts.

The US Energy Transition and Latin America's Mineral Resource Endowment

In recent years, Latin America's prospects to deepen its engagement in global supply chains, especially those associated with critical minerals,[4] have become more plausible against the backdrop of major geopolitical events and their implications. The trade dispute that the US and China[5]

[4] Critical minerals refer to "any non-fuel mineral that has a risk of a supply chain disruption, serves an essential function in one or more energy technologies, including technologies that produce, transmit, store, and conserve energy", and whose shortage represents a potential hazard to US economic security (Congress of the United States, 2021: 1381).

[5] The mutual imposition of tariffs has affected bilateral trade flows valued at approximately 450 billion USD (Fajgelbaum et al., 2023).

have sustained since 2018–2019 (Bown & Kolb, 2023) exposed that strategic supply chains serving the US industrial apparatus were somewhat vulnerable. The Biden administration (2021–2025) soon learnt that, concerning critical minerals, China's grip on global supply chains represents a risk to US decarbonisation plans and economic security. Another event worth considering is the Russian invasion of Ukraine in 2022. This conflict, which compelled the international community to impose economic and trade sanctions on Russia, validated the extent of Russia's influence as a producer and exporter of commodities like natural gas.[6] In this instance, the natural gas price spikes and supply disruptions that Europe[7] endured (Zettelmeyer et al., 2022), and their consequences, were a grim reminder of how important is to ease reliance on hydrocarbons (imports) and accelerate the deployment of clean energy technologies, whose expansion is also linked to the supply of critical minerals.

Hence, Latin American countries, as holders of considerable mineral reserves and leading producers of lithium and copper, among other minerals (as Fig. 2.1 illustrates), are well positioned to play a larger-than-expected role in both the energy transition (Bernal et al., 2023) and the global ecosystems of production and supply of minerals (Price, 2022). However, it remains to be seen whether the region's countries are poised to take advantage of this economic landscape—and whether they can overcome its challenges—since one of the underlying requirements is to put in place the right institutional and policy frameworks. As the subsequent sections elaborate, these include frameworks aimed at safeguarding and preserving the rights of local stakeholders such as Indigenous and other local frontline communities.

Concerned about supply chain vulnerabilities impaired by the trade tensions with China and even the outbreak of COVID-19, the US government issued an Executive Order in February 2021 to review and identify the existing risks affecting US supply chains (The White House, 2021a).

[6] As of 2021, Russia was the world's largest natural gas exporter (EIA, 2022). Based on the value of exports, Russia was also the world's second largest exporter of crude oil and the world's largest exporter of nitrogen fertilisers in that year (UN Comtrade, 2021). Natural gas is the most important raw material for the making of nitrogen fertilisers, which are the world's most demanded group of nutrients (Our World in Data, 2020).

[7] As of 2021, Europe accounted for 74% of Russia's exports of natural gas (EIA, 2022).

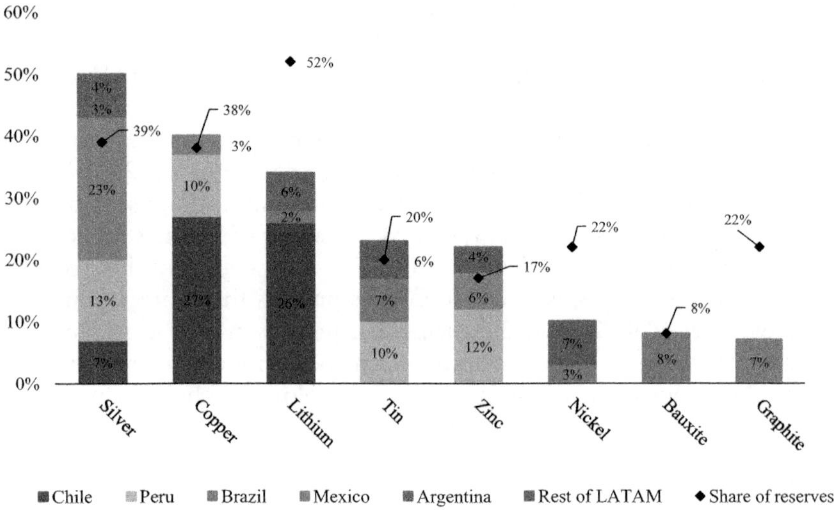

Fig. 2.1 Latin America's share of global production and reserves of selected minerals, 2021 (IEA, 2022)

The resultant report, presented in June 2021, resolved that strategic supply chains such as critical minerals and large-capacity batteries[8] are geographically concentrated, a feature that makes US companies vulnerable to potential supply disruptions. Notably concerning is the role of China, which despite boasting a limited advantage in terms of mineral reserves and production, its influential position stems from securing supply while developing robust mineral refining capabilities at home. For example, as for global lithium and cobalt refining, both of which are essential for manufacturing high-capacity batteries,[9] China's share stands at 60% and 80%, respectively—a condition that is perceived as a risk to the future of the electric vehicle (EV) industry in the US (The White House, 2021b).

[8] Large-capacity batteries play a pivotal role in reducing the fossil fuel footprint of power generation and in electrifying transportation, notably through electric vehicles. The report illuminates four strategic supply chains: semiconductor manufacturing and advanced packaging, large-capacity batteries, critical minerals and materials, and pharmaceuticals along with advanced pharmaceutical ingredients (The White House, 2021a).

[9] Nickel, especially Class 1, or high-purity nickel, is another critical mineral for electric vehicles and energy storage batteries (The White House, 2021b).

Down the road, the urge to secure mineral supply will build up as the world's efforts to reduce emissions and decarbonise energy systems gain momentum. In that regard, demand for critical minerals is anticipated to increase by up to six times between 2020 and 2040 (IEA, 2021).[10] Against this backdrop, it became evident for the US that its supply chains need to be more globalised, resilient, and less dependent on China. To achieve so, President Biden signed into law two important pieces of legislation, namely, the Infrastructure Investment and Jobs Act (IIAJ)[11] on November 15, 2021, and the Inflation Reduction Act (IRA) on August 16, 2022, earmarking considerable resources—in the form of tax incentives, loans, and grants—to boost domestic capabilities in clean energy technologies, including electric vehicles (EVs) and batteries.

The question arises as to whether the effects of these legislations will reverberate outside the US. The sheer size of resources compromised through the IIAJ and IRA, coupled with the conditions attached, suggests that both could bear implications for Latin America. To address energy security and combat climate change, the IRA alone will invest 369 billion USD over a ten-year span. Moreover, with the US government's goal for EVs[12] to constitute 50% of all sales by 2030 (The White House, 2023a), the IIAJ contemplates to roll out a multi-billion-dollar scheme to secure the supply of critical minerals, expand domestic capacity to process them, and advance battery manufacturing. The IIAJ also seeks to invest up to 7.5 billion USD to deploy around 500,000 EV chargers all over the country with the aim to accelerate the adoption of EVs (The White House, 2021c). These legislations, together with the CHIPS and Science Act—which centres on building capacity to manufacture semiconductors for EVs and chargers—involve an investment of around 135 billion USD just to strengthen the US electric vehicle chain (The White House, 2022).

In addition to the previously mentioned policies, primarily aimed at the supply side of EVs, the IRA also endeavours to stimulate demand for EVs, notably through consumer tax credits that can reach up to 7500 USD. However, to claim the full credit on EV purchases, buyers must meet specific requirements related to battery components and critical

[10] Based on a scenario where the world achieves net-zero emissions by 2050 (IEA, 2021). In the IEA Sustainable Development Scenario (SDS), which is consistent with meeting the Paris Agreement Goals, demand for critical minerals will quadruple by 2040 (IEA, 2021).

[11] More commonly known as the Bipartisan Infrastructure Deal.

[12] It includes battery electric, fuel cell, and plug-in hybrid vehicles.

minerals. For example, Section 13401, paragraph (1)(G) of the IRA mandates that for vehicles purchased in or after 2023, the final assembly must occur in North America, including Mexico (The White House, 2023b). This rule also pertains to battery manufacturing, with the eligible percentage of domestically produced component value for a 3750 USD tax credit increasing annually, starting at 50% in 2023. By 2029, all battery components must be manufactured within North America.

Regarding critical minerals, to qualify for an additional 3750 USD in credits in 2023, at least 40% of the value of the battery's minerals (or 80% in 2027) must be mined or processed either in the US or in a country with which the US has an FTA (US Department of the Treasury, 2023). In Latin America, the US has FTAs with 11 countries, including Guatemala and Mexico, as Table 2.1 indicates.

Put together, these policies are poised to trigger demand for critical minerals in the upcoming years. Considering the US government's endeavour to diversify and strengthen supply chains and the stipulations of the IRA outlined above, it is plausible that US firms will shift their focus to Latin America's extractive and manufacturing sector, prompted by the region's rich resource endowment. The IRA will likely extend its reach beyond typical participants like Mexico, a prominent car manufacturer,

Table 2.1 US FTA with Latin American countries (Office of the United States Trade Representative, n.d.)

Partner	*Entry into force*	*Main metallic minerals*
Chile	2004	Copper, gold, iron, manganese, molybdenum, silver
Colombia	2012	Gold, nickel
Costa Rica	2009	Bauxite, copper
Dominican Republic	2007	Bauxite, copper, gold, nickel, silver, zinc
El Salvador	2006	Gold, silver
Guatemala	2006	Copper, gold, iron, lead, nickel, silver
Honduras	2006	Copper, gold, lead, silver, zinc
Mexico	1994	Copper, gold, iron, lead, silver, zinc
Nicaragua	2006	Copper, gold, iron, lead, silver, zinc
Panama	2012	Bauxite, copper, gold, iron, manganese, molybdenum, silver
Peru	2009	Copper, gold, iron, lead, molybdenum, silver, tin, zinc

and Chile, a premier supplier of minerals; Guatemala serves as a prime example.

Beyond lithium and cobalt, nickel is vital for EV and energy storage battery chains. Guatemala, home to the world's tenth largest reserves[13] (Church & Crawford, 2020), amplified its production from 2.4 thousand metric tons in 2012 to 64.9 thousand metric tons in 2021 (Statista, 2021). For the US, nickel doesn't face the geographical diversification challenges inherent to lithium and cobalt in the sense that a larger number of countries participate in the value chain. However, in the wake of escalating future demand, the predominance of mining and refining in non-FTA countries—75.1% and 86.9%, respectively—may pose supply concerns in the future (S&P Global, 2023), given that IRA tax credits are limited to FTA partners like Guatemala. This raises the question of whether Guatemala, with its untapped nickel reserves, could emerge as a key player in the US battery and EV supply chains. Despite numerous potential obstacles, given the projected demand for EVs and batteries in the US, geographical proximity, and America's pursuit of resilient supply chains, dismissing this possibility would be premature.

Mapping Participation across Energy Transition Governance

The participation of local frontline communities in decision-making related to mineral extraction projects that might impact them has been a central concern in energy transition governance (Church & Crawford, 2020: 299). This section aims to provide a non-exhaustive overview of how the international human rights and trade and investment frameworks address such participation, as the increasing number of mineral extraction-related court cases has inevitably brought these legal domains into the realm of energy transition governance. While these fields differ in the depth of what they consider participation—human rights law provides comprehensive guidance to tackle power imbalances in natural resource extraction, whereas trade and investment law occasionally refers to or even neglects the concept—they both agree on how to materialise it. In essence, as the next paragraphs show, both view the confines of nation-states as the primary arena for ensuring participation, mainly through domestic regulation.

[13] The Peterson Institute for International Economics notes, "Guatemala may hold some of the largest nickel reserves in the world, but mining there has faced stiff opposition from the indigenous community" (Leruth et al., 2022: 15).

The Rights to Public Participation and to Free, Prior, and Informed Consent

As the human rights framework builds upon recognising the "inherent dignity of all individuals" (United Nations, 1948), it aims to equalise asymmetrical power dynamics where states and businesses dominate. It does this by valuing and amplifying the voices of local communities, primarily through promoting and ensuring their participation in environmentally and socially impactful processes. Though fundamental instruments in both the universal and Inter-American Human Rights Systems (IAHRS) do not explicitly enshrine the right to public participation or the right to FPIC, these rights have been elucidated through the interpretative work of treaty bodies over time, thereby clarifying the distinctive attributes of each.

Rooted principally in environmental considerations, the right to participation transcends the scope of FPIC, encompassing the involvement of all concerned citizens in matters related to the environment (Barragán-Terán & Erazo Robles, 2021: 158). The Rio Declaration of 1992 constituted the first acknowledgement of the principle of public participation in environmental matters (Principle 10). Following this landmark, the principle has permeated other multilateral environmental agreements as well as various human rights instruments. These include the Aarhus Convention[14] (adopted in 1998 and applicable to the UN Economic Commission for Europe region) and the Escazú Agreement[15] (adopted in 2018 and applicable to the Latin American countries). The latter specifically enjoins states to "facilitate access to environmental information" (article 5.3) and "to implement open and inclusive participation in environmental decision-making processes" (article 7).

The essential character of promoting and ensuring public participation has also been underscored by several treaty bodies and special procedures within the human rights system. In the 2013 inaugural report of the Independent Expert on Human Rights and the Environment, it was articulated that facilitating public participation in environmental decision-making constituted a procedural obligation incumbent upon states under human rights law (UN Human Rights Council, 2013: para. 29). Most

[14] This chapter refers to the Convention on Access to Information, Public Participation in Decision-Making, and Access to Justice in Environmental Matters as the Aarhus Convention.

[15] The Escazú Agreement is the name given to the Regional Agreement on Access to Information, Public Participation, and Justice in Environmental Matters in Latin America and the Caribbean.

recently, this duty was further solidified with the recognition of a clean, healthy, and sustainable environment as a human right by the UN General Assembly in 2022 (UN General Assembly, 2022a: 3).

Although originating from the broader right to public participation, the FPIC is *rationae personae* right, as it specifically pertains to Indigenous peoples (Puig, 2021: 39). It extends beyond the environmental challenges faced by these communities to encompass their unique social and cultural circumstances (Barragán-Terán & Erazo Robles, 2021: 158). The 1989 Indigenous and Tribal Peoples Convention of the International Labour Organization (ILO Convention 169) acknowledges these populations' right to occupy their lands and consequently stipulates their free and informed consent as a prerequisite to any relocation (article 16). Additionally, the UN Declaration on the Rights of Indigenous Peoples (UNDRIP) of 2007 affirms Indigenous peoples' right to self-determination, compelling states to obtain their FPIC prior to approving any project that may affect their lands or territories (article 32).

Case law that has further elucidated the scope of the right to FPIC has been particularly prolific within the IAHRS (Morales Antoniazzi & Barraco, 2020). On numerous occasions, the Inter-American Court of Human Rights (IACtHR) has recognised this right as deriving from property rights, referencing UNDRIP, and more recently even deeming it a principle of international law (de Casas, 2019: 250). Notably, in the case of Kichwa Indigenous People of Sarayaku v. Ecuador (2012), this court found the state responsible for violating the right to consultation, among other rights. In the reparations, the court provided core standards concerning the nature and scope of the required consultation. Specifically, the court mandated the state to "adopt the necessary legislative, administrative, or any other type of measures to give full effect, within a reasonable time, to the right to prior consultation of the indigenous and tribal peoples and communities", by sufficiently and adequately *regulating* it through appropriate norms for its practical implementation (para. 341.4).

Finally, in any recount, even a minimal one, of the standards pertaining to the extraction of critical minerals from a human rights law standpoint, it is imperative to include consideration of the UN Principles on Business and Human Rights (UN Principles). Endorsed by the Human Rights Council in 2011, they set out specific measures to prevent and mitigate the potential human rights impacts of business operations. Particularly relevant for our overview is Principle 18. This principle mandates that businesses conduct in-depth consultations with stakeholders who might be affected, serving as a core element of their human rights due diligence process. In embracing the wider concept of the right to participation, it can also be interpreted to include the right to FPIC (Barragán-Terán & Erazo Robles, 2021).

Given its orientation towards the nation-state, this legal domain assumes that governments, primarily through their regulatory capacities, have the necessary authority in every situation to equitably navigate the complex power dynamics underpinning diverse interests, even those that go beyond traditional state boundaries.

Trade and Investment Law's Pathway to Participation

The growing awareness of the social and environmental consequences of cross-border trade and investment has prompted most recent FTAs to increasingly incorporate considerations aimed at addressing such impacts (Alam, 2015: 303; Luke, 2019: 153; Luttrell, 2019: 274). Specifically, general "exception clauses" related to the environment in such agreements permit member states to adopt or maintain measures necessary to protect human health or the environment[16] (Fernández Masiá, 2020: 124; Gantz, 2020: 6; Viñuales, 2012) without incurring liability towards investors. These, along with the (seldom) application of human rights standards in trade and investment dispute settlement (Brown & Cucinotta, 2019; Hirsch, 2021: 129; Steininger, 2018), have been recently applauded as progressive measures that were unthinkable in earlier times (Desierto, 2022). However, as this chapter elaborates, while offering a glimmer of hope, these steps have been insufficient for making participation a reality, particularly the right to FPIC.

The integration of exception clauses related to the environment but not an explicit acknowledgement of a human rights language has a restrictive implication for participation and FPIC (Luke, 2019: 154; Puig, 2021:

[16] For example, the 2018 Canada-Mexico-US FTA (USMCA), which replaced the 1992 North American Free Trade Agreement (NAFTA), improved protections for the environment. These enhanced provisions address conservation in areas such as fisheries subsidies, shark finning, invasive species, and cooperation on marine litter and air quality (Chapter 24). Furthermore, its exception clause (reinstated from NAFTA) sets out that "[i]n the event of any inconsistency between a Party's obligations under this Agreement and its respective obligations under the following multilateral environmental agreements [...] a Party's obligations under this Agreement shall not preclude the Party from taking a particular measure to comply with its obligations under the covered agreement, provided that the primary purpose of the measure is not to impose a disguised restriction on trade" (Article 1.3, Para. 1). For a commentary on these USMCA provisions, please see Gantz (2020: 8).

78). Notably, as such clauses justify restrictions on trade and investment *only when they strictly pertain to environmental protection* (Brown & Cucinotta, 2019), domestic environmental regulation becomes the *sole avenue* for host states to implement participation and FPIC without rising liability towards investors. Environmental Impact Assessments (EIAs) as most countries' cornerstone of such regulatory compendium thus constitute a central mechanism for national governments to achieve this aim (Craik, 2019: 896). However, because often EIAs are inclined towards a broader public participation approach and do not expressly incorporate human rights standards themselves, they fail to render extractive projects incompatible with the realisation of Indigenous peoples' FPIC without infringing FTA provisions.

As the next section illustrates through the cases of Guatemala and Mexico, environmental provisions within FTAs are still an insufficient means for fulfilling the right to public participation, particularly the right to FPIC. By confining their realisation to the hyper-legalistic realm of domestic regulation (Brown & Cucinotta, 2019: 189; Fernández Masiá, 2020: 124), mainly, but not limited to EIAs, international trade and investment law's pathways to participation still present an illusory facade for countering transnational power asymmetries converging in extractive projects. Similarly, the human rights framework, by primarily viewing nation-states as the primary entities responsible for and driving participation, shares this shortcoming.

Participation Rights, Domestic Courts, and Investment Arbitration: Mutual Losses in Guatemala and Mexico

Since the 1980s, most Latin American states have enacted structural reforms aimed at promoting foreign trade and investment, scaling-up extractive and industrial activities, and liberalising their internal markets (Alviar García, 2022; Castellanos Giracca & Castellanos Howell, 2020; Cedillo, 2017). In line with this trend, the Guatemalan Constitution, adopted in 1985, sought to promote accession to FTA and foreign investment by setting out the duty of the state to protect the formation of capital, savings, and investment (article 119 k) (Castellanos Giracca & Castellanos Howell, 2020: 422). The same is true for Mexico: while the Mexican Constitution dates back to 1917, during the past decades it has undergone significant amendments to accommodate the neoliberal economic paradigm (Pou Giménez, 2022). At the international level, this country joined the General Agreement on Tariffs and Trade (GATT) in 1986, followed by Guatemala five years later in 1991. Currently, Mexico is state party to 14 FTA in force, including the Canada-Mexico-United States Agreement (USMCA),[17] while Guatemala is a party to nine, encompassing the Central America–Dominican Republic–the US (DR-CAFTA)[18] (Organization of American States, n.d.-a).

Parallel to such legal reforms closely tied to the countries' economic transformation, Guatemala and Mexico also progressively reconfigured their legal *corpus* pertaining to human rights and the environment (Pou Giménez, 2022; Sieder, 2007: 8). These efforts, at least in theory, aim to overcome historical exclusion and protect the environment, incorporating provisions relevant to the right to participation and FPIC. For example, the Guatemalan Constitution enshrines a comprehensive list of fundamental rights and guarantees under Title II. Furthermore, it gives human rights treaties precedence over domestic law (article 46), thereby emboldening national courts as interpreters of such convergence. Notably, the constitution mandates the state to ensure social participation in environmental planning, management, and oversight (article 46). In 2011, Mexico enacted a major constitutional reform, expanding the recognition and precedence of human rights to those enshrined in international treaties. While this bolstered the legal and institutional frameworks for the protection of human dignity across all branches of the state, the Mexican Constitution does not explicitly recognise public participation. However, it does so indirectly through its article 4, which underpins the General Law of Ecological Balance and Environmental Protection (LGEEPA).[19]

[17] The USMCA was signed on November 30, 2018, and entered into force on July 1, 2020. It replaced the North America Free Trade Agreement (NAFTA), which was signed on December 17, 1992, and took effect on January 1, 1992.

[18] The DR-CAFTA was signed by Guatemala on August 5, 2004, and entered into force for the country on July 1, 2006.

[19] Article 4, para. 5 of the Mexican Constitution establishes that: "Every person has the right to a healthy environment for their development and well-being. The State will guarantee respect for this right. Environmental damage and deterioration will generate responsibility for whoever causes it, in terms of what is provided by law".

In connection with this provision, article 1 of LGEEPA states that: "This law is regulatory of the provisions of the Political Constitution of the United Mexican States that refer to the preservation and restoration of ecological balance, as well as to environmental protection, in the national territory and the areas over which the nation exercises its sovereignty and jurisdiction".

Within this framework, participation in environmental decision-making relative to extractive projects is regulated via the EIA.[20]

Regarding the specific right to FPIC, neither the Guatemalan nor the Mexican Constitution provides an explicit reference. However, both countries have ratified the ILO Convention 169.[21] As a result of this ratification and the open human rights clauses in such countries' constitutions, this treaty becomes integral to their *corpus iuris*. In addition to this pivotal international instrument, Guatemala and Mexico are both signatories to the nine core UN international human rights instruments,[22] as well as several Inter-American treaties;[23] but only Mexico has ratified the Escazú Agreement.[24] Moreover, each country has ratified over 20 multilateral environmental agreements, including the Convention on Biological

[20] Article 34, para. 3 of LGEEPA sets out that "The [Ministry of the Environment and Natural Resources], upon request of any person from the concerned community, *may* conduct a public consultation [...]" (emphasis added by the authors).

[21] Guatemala ratified this Convention on June 5, 1996; Mexico had done so six years earlier, on September 5, 1990 (International Labour Organization, n.d.).

[22] These treaties include the International Covenant on Economic, Social and Cultural Rights (ICESCR, 1966), the International Covenant on Civil and Political Rights (ICCPR, 1966), the International Convention on the Elimination of All Forms of Racial Discrimination (ICERD, 1965), the Convention on the Elimination of All Forms of Discrimination Against Women (CEDAW, 1979), the Convention Against Torture and Other Cruel, Inhuman or Degrading Treatment or Punishment (CAT, 1984), the Convention on the Rights of the Child (CRC, 1989), the International Convention on the Protection of the Rights of All Migrant Workers and Members of Their Families (ICMW, 1990), the Convention on the Rights of Persons with Disabilities (CRPD, 2006), and the International Convention for the Protection of All Persons from Enforced Disappearance (CED, 2006) (UN Office of Legal Affairs, n.d.).

[23] Both Guatemala and Mexico have adopted the American Convention on Human Rights (ACHR, 1969); the Additional Protocol to the American Convention on Human Rights in the Area of Economic, Social, and Cultural Rights, commonly known as the Protocol of San Salvador (1988); the Inter-American Convention to Prevent and Punish Torture (1985); the Inter-American Convention on Forced Disappearance of Persons (1994); and the Inter-American Convention on the Prevention, Punishment, and Eradication of Violence against Women, known as the Convention of Belém do Pará (1994) (Organization of American States, n.d.-b).

[24] Guatemala signed this treaty on September 27, 2018, but had not ratified it as of the time of writing. Mexico signed it on the same date and later ratified it on January 22, 2021 (CEPAL, n.d.).

Diversity, the United Nations Framework Convention on Climate Change, and its legally binding Paris Agreement (United Nations, n.d.).[25]

Such simultaneous yet diverging trajectories of trade and investment and human rights frameworks in Guatemala and Mexico have created a paradoxical backdrop for the energy transition. On one hand, the countries' neoliberal inclinations, combined with the increasing demand for minerals critical for the US decarbonisation plans, provide fertile ground for both national and transnational economic interests to promote large-scale investment as a significant business opportunity. On the other hand, both countries have an obligation to protect and fulfil the rights of the most vulnerable—a primary concern for those directly affected and a cause championed by transnational advocacy networks. These seemingly opposing currents of interest are intertwined in a complex web of power dynamics that operate both within and beyond the countries' national boundaries (Lachapelle et al., 2017; Newell, 2019; Pastukhova & Westphal, 2020). Foreign investors, predominantly based in high-income countries like the US, along with national elites in host countries, perpetuate and deepen the socio-economic hierarchies inherent in the current global ecological crisis, leaving historically marginalised groups such as Indigenous peoples, *campesinos*, individuals of African descent, and other ethnic minorities voiceless (UN General Assembly, 2023: para. 8; 2022b).

Participation rights act as a counterbalance to these highly asymmetrical power dynamics. As they serve as the structure of the procedural dimension of climate justice, they help tilt the pendulum towards populations that have historically been discriminated against by racial capitalism (Gonzalez, 2021; UN General Assembly, 2023: para. 8; 2022b: para. 82). These groups, despite not having contributed to the current climate crisis nor holding either state or investment power, are most likely to bear the socio-environmental burden of the extractive activities of the US energy transition. Although participation rights are entrenched in the legal *corpus* of Guatemala and Mexico relating to human rights and the environment,

[25] Guatemala ratified the Convention on Biological Diversity (CBD) on July 10, 1995, after signing it on June 13, 1992. The country also ratified the United Nations Framework Convention on Climate Change (UNFCCC) on December 15, 1995, following its signature on the same date in 1992, and the Paris Agreement on January 25, 2017, which it had signed on April 22, 2016. Mexico, on the other hand, ratified the CBD on March 11, 1993, having signed it on June 13, 1992. It signed the UNFCCC on June 13, 1992, and ratified it on March 11, 1993. The country ratified the Paris Agreement on September 21, 2016, following its signature on April 22, 2016.

substantial scholarship and institutional reports indicate that these rights have not been effectively realised in practice—as the following paragraphs will detail. As a result, affected communities increasingly seek rights-based recourse in domestic courts to protect themselves against their lack of participation in the authorisation processes of extractive enterprises, which lead to harmful socio-environmental effects (Heffron, 2022; Sieder, 2007; Schilling-Vacaflor, 2019; Spalding, 2023).

The judicialisation of participation rights, prevalent in both Guatemala and Mexico, tends to emerge once grievances between involved parties become tangible. This often sits on a lose-lose logic for both the affected communities and the states: domestic rulings that favour local communities but oppose the interests of foreign investors can be overturned by investment arbitration initiated by these investors, possibly saddling the state with substantial financial liabilities. The ensuing analysis reveals how the lose-lose trajectory of judicialising participation cannot be effectively countered by solely relying on domestic regulation. While this approach is necessary, it proves insufficient when navigating the intricate web of transnational power dynamics.

Judicialising Participation in Guatemala: The Need to Overcome Regulatory Failures

With 44% of its population comprising Indigenous peoples (Us et al., 2021: 10) and significant reserves of minerals (as section "The US Energy Transition and Latin America's Mineral Resource Endowment" outlines), Guatemala has witnessed prolonged tensions between Indigenous groups and extractive industries (either directly or indirectly owned by foreign entities) (*Observatorio de Conflictos Mineros en América Latina*, n.d.). Indigenous communities have increasingly resorted to both national and international human rights tribunals to address the socio-environmental challenges arising from such corporate exploitation activities, with their right to FPIC forming a central argument (Spalding, 2023; Sieder, 2007). This reactive approach has emerged in response to the issuance of mining licences without securing the participation and FPIC from the affected populations, and without an adequate assessment of environmental impacts (UN Human Rights Council, 2018a: para. 62). Studies indicate that the execution of the EIA—the primary policy tool through which the Guatemalan government seeks to ensure the participation and consultation of Indigenous populations affected by major projects, including critical mineral extraction—is inadequate, as it does not align with human rights standards (Aguilar-Støen & Hirsch, 2015: 474; CIDH, 2017: 106; Spalding, 2023: 108).

Consequently, mining projects in Guatemala often start on, and progress over, challenging terrains that inevitably give rise to conflicts between impacted local communities and corporations. This leads to the vindication of rights to participation and FPIC increasingly being pursued *ex-post*, once the grievances are tangible, through human rights judicial pathways. Domestically, between 1996 and 2018, Guatemala's Constitutional Court delivered over 55 rulings on consultations involving Indigenous peoples (UN Human Rights Council, 2018a: para. 65). Internationally, in 2020, the Inter-American Commission on Human Rights (IACHR) presented a case to the IACtHR concerning the state's failure to implement FPIC while approving mining licences, adversely impacting the Maya Q'eqchi' Indigenous Community of Agua Caliente (CIDH, 2020). Although the case remains unresolved at the time of this writing, its claims echo broader concerns about lapses in the country's EIA process.

While human rights tribunals may tilt the justice pendulum in favour of local community claimants, these judicial avenues do not represent the final verdict. Notably, in cases involving foreign investors as defendants, they retain the option to resort to investor-state dispute settlement mechanisms (Lozada Gómez, 2020; UN General Assembly, 2023: para. 24). To be precise, Article 10.16 of the DR-CAFTA enables them to challenge the Guatemalan state for breaching its obligations under such an agreement. Hence, Guatemala has faced challenges in nine cases before the International Centre for Settlement of Investment Disputes (ICSID), two of which, *Energía y Renovación v. Guatemala* (2021) and *Kappes v. Guatemala* (2018), are linked to unrest among Indigenous peoples due to a lack of participation and consultation (UNCTAD, n.d.).

The *Kappes v. Guatemala* case is particularly relevant to this chapter's analysis due to its exclusive focus on the mining sector and the involvement of a US-based investor. Its significance is underscored by its explicit challenge to the procedural underpinnings of participation rights rulings made by domestic tribunals. Two shareholders of US nationality, holding interests in *Exploraciones Mineras de Guatemala* (Exmingua) and *Minerales KC Guatemala*—both mining enterprises—acted as claimants. In 2018, they submitted a dispute against the Guatemalan state, alleging that it violated the principles of national treatment, most-favoured-nation treatment, and the minimum standard of treatment as outlined in the

DR-CAFTA. These alleged violations arose from the claimants' characterisation of actions by the Guatemalan Constitutional Court and the Ministry of Energy and Mines (MEM) as arbitrary and discriminatory, especially in response to the suspension of their mining licences due to *amparos* stemming from an alleged failure to consult with local communities. The claimants contended that the respondents provided preferential treatment to similar mining projects run by domestically owned enterprises. They pointed to the *Oxec* and *Escobal* lawsuits as evidence that the Constitutional Court showed a more expedient and favourable disposition, allowing those operations to proceed while the MEM facilitated consultations. Conversely, claimants argued that in the *amparo* filed by local communities against *Exmingua and Minerales KC*, the court chose to suspend the mining companies' operations, adversely affecting their economic interests.

The Guatemalan state rebutted such allegations solely on procedural grounds. It contended that the tribunal should dismiss the claims due to a failure to meet the notice requirement under DR-CAFTA. While the case remains pending, it underscores the complexities of judicialising participation rights. Notably, the case's focus on the jurisdictional function of the Guatemalan state still requires the respondent state to rely on the adequacy of its domestic environmental regulations (Aldana & Abate, 2016: 665). Given that Chapter Seventeen of the DR-NAFTA permits state parties to establish their own environmental protection levels, having a "thick" environmental regulation in place is deemed essential in defending courts' non-discriminatory assessments of whether mining companies' EIAs have been lawfully granted. In other words, because DR-NAFTA allows unilateral regulation mainly in environmental matters, Guatemala can only demonstrate that its courts impartially evaluate the adherence of all mining companies—national or foreign—to participation rights primarily through a clearly defined and robust environmental framework that upholds these rights.

From this perspective, the latent threat of foreign investors resorting to investment arbitration—even after local communities secure victories in domestic courts—underlines the perception that regulating participation rights domestically, primarily through the environmental domain, is essential for realising these rights. Ultimately, having participation and FPIC robustly and explicitly incorporated within EIA procedures equips defendant states to shield themselves against unfavourable arbitration awards. Without such substantial means of defence, as seen in this case, Guatemala might find itself compensating companies like *Exmingua* and *Minerales*

KC sums surpassing USD 350,500,000—an amount roughly equivalent to the financial expenditure required to establish up to 17 fully equipped healthcare facilities in the country (UN General Assembly, 2023: para. 32).[26]

Participation before Mexican Courts: A Matter Beyond the Domestic

Mexico stands at a crossroads of immense mineral wealth and escalating socio-environmental tensions. According to the *Observatorio de Conflictos Mineros en América Latina* (n.d.), the nation has borne witness to 58 documented mining conflicts, a figure unmatched by any other Latin American country—Chile trails behind with 49 conflicts, and Peru with 46. Disturbingly, many of these mining disputes unfold within the territories of Indigenous communities (Serrano Carreto, 2006: 10), a significant note given that Mexico embraces the largest Indigenous demographic in the Americas, accounting for 16.9 million people or 15.1% of its population (IWGIA, n.d.). The striking juxtaposition of this country's mineral wealth—as section "The US Energy Transition and Latin America's Mineral Resource Endowment" outlined—with the stark challenges its Indigenous and environmental activists face paints a complex picture. The nation is, regrettably, renowned as one of the most dangerous places for human rights defenders (IWGIA, n.d.).

The rights to participation and to FPIC of populations impacted by mineral extraction, particularly Indigenous communities, are central to Mexico's socio-environmental conflicts. This was underscored during the 2017 visit to the country of Ms. Victoria Tauli Corpuz, then the UN Special Rapporteur on the rights of Indigenous Peoples. She identified this concern among the top five issues developing concurrently with Mexico's drive to boost foreign investment in mining over the past 20 years (UN Human Rights Council, 2018b: para 31). While the rights to participation and the Indigenous peoples' FPIC are both broadly anchored in the National Constitution and multiple secondary legislation,

[26] The estimated cost for a healthcare facility in Zacapa, Guatemala, is 150 million Guatemalan Quetzals, equivalent to around 19 million USD as of the exchange rate on September 25, 2023. This cost was reported by Bnamericas (2019), based on a public announcement made by the Guatemalan Institute for Social Security relative to its healthcare investment.

translating these written provisions into tangible reality has emerged as a significant challenge (CNDH, 2016: para. 38; UN Human Rights Council, 2018b: para 33).

The judiciary has been increasingly invoked to bridge this enforcement gap when disputes arise between affected communities and extractive initiatives. The Mexican Supreme Court has delivered seven landmark judgments to date concerning FPIC in relation to environmental issues, including those posed by mining projects (SCJN, n.d.). Notably, in most of these cases, the country's apex court has halted the projects due to failures of the Executive branch to meet the essential conditions ensuring that participation and FPIC are realised (CNDH, 2016: para. 105; UN Human Rights Council, 2018b: para 35). Such discrepancy between what the law mandates and what the government actually conducts is often predominantly ascribed by institutional commentators to the inadequacy of specific domestic regulations on participation and FPIC, especially in their alignment with international human rights standards.[27]

Consider, for example, Mexico's EIA as delineated in the LGEEPA and its related regulations. Instead of viewing participation as a foundational state duty, it is treated as discretionary.[28] More specifically, LGEEPA's regulatory framework on EIA provides no explicit guidelines on this right's implementation and pays little regard to its essential enabling conditions such as transparency and access to information.[29] Consequently, the Ministry of the Environment (the agency responsible for

[27] For example, in its General Recommendation 27/2016, the National Human Rights Commission emphasised the importance of state or federal legislation on FPIC, stipulating that they must contain at least the standards provided in international treaties on the human rights of Indigenous peoples (CNDH, 2016: para. 97). Consequently, it recommended that the General Congress study, discuss, and vote on legislation regarding FPIC. In the same vein, several legislatures and almost all national political parties have asserted that the remedy for the arbitrariness in which consultation processes occur in the country is the enactment of a law of general scope. Parliamentary members from different political parties have submitted proposals to the joint committees of Governance and Indigenous Affairs, reviewed by the Budget and Public Account Committee. However, their processing has not yet resulted in the enactment of a law (Fundación para el Debido Proceso and Oxfam Mexico, 2018: p. 50).

[28] Please, see footnote 20 above.

[29] For instance, Article 43 of the LGEEPA's Regulation on Environmental Impact Assessment bestows discretion upon the Ministry of the Environment and Natural Resources to decide whether or not to organise a public information meeting in instances involving works or activities with the potential to cause serious ecological imbalances or damage to public health or ecosystems.

implementing the EIA) is given leeway, sometimes sidestepping the enforcement of such rights and approving potentially harmful, polluting, or extractive projects—projects that local communities might have otherwise voiced concerns about, advocating for a different approach to their execution.

Given this course of action, tensions frequently arise once projects become operational and finalised, often leading to investment arbitration. In three out of the four investment dispute cases lodged against Mexico—cases either directly or indirectly related to participation rights—the adjudication has invariably been unfavourable to the nation. In each of these instances—which are not confined to mining but encompass polluting industries—communities have underscored the Ministry of the Environment's authorisation of the investors' EIA without conducting consultation. This omission has prompted municipalities to withhold local permits from investors, leading companies to claim "indirect expropriation" in breach of Mexico's FTA with both the US and Spain. While a comprehensive dissection of these cases—namely, *Abengoa v. Mexico* (2009), *TecMed v. Mexico* (2000), and *Metaclad v. Mexico* (1997)—would be beyond the scope of this chapter, it is important to underline a recurring theme: both the Mexican state and the investors' neglect of participation rights have been consistently disregarded by the ICSID proceedings. For this tribunal, the mere issuance of an EIA by the Ministry of Environment stands as irrefutable proof of the investors' commitment to addressing environmental issues and operating within legal parameters. The *TecMed v. Mexico* case, which was brought by a Spanish company to the ICSID in 2000 under the Mexico-Spain BIT, provides an example. Here, the tribunal concluded that the municipality's denial of local permits to the Spanish firm for running a landfill of hazardous industrial waste was more influenced by "socio-political concerns"—namely community opposition—than by "the danger that the [project] may pose to public health, ecological balance or the environment" (2003: para. 129).

Of course, addressing domestic regulatory failures by further incorporating the right to participation and FPIC in explicit alignment with international human rights standards may complicate the Ministry of the Environment's approval process for EIAs when consultations are bypassed. However, while this approach is undeniably necessary, it appears to be insufficient on its own. As evidenced by the fourth case levied against Mexico regarding participation rights, the Mexican government, despite rejecting projects based on the outcomes of consultations, has nonetheless not been spared from investment arbitration.

In 2019, the US company Odyssey initiated an arbitration claim under the 1992 NAFTA. It alleged breaches of the national treatment, minimum standard of treatment, and expropriation NAFTA articles resulting from the Ministry of the Environment's refusal to authorise a project related to the extraction of the largest phosphorite ore deposit in the Americas located in Mexico's Ulloa Gulf (2019: para. 123). Odyssey argued that this rejection was driven by biased political motivations. It highlighted that during the same period, the Ministry approved several environmentally contentious projects owned by Mexican companies. Conversely, the respondent state asserted its sovereign right to protect and regulate its environment (2021: para. 11). The country invoked the right to a clean environment and justified its denial of the EIA permit on various grounds. One of these was the outcome of the 2015 consultation proceedings for the claimant's EIA, which saw the participation of over 500 individuals, encompassing a diverse set of actors—from NGOs and local residents to fishermen and international organisations (2021: para. 269).

As of this writing, a resolution to this case is still pending. However, it is crucial to acknowledge that the ICSID's decision, whatever it may be, could potentially take precedence over the right to participation that the Mexican government aimed to realise through the aforementioned consultation. In this context, much like in the Guatemalan case, the arbitral award holds significant potential to either bolster or overshadow such fundamental rights.

The Role of Courts in Transitioning *All* Together

In discussing the cases from Guatemala and Mexico, this chapter has pointed out the complexities of judicialising participation rights amidst the transnational interests and power dynamics inherent in the US-centric energy transition. As affected communities increasingly turn to domestic courts as a last and influential recourse to assert their participation rights, it is becoming evident that such pathway may neither be the last nor the decisive alternative. The predominant neoliberal canon, along with the growing demand for critical minerals, heightens the prominence of international trade and investment law, making the turn to investment arbitration an increasingly latent threat. The *Kappes v. Guatemala* case exemplifies how, even when local communities successfully approach domestic courts to challenge their lack of consultation in the authorisation of resource extraction projects, their hard-fought victories are questioned and might be overturned through investment arbitration. Furthermore, Mexico's *Abengoa, TecMed*, and *Metaclad* cases emphasise that the realisation of participation rights is not necessarily upheld by investment tribunals unless these rights and their implementation statutes are clearly enshrined within a country's environmental regime. Given these experiences, one cannot deny the importance of countries implementing a robust environmental framework that incorporates both participation rights and FPIC. Not only is this foundational for guiding executive branches bridge the persistent gap between the written law and its practical application, but, crucially, it also serves as a defence mechanism for states in investor-state dispute settlement.

The *Odyssey v. Mexico* case presents a counter-factual scenario that shows how domestic environmental regulation, while essential, is not sufficient on its own. Even when public consultation is conducted, the lose-lose logic inherent in resorting to investment tribunals remains a latent risk for both local communities and states. Foreign investors, unsatisfied with consultation outcomes that reject their projects, can still find refuge in legal venues where the procedural dimension of environmental justice promoted by participation rights is not a foundational principle (UN General Assembly, 2023: para. 12). Importantly, the absence of human rights language in FTAs grants investment tribunals significant discretion when determining how they consider such legal principles in their awards (Steininger, 2018). While there have been instances where such tribunals invoke human rights laws to uphold the public interest actions of host states, such references are more often the exception than the rule[30] (Broude & Henckels, 2021: 101).

[30]The *Urbaser v. Argentina* case (2016) before the ICSID is the first to involve a Latin American country where an arbitration tribunal, citing international human rights instruments, determined that obligations such as the right to water can be directly imposed on international corporations. Urbaser, a shareholder in a concessionaire responsible for water supply and sewerage services, faced financial losses due to Argentina's emergency measures against its financial crisis, leading to the concessionaire's insolvency. Consequently, Urbaser initiated arbitral proceedings against Argentina. In response, Argentina filed a counterclaim, alleging that the concessionaire's failure to provide necessary investment in supply services resulted in violations of the human right to water (Pineda Rios, 2022: 89; Schacherer, 2018). This single case pales in comparison to the number of investment claims seen lodged against countries in Latin America (Bottini & Méndez Bräutigam, 2019: 408; Fernández Masiá, 2020: 115; ICSID, 2023: 12).

It is important to note that references in FTAs to environmental and public health objectives, framed as exception clauses to mitigate liabilities towards investors, do not *de facto* align with human rights law. From the cases analysed, it is evident that such clauses can even limit the latitude for states and investment tribunals to interpret participation rights in a protective manner. These clauses confine states to implement their human rights obligations predominantly, if not exclusively, through specific domestic environmental regulations[31] (Luke, 2019: 154; Viñuales, 2012), as previously discussed. Consequently, they narrow the interpretative focus of investment tribunals to the *minutiae* of such regulations. This implies that, without such regulations, tribunals are hindered from addressing blatant participation injustices committed by either investors or states (Bottini & Méndez Bräutigam, 2019: 438; Brown & Cucinotta, 2019: 197).

The US-led neoliberal tilt of our era suggests that the energy transition will face contrasting currents of interests woven into a complex and highly unequal matrix of power dynamics. As these dynamics operate both within Latin American countries and beyond, the transnational polities of the energy transition require transnational responses. Based on this premise, the chapter has consistently argued that harnessing the environmental and climate justice potential of participation rights requires *both* human-rights-oriented domestic environmental frameworks and strong global solidarities. The importance of this dual approach is underscored by the insufficiency of the individual components: while each component—robust rights-based EIAs and global solidarities—is necessary, neither is adequate on its own.

For instance, Mexico's recent amendment of its mining regulatory framework to strengthen the implementation of Indigenous communities' FPIC is a case in point. On May 9, 2023, changes to the Mining Law mandated that the Ministry of Economy must involve the National Institute of Indigenous Peoples in consultations, which must be funded by corporations seeking mining rights (Secretaría de Gobernación, 2023). However, these reforms did not extend to the LGEEPA's EIA, the regulatory instrument that operationalises the USMCA's exceptions for

[31] Although focused on a state-to-state dispute and not directly related to environmental concerns, Phillip Paiement (2018) offers a comprehensive illustration of the inherent limitations within investment arbitration in enforcing labour rights. His meticulous analysis of the first conflict between the US and Guatemala under the DR-CAFTA illuminates how the failure of arbitration tribunals to uphold labour rights predominantly stems from their often narrow and rigid interpretations of domestic labour laws (Paiement, 2018: 689).

imposing regulations on foreign investors (Article 1.3, Para. 1).[32] The effectiveness of such legislative changes thus remains to be seen, particularly in light of the pro-investor and hyper-legalistic interpretations typical of ICSID proceedings—as earlier exemplified in the *TecMed v. Mexico* case. Here is then where the role of global solidarity becomes crucial. While Canada, a USMCA party, negotiated its exclusion from the investor-state dispute settlement model in the agreement, Mexico did not (UN General Assembly, 2023: para. 19). That is, for the most part, Canadian and US investors in Mexico can still resort to investment arbitration in the face of grievances (Gantz, 2020: 48). Imagine a united stance between Mexico and Canada. This could not only eliminate the investor-state dispute settlement mechanism between Mexico and Canada but also create a counterbalance at the negotiating table with the US, thereby agreeing on a more equitable dispute settlement mechanism within the investment framework. Such a development would significantly support not only Mexico's but also Canada's and US's current national efforts towards realising participation rights.

Reimagining both domestic environmental regulations and international investment law concurrently and complementarily is crucial for counteracting the inequalities that the current racial capitalist trade and investment systems perpetuate and deepen to the detriment of historically marginalised populations. Aligning both fields with the principles and language of human rights can serve as a safeguard against the economic and political pressures that host states face from both internal and external wealthier economic actors. Such an alignment precludes the creation of "freeways" where private interests are allowed to override the public good, thus jeopardising the environment, the climate system, and the most vulnerable.

Concluding Remarks

As local communities' increasing reliance on national courts to safeguard their participation rights encounters the lose-lose logic embedded in the current interface between human rights and the trade and investment system, the proposition of a novel approach to international trade and investment law no longer appears "radical". Instead, it resonates with what the critical legal scholar Roberto Unger termed as "ad-hoc corrective

[32] Please, see footnote 18 above.

interventions", as it stands at an intermediate space of solution that avoids *extreme* responses (Unger, 1983: 626).

Viewed within Unger's parameters, the global system faces a choice: allow inequalities to accumulate unrestrictedly by not addressing the injustices of the current investment arbitration architecture or act immediately to correct them by dismantling the system entirely (Unger, 1983: 626). Extreme reactions correspond to either approach, exemplified by the increasing dispossession and displacement of racial minorities from their territories (UN General Assembly, 2022a), the withdrawal of some Latin American countries from the ICSID, or even the termination of their FTAs (Goodwin & Paiement, 2021: 86). Fortunately, recent steps towards avoiding extremes have materialised. These include the pursuit of more balanced FTA models by states such as The Netherlands (Goodwin & Paiement, 2021: 87), high-level political calls for a comprehensive reform of investor-state dispute settlement, focusing on the protection of human rights and the environment (UN General Assembly, 2023: para. 75), and the implementation of "The Hague Rules on Business and Human Rights". The potential of these measures is particularly promising. The European Union, for example, a key trade and investment partners to several Latin American countries, has intensified efforts to agreeing on new FTA and arbitration models, which are foreseen to unleash effective cross-regional solidarities in the pursuit of human rights protection (Pineda Rios, 2022).

Just consider the recent trend of integrating environmental and labour concerns into FTAs (Alam, 2015: 303; Luke, 2019: 153; Luttrell, 2019: 274). Mere decades ago, such amalgamation was deemed both theoretically and practically untenable (Desierto, 2022). However, these integrations have since become commonplace, fuelled by the growing awareness of the socio-environmental repercussions of cross-border trade and investment and by national political actors championing transnational solidarities. For example, the US Congress played a pivotal role in incorporating labour and environmental obligations into the country's FTAs (Brookfield & Di Pierdomenico, 2022: 232).

Similarly, in the context of the energy transition, the intricate power dynamics spanning transnationally have laid bare the limits of the judicial approach confined solely to the nation-state. This situation represents a significant bottleneck for environmental and climate justice within the current trade and investment system, which, if left unreformed, could increasingly intensify injustices and lead to global distrust—and thus nationalistic encapsulation—thereby obstructing the cooperative trade that energy transition requires. Therefore, paying special attention to participation rights in investment arbitration is not a radical response but rather a sine qua non for transitioning *all* together.

Acknowledgments This chapter is made possible by the TransLitigate project (ERC-2021-STG 101039648), funded by the European Union. Views and opinions expressed are however those of the author(s) only and do not necessarily reflect those of the European Union or European Research Council. Neither the European Union nor the granting authority can be held responsible for them.

References

Aguilar-Støen, M., & Hirsch, C. (2015). Environmental Impact Assessments, Local Power and Self-Determination: The Case of Mining and Hydropower Development in Guatemala. *The Extractive Industries and Society, 2*(3), 472–479. https://doi.org/10.1016/j.exis.2015.03.001

Alam, S. (2015). Trade and the Environment: Perspectives from the Global South. In C. G. Gonzalez, J. Razzaque, S. Alam, & S. Atapattu (Eds.), *International Environmental Law and the Global South* (pp. 297–316). Cambridge University Press. https://doi.org/10.1017/CBO9781107295414.015

Aldana, R., & Abate, R. S. (2016). Banning Metal Mining in Guatemala. *Vermont Law Review, 040*, 597–671.

Alviar García, H. (2022). Foreign Capital and Constitutions. In C. H. Mendes, R. Gargarella, & S. Guidi (Eds.), *The Oxford Handbook of Constitutional Law in Latin America* (p. 0). Oxford University Press. https://doi.org/10.1093/oxfordhb/9780198786900.013.22

Anglés Hernández, M., & Rovalo Otero, M. (2019). Mexico. In E. Lees & J. E. Viñuales (Eds.), *The Oxford Handbook of Comparative Environmental Law* (p. 336-C18.P59). Oxford University Press. https://doi.org/10.1093/law/9780198790952.003.0013

Barragán-Terán, D., & Erazo Robles, D. (2021). El Acuerdo de Escazú y el Convenio 169 de la OIT: Sinergias y desafíos para garantizar el derecho a la consulta en materia ambiental en Ecuador. In G. A. Rodríguez (Ed.), *Identificación de problemáticas para el tratamiento de los conflictos ambientales* (pp. 155–182). Grupo Editorial Ibáñez.

Bernal, A., Husar, J., & Bracht, J. (2023, April 7). Latin America's opportunity in critical minerals for the clean energy transition. *International Energy Agency*. Retrieved from https://tinyurl.com/58fs5jk2

Bnamericas. (2019, January 17). Guatemala revela planes para construir nuevos hospitales. *Bnamericas.* Retrieved from https://www.bnamericas.com/es/noticias/guatemala-revela-planes-para-construir-nuevos-hospitales

Bottini, G., & Méndez Bräutigam, E. (2019). Investment Claims Against Latin American States: Environmental Protection and the Applicable Law. In K. Miles (Ed.), *Research Handbook on Environment and Investment Law* (pp. 408–439). Edward Elgar Publishing. https://doi.org/10.4337/9781784714635.00024

Bown, C., & Kolb, M. (2023, June 1). Trump's trade war timeline: An up-to-date guide. *Peterson Institute for International Economics.* Retrieved from https://www.piie.com/sites/default/files/documents/trump-trade-war-timeline.pdf

Brookfield, R., & Di Pierdomenico, L. (2022). North American Trade: NAFTA and US-Caused Global Trade Tensions. In D. Bethlehem, D. McRae, R. Neufeld, & I. Van Damme (Eds.), *The Oxford Handbook of International Trade Law (2e)* (p. 211-C8.P85). Oxford University Press. https://doi.org/10.1093/oxfordhb/9780192868381.013.9

Broude, T., & Henckels, C. (2021). Not All Rights are Created Equal: A Loss–Gain Frame of Investor Rights and Human Rights. *Leiden Journal of International Law, 34*(1), 93–108. https://doi.org/10.1017/S092215652000062X

Brown, C., & Cucinotta, D. (2019). Treatment Standards in Environment-Related Investor-State Disputes. In K. Miles (Ed.), *Research Handbook on Environment and Investment Law*. Edward Elgar Publishing. https://doi.org/10.4337/9781784714635.00014

Castellanos Giracca, M., & Castellanos Howell, Á. R. (2020). Marco legal sobre las inversiones extranjeras y arbitraje de inversiones en Guatemala. In C. Esplugues Mota (Ed.), *Tratado de inversiones extranjeras y arbitraje de inversiones en Iberoamérica* (pp. 421–441). Tirant lo Blanch.

Cedillo, E. (2017). The Impact of Mexico's 2011 Human Rights Constitutional Amendment on Arbitral Practice: A View from Local Actors. In *Local Engagement with International Economic Law and Human Rights* (pp. 105–124). Edward Elgar Publishing. Retrieved from https://www.elgaronline.com/display/edcoll/9781785367182/9781785367182.00012.xml

Church, C., & Crawford, A. (2020). Minerals and the Metals for the Energy Transition: Exploring the Conflict Implications for Mineral-Rich, Fragile States. In M. Hafner & S. Tagliapietra (Eds.), *The Geopolitics of the Global Energy Transition* (pp. 279–304). Springer International Publishing. https://doi.org/10.1007/978-3-030-39066-2_12

Comisión Económica para América Latina (CEPAL). (n.d.). Acuerdo Regional sobre el Acceso a la Información, la Participación Pública y el Acceso a la Justicia en Asuntos Ambientales en América Latina y el Caribe. *Observatorio del Principio 10 en América Latina y el Caribe*. Retrieved from https://observatoriop10.cepal.org/es/tratado/acuerdo-regional-acceso-la-informacion-la-participacion-publica-acceso-la-justicia-asuntos

Comisión Interamericana de Derechos Humanos (CIDH). (2017). Situación de los derechos humanos en Guatemala (OEC/Ser.L/V/II Doc. 208/17). Retrieved from https://www.oas.org/es/cidh/informes/pdfs/Guatemala2017-es.pdf

Comisión Interamericana de Derechos Humanos (CIDH). (2020, September 8). CIDH presenta caso sobre Guatemala ante la Corte Interamericana. *Organization of American States*. Retrieved from https://www.oas.org/es/cidh/prensa/comunicados/2020/211.asp

Comisión Nacional de los Derechos Humanos (CNDH). (2016, July 11). Recomendación General No. 27/2016 sobre el derecho a la consulta previa de los pueblos y comunidades indígenas de la República Mexicana. Mexico. Retrieved from https://www.cndh.org.mx/sites/default/files/doc/Recomendaciones/generales/RecGral_027.pdf

Congress of the United States. (2021). Consolidated Appropriations Act, 2021. H.R. 133, 166th Congress. Retrieved from https://www.congress.gov/116/bills/hr133/BILLS-116hr133enr.pdf

Couso, J. A. (2006). The Changing Role of Law and Courts in Latin America: From an Obstacle to Social Change to a Tool of Social Equity. In R. Gargarella, P. Domingo, & T. Roux (Eds.), *Courts and Social Transformation in New Democracies* (pp. 61–79). Ashgate Publishing Limited.

Craik, N. (2019). The Assessment of Environmental Impact. In E. Lees & J. E. Viñuales (Eds.), *The Oxford Handbook of Comparative Environmental Law* (pp. 876–900). Oxford University Press. https://doi.org/10.1093/law/9780198790952.003.0039

de Casas, C. I. (2019). The Corporate Responsibility to Respect Consultation Rights in the Americas: How the Inter-American System Can Better Promote Free, Prior, and Informed Consent. In I. Feichtner, M. Krajewski, & R. Roesch (Eds.), *Human Rights in the Extractive Industries: Transparency, Participation, Resistance* (pp. 247–279). Springer International Publishing. https://doi.org/10.1007/978-3-030-11382-7_10

De Castro, F., Hogenboom, B., & Baud, M. (2016). Environment and Society in Contemporary Latin America. In F. De Castro, B. Hogenboom, & M. Baud (Eds.), *Environmental Governance in Latin America* (pp. 1–28). Palgrave Macmillan. https://doi.org/10.1057/978137505729

Desierto, D. A. (2022). Coming Full Circle on Human Rights in the Global Economy: International Economic Law Tools to Realize the Right to Development. *Loyola University Chicago International Law Review, 18*(1), 1–30.

EJAtlas. (n.d.). *Global Atlas of Environmental Justice*. Retrieved September 20, 2023, from https://ejatlas.org/

Energy Information Administration (EIA). (2022, March 14). Europe is a key destination for Russia's energy exports. *Today in Energy*. Retrieved from https://www.eia.gov/todayinenergy/detail.php?id=51618

Fajgelbaum, P., Goldber, P., Kennedy, P., Khandelwal, A., & Taglioni, D. (2023). *The US-China Trade War and Global Reallocations*. Working Paper 29562 (revised version). National Bureau of Economic Research (NBER). Retrieved from https://www.nber.org/system/files/working_papers/w29562/w29562.pdf

Fernández Masiá, E. (2020). La situación del arbitraje de inversiones en Latinoamérica: Base jurídica, evolución y situación presente. In C. Esplugues Mota (Ed.), *Tratado de inversiones extranjeras y arbitraje de inversiones en Iberoamérica* (pp. 111–136). Tirant lo Blanch.

Fundación para el Debido Proceso (DPLF) and Oxfam Mexico. (2018). *Implementación de la consulta y consentimiento previo, libre e informado*.

Gantz, D. A. (2020). *An Introduction to the United States-Mexico-Canada Agreement: Understanding the New NAFTA*. Edward Elgar Publishing. https://doi.org/10.4337/9781839105326

Gonzalez, C. G. (2015). Human Rights, Environmental Justice, and the North-South Divide. In A. Grear & L. J. Kotzé (Eds.), *Research Handbook on Human Rights and the Environment* (pp. 449–472). Edward Elgar Publishing.

Gonzalez, C. (2021). Racial Capitalism, Climate Justice, and Climate Displacement. *Oñati Socio-Legal Series, 11*. https://doi.org/10.35295/osls.iisl/0000-0000-0000-1137

Goodwin, M., & Paiement, P. (2021). Litigating the Crisis: Towards a Rebalancing of the Rights of Investors Versus Public Interest? In E. Aarts, H. Fleuren, M. Sitskoorn, & T. Wilthagen (Eds.), *The New Common: How the COVID-19 Pandemic is Transforming Society* (pp. 83–90). Springer International Publishing. https://doi.org/10.1007/978-3-030-65355-2

Heffron, R. J. (2022). Applying Energy Justice into the Energy Transition. *Renewable and Sustainable Energy Reviews, 156*, 111936. https://doi.org/10.1016/j.rser.2021.111936

Hirsch, M. (2021). Social Movements, Reframing Investment Relations, and Enhancing the Application of Human Rights Norms in International Investment Law. *Leiden Journal of International Law, 34*(1), 127–154. https://doi.org/10.1017/S0922156520000643

International Centre for Settlement of Investment Disputes (ICSID). (2023). *The ICSID Caseload—Statistics* (Issue 2023-1). World Bank Group.

International Energy Agency (IEA). (2022, October 24). *Latin America's share in the production and reserves of selected minerals, 2021*. https://tinyurl.com/9c8n3tp3

International Labour Organization. (n.d.). Ratifications of C169—Indigenous and Tribal Peoples Convention, 1989 (No. 169). *Normlex*. Retrieved from https://

www.ilo.org/dyn/normlex/en/f?p=NORMLEXPUB:11300:0::NO::P11300_INSTRUMENT_ID:312314

International Work Group for Indigenous Affairs (IWGIA). (n.d.). *Indigenous Peoples in Mexico*. Retrieved from https://www.iwgia.org/en/mexico.html

Kichwa Indigenous People of Sarayaku v. Ecuador. (2012). Inter-American Court of Human Rights. Retrieved from https://corteidh.or.cr/docs/casos/articulos/seriec_245_ing.pdf

Lachapelle, E., MacNeil, R., & Paterson, M. (2017). The Political Economy of Decarbonisation: From Green Energy 'Race' to Green 'Division of Labour'. *New Political Economy, 22*(3), 311–327. https://doi.org/10.1080/13563467.2017.1240669

Leruth, L., Mazarei, A., Régibeau, P., & Renneboog, L. (2022, August). Green energy depends on critical minerals. Who controls the supply chains? *Paterson Institute for International Economics Working paper 22-12*. Retrieved from https://www.piie.com/sites/default/files/documents/wp22-12.pdf

Lozada Gómez, M. (2020). Y los jueces nacionales, ¿qué? La omisión de los recursos internos como un rasgo imperialista del mecanismo de soluci6n de controversias en materia de protecci6n internacional a la inversión extranjera. In P. A. Acosta Alvarado, L. Betancur Restrepo, & E. Prieto Rios (Eds.), *Derecho internacional: Investigación, estudio y enseñanza.Tomo 3. Economía, medio ambiente y desarrollo frente al derecho internacional* (pp. 123–154). Universidad del Rosario. https://doi.org/10.12804/tj9789587844221

Luke, E. (2019). Environment and Human Rights in an Investment Law Frame. In K. Miles (Ed.), *Research Handbook on Environment and Investment Law*. Edward Elgar Publishing. https://doi.org/10.4337/9781784714635.00012

Luttrell, S. (2019). Green Multilateralism: 'Mega FTAs' and the Changing Interface Between Environmental Regulation and Investment Protection. In K. Miles (Ed.), *Research Handbook on Environment and Investment Law*. Edward Elgar Publishing. https://doi.org/10.4337/9781784714635.00017

Martinez-Alier, J., & Walter, M. (2016). Social Metabolism and Conflicts over Extractivism. In F. De Castro, B. Hogenboom, & M. Baud (Eds.), *Environmental Governance in Latin America* (pp. 58–85). Palgrave Macmillan. https://doi.org/10.1057/978137505729

Morales Antoniazzi, M., & Barraco, M. (2020). Aproximación al ius commune: Interamericanización en los derechos de participación y defensa del medio ambiente. In H. Jiménez Guanipa & M. Luna Leal (Eds.), *Crisis climática, transición energética y derechos humanos. Tomo II: Protección del medio ambiente, derechos humanos y transición energética* (pp. 15–42). Fundación Heinrich Böll.

Newell, P. (2019). Trasformismo or Transformation? The Global Political Economy of Energy Transitions. *Review of International Political Economy, 26*(1), 25–48. https://doi.org/10.1080/09692290.2018.1511448

Observatorio de Conflictos Mineros en América Latina. (n.d.). *Conflictos Mineros en América Latina*. Retrieved from https://mapa.conflictosmineros.net/ocmal_db-v2/

Odyssey v. Mexico. (2019). *Notice of Arbitration* (Case No. UNCT/20/1). International Centre for Settlement of Investment Disputes.

Odyssey v. Mexico. (2021). *Counter Memorial* (Case No. UNCT/20/1). International Centre for Settlement of Investment Disputes.

Office of the United States Trade Representative (USTR). (n.d.). *Free Trade Agreements*. Retrieved from https://ustr.gov/trade-agreements/free-trade-agreements

Organization of American States. (n.d.-a). Information by Country. *Foreign Trade Information System*. Retrieved from http://www.sice.oas.org/ctyindex/GTM/GTMagreements_e.asp

Organization of American States. (n.d.-b). Basic Documents in the Inter-American System. *Inter-American Commission on Human Rights*. Retrieved from https://www.oas.org/en/IACHR/jsForm/?File=/en/iachr/mandate/basic_documents.asp

Our World in Data. (2020). *Fertilizer Use per Hectare of Cropland, 2020*. Retrieved from https://tinyurl.com/2t3xbeeh

Paiement, P. (2018). Leveraging Trade Agreements For Labor Law Enforcement: Drawing Lessons From the US-Guatemala CAFTA Dispute. *Georgetown Journal of International Law, 49*, 675–692.

Parra-Vera, Ó. (2016). The Protection of Social Rights. In J. F. Gonzalez-Bertomeu & R. Gargarella (Eds.), *The Latin American Casebook. Courts, Constitutions, and Rights* (pp. 147–171). Routledge.

Pastukhova, M., & Westphal, K. (2020). Governing the Global Energy Transformation. In M. Hafner & S. Tagliapietra (Eds.), *The Geopolitics of the Global Energy Transition* (pp. 341–364). Springer International Publishing. https://doi.org/10.1007/978-3-030-39066-2_15

Pineda Rios, D. (2022). El arbitraje internacional como mecanismo para desarrollar el tercer pilar, la reparación. Experiencias cruzadas entre la Unión Europea y América Latina. In D. Iglesias Márquez & A. L. Walter de Santana (Eds.), *Derechos humanos y empresas: Retos y debates multidisciplinarios en Latinoamérica* (pp. 87–98). Tirant Lo Blanch.

Pou Giménez, F. (2022). The Constitution of Mexico. In C. H. Mendes, R. Gargarella, & S. Guidi (Eds.), *The Oxford Handbook of Constitutional Law in Latin America* (p. 203-C9.P60). Oxford University Press. https://doi.org/10.1093/oxfordhb/9780198786900.013.10

Price, J. (2022, August 16). Why the US Inflation Reduction Act could benefit both mining and energy in Latin America. *Americas Market Intelligence*. Retrieved from https://tinyurl.com/cxjwbapw

Puig, S. (2021). *At the Margins of Globalization: Indigenous Peoples and International Economic Law*. Cambridge University Press. https://doi.org/10.1017/9781108596503

S&P Global. (2023). *Inflation Reduction Act: Impact on North America Metals and Minerals Market*. Final Report. Retrieved from https://cdn.ihsmarkit.com/www/prot/pdf/0823/Impact-IRA-Metals-Minerals-Report-FINAL-August2023.pdf

Schacherer, St. (2018, October 18). Urbaser v. Argentina. *Investment Treaty News. International Institute for Sustainable Development*. Retrieved from https://www.iisd.org/itn/en/2018/10/18/urbaser-v-argentina/

Schilling-Vacaflor, A. (2019). Norm Contestation and (Non-)Compliance: The Right to Prior Consultation and FPIC in the Extractive Industries. In I. Feichtner, M. Krajewski, & R. Roesch (Eds.), *Human Rights in the Extractive Industries: Transparency, Participation, Resistance* (pp. 311–337). Springer International Publishing. https://doi.org/10.1007/978-3-030-11382-7_12

Seck, S. L. (2015). Transnational Corporations and Extractive Industries. In C. G. Gonzalez, J. Razzaque, S. Alam, & S. Atapattu (Eds.), *International Environmental Law and the Global South* (pp. 380–398). Cambridge University Press. https://doi.org/10.1017/CBO9781107295414.019

Secretaría de Gobernación. (2023, May 8). Decreto por el que se reforman, adicionan y derogan diversas disposiciones de la Ley Minera, de la Ley de Aguas Nacionales, de la Ley General del Equilibrio Ecológico y la Protección al Ambiente y de la Ley General para la Prevención y Gestión Integral de los Residuos, en materia de concesiones para minería y agua. *Diario Oficial de la Federación*. Retrieved from https://www.dof.gob.mx/nota_detalle.php?codigo=5688050&fecha=08/05/2023#gsc.tab=0

Serrano Carreto, E. (Ed.). (2006). *Regiones indígenas de México*. Comisión Nacional para el Desarrollo de los Pueblos Indígenas.

Sieder, R. (2007). The Judiciary and Indigenous Rights in Guatemala. *International Journal of Constitutional Law, 5*. https://doi.org/10.1093/icon/mom007

Spalding, R. J. (2023). Mining, Maybe: Intermittent Mining in Guatemala. In R. J. Spalding (Ed.), *Breaking Ground: From Extraction Booms to Mining Bans in Latin America*. Oxford University Press. https://doi.org/10.1093/oso/9780197643150.003.0004

Statista. (2021). *Mine Production of Nickel in Guatemala from 2012 to 2021*. Retrieved from https://tinyurl.com/2f34m5x4.

Steininger, S. (2018). What's Human Rights Got To Do With It? An Empirical Analysis of Human Rights References in Investment Arbitration. *Leiden Journal of International Law, 31*(1), 33–58. https://doi.org/10.1017/S0922156517000528

Suprema Corte de Justicia de la Nación (SCJN). (n.d.). *Sentencias relevantes en materia de Derechos Humanos: Personas, pueblos y comunidades indígenas.* Retrieved from https://www.scjn.gob.mx/derechos-humanos/buscadores-juridicos/sentencias-relevantes-en-materia-de-derechos-humanos/1299?field_tema_value=consulta&field_sinopsis_value=ambiente&field_numero_de_expediente_value=

TecMed v. Mexico. (2003). *Award* (Case No. ARB (AF)/00/2). International Centre for Settlement of Investment Disputes.

The White House. (2021a). Executive Order on America's Supply Chains. *Briefing Room—Presidential Actions.* Retrieved from https://tinyurl.com/3dhre7bd

The White House. (2021b). *Building Resilient Supply Chains, Revitalizing American Manufacturing, and Fostering Broad-Base Growth.* Retrieved from https://tinyurl.com/56vrffde

The White House. (2021c, November 6). Fact Sheet: The Bipartisan Infrastructure Deal. *Briefing Room—Statements and Releases.* Retrieved from https://tinyurl.com/2r7k26v8

The White House. (2022, October 19). Fact Sheet: Biden-Harris Administration Driving US Battery Manufacturing and Good-Paying Jobs. *Briefing Room—Statements and Releases.* Retrieved from https://tinyurl.com/y8rmb3zd

The White House. (2023a, April 17). Fact Sheet: Biden-Harris Administration Announces New Private and Public Sector Investments for Affordable Electric Vehicles. *Briefing Room—Statements and Releases.* Retrieved from https://tinyurl.com/bdz8a4u5

The White House. (2023b, January). *Building a Clean Energy Economy: A Guidebook to the Inflation Reduction Act's Investments in Clean Energy and Climate Action.* Retrieved from https://tinyurl.com/bdz5mr9n

UN Comtrade. (2021). Exports of Crude Oil and Nitrogen Fertilizers Per Country, 2021. *UN Comtrade Database.* Retrieved from https://comtradeplus.un.org/TradeFlow

UN Conference on Trade and Development (UNCTAD). (n.d.). Guatemala. *Investment Dispute Settlement Navigator.* Retrieved from https://investmentpolicy.unctad.org/investment-dispute-settlement/country/86/guatemala

UN General Assembly. (2022a). *The Human Right to a Clean, Healthy and Sustainable Environment (A/76/L.75).* Retrieved from https://digitallibrary.un.org/record/3982508?ln=en

UN General Assembly. (2022b). *Contemporary Forms of Racism, Racial Discrimination, Xenophobia and Related Intolerance (A/77/549).*

UN General Assembly. (2023). *Paying Polluters: The Catastrophic Consequences of Investor-State Dispute Settlement for Climate and Environment Action and Human Rights (A/78/168).*

UN Human Rights Council. (2013). *Report of the Independent Expert on the Issue of Human Rights Obligations Relating to the Enjoyment of a Safe, Clean, Healthy*

and Sustainable Environment, John H. Knox (A/HRC/25/53). Retrieved from https://www.ohchr.org/en/special-procedures/sr-environment/annual-thematic-reports
UN Human Rights Council. (2018a). *Report of the Special Rapporteur on the Rights of Indigenous Peoples on Her Visit to Guatemala (A/HRC/39/17/Add.3)*. Retrieved from https://www.ohchr.org/en/special-procedures/sr-indigenous-peoples/country-visits
UN Human Rights Council. (2018b). *Report of the Special Rapporteur on the Rights of Indigenous Peoples on Her Visit to Mexico (A/HRC/39/17/Add.2)*. Retrieved from https://www.ohchr.org/en/special-procedures/sr-indigenous-peoples/country-visits
UN Office of Legal Affairs. (n.d.). Multilateral Treaties Deposited with the Secretary General—Chapter IV: Human Rights. *UN Treaty Collection*. Retrieved from https://treaties.un.org/Pages/Treaties.aspx?id=4&subid=A&clang=_en
Unger, R. M. (1983). The Critical Legal Studies Movement. *Harvard Law Review, 96*(3), 561–675. https://doi.org/10.2307/1341032
United Nations. (1948). Universal Declaration of Human Rights. United Nations General Assembly. Retrieved from https://www.un.org/en/about-us/universal-declaration-of-human-rights
United Nations. (n.d.). Access Information on Multilateral Environmental Agreements—Parties. *InforMEA*. Retrieved from https://www.informea.org/en/countries
US Department of the Treasury. (2023, March 21). Treasury Releases Proposed Guidance to New Clean Vehicle Credit to Lower Costs for Consumers, Build US Industrial Base, Strengthen Supply Chains. *Press Release*. Retrieved from https://home.treasury.gov/news/press-releases/jy1379
Us, H., Mendoza, C., & Guzmán, V. (2021). *Pueblos indígenas en Guatemala: desafíos demográficos, lingüísticos y socioeconómicos*. Banco Interamericano de Desarrollo.
Viñuales, J. E. (2012). *Foreign Investment and the Environment in International Law*. Cambridge University Press. https://doi.org/10.1017/CBO9780511902567
von Bogdandy, A. (2017). Ius Constitutionale Commune en América Latina: Observations on Transformative Constitutionalism. In A. von Bogdandy, E. Ferrer Mac-Gregor, M. Morales Antoniazzi, & F. Piovesan (Eds.), *Transformative Constitutionalism in Latin America. The Emergence of a New Ius Commune* (pp. 27–48). Oxford University Press.
Walter, M., & Urkidi, L. (2016). Community Consultations: Local Responses to Large-Scale Mining in Latin America. In F. De Castro, B. Hogenboom, & M. Baud (Eds.), *Environmental Governance in Latin America* (pp. 287–325). Palgrave Macmillan. https://doi.org/10.1057/978137505729

Yamin, A. E. (2022). The Right to Health. The Potential and Limits of Catalysing Systemic Change through the Courts. In C. Hübner Mendes, R. Gargarella, & S. Guidi (Eds.), *The Oxford Handbook of Constitutional Law in Latin America* (pp. 758–781). Oxford University Press.

Zettelmeyer, J., Tagliapietra, S., Zachmann, G., & Heussaff, C. (2022). Beating the European Energy Crisis. *International Monetary Fund*. Retrieved from https://tinyurl.com/y4j4752m

CHAPTER 3

The Right to a Decent Standard of Living: A Practical Tool to Guarantee Basic Needs at the End of the Fossil Fuel Era

Ana García-Juanatey, Bettina Steible, and Luca Volpi

Introduction

At present, despite evidence that exceeding 1.5 °C global warming could result in surpassing various climate tipping points (Armstrong McKay et al., 2022), everything indicates that humanity will go on burning fossil fuels as long as possible, unchaining devastating impacts on human communities around the world. The current outlook is that the rise in global temperature will exceed 1.5 °C during the twenty-first century and that limiting warming below 2 °C will be very challenging at this point (IPCC,

A. García-Juanatey (✉)
CEI International Affairs, University of Barcelona, Barcelona, Spain
e-mail: ana.garcia@ceibcn.com

B. Steible
Universitat Oberta de Catalunya, Barcelona, Spain

L. Volpi
Societat Orgànica, Barcelona, Spain

D. Iglesias Márquez et al. (eds.), *Legal Challenges at the End of the Fossil Fuel Era*, Global Issues,
https://doi.org/10.1007/978-3-031-61766-9_3

2023). This will trigger devastating humanitarian impacts, especially for vulnerable communities living in the Global South. According to UN General Secretary António Guterres, "the climate crisis is the biggest threat to our survival as a species and is already threatening human rights around the world" (2020).

Moreover, climate change is not the only systemic problem that humanity is facing, as there are other interlinked threats—such as biodiversity loss, resource depletion and pollution, among others—which are also closely related to the use of fossil fuels. In fact, climate change is described by some to be not "the problem" but just a "symptom of a much larger dysfunction" (Hagens, 2020). In line with this view, the role played by the intensive and growing use of fossil fuels is of crucial importance to explain different overlapping crises that impinge upon human society. At present, industrial societies based on infinite growth are deeply dependent on fossil fuels. Consequently, the use of these fuels has deteriorated the natural world and altered biophysical processes that sustain life on Earth, as it is the main cause behind the current climate change and it has an important role in the surpassing of other planetary boundaries, such as biodiversity loss and others (Rockström et al., 2009).

On the other hand, another side of this intensive use of fossil fuels is its progressive depletion, which will have an enormous impact on human economic and commercial structures, on the current standards of living and, also, on human rights. Even though estimates of remaining oil and natural gas vary widely (Mohr et al., 2015), it is a well-established fact that the easily accessible high-quality oil has mostly been already extracted (Masnadi & Brandt, 2017). Since welfare and, more in general, human wellbeing have been built on fossil fuel-dependent infrastructures, the growing scarcity of these fuels—that will generate a tendency to higher and more fluctuating prices—will have direct and indirect impacts on human rights, as it affects the availability of goods and services necessary to guarantee human dignity (access to sufficient and adequate food, decent housing, etc.). Furthermore, despite dominant narratives about an ongoing energy transition that will achieve net zero emissions in the near future, there is growing evidence that the current transition will not be able to shift our dependence on fossil fuels in key sectors (Smil, 2010, 2022). Therefore, and to put it short, we are at a point where "multiple interrelated risks all point to an impending, imposed reduction in energy/material throughput in coming decades" (Hagens, 2020: 12).

Against this backdrop, this chapter aims to discuss these vital challenges through the lenses of a human right of renewed relevance in the current context: the right to an adequate standard of living. This social and economic right, recognised in Article 25 of the Universal Declaration of Human Rights (UDHR) and Article 11 of the International Covenant on Economic, Social and Cultural Rights (ICESCR), is gaining special momentum, as the consequences of climate change and the fossil fuels depletion will significantly alter the possibilities of satisfying basic material needs, such as food, water, and housing. Therefore, the primary purpose of this chapter is to examine the role of the right to an adequate standard of living in guaranteeing basic needs depending on energy in the current context. Thus, our premise is that the absolute standards implied in this right could help to navigate these challenges in the next few years, giving priority to the satisfaction of basic needs, such as access to water, food, and shelter. In addition, this work also discusses the limitations of the human rights framework to integrate the needs of future generations as well as the concept of planetary boundaries.

This chapter starts by defining the right to an adequate standard of living, its content, and the corresponding state obligations (section "Introduction"). In this respect, the current Anthropocene era, characterised by the impact of human beings on Earth, challenges the notion of 'adequate standard', which needs to be reframed in light of society's fossil fuel dependence and planetary boundaries (section "The Right to an Adequate Standard of Living"). Lastly, this chapter concludes with a proposal to adapt the right to an adequate standard of living to these challenges (section "An Adequate Standard of Living at the End of the Fossil Fuel Era").

The Right to an Adequate Standard of Living

Formulated by US President Franklin Roosevelt as "freedom from want" in his address to the United States Congress in 1941, the right to an adequate standard of living was first recognised at the international level in the UDHR. The latter refers to Roosevelt's concept "as the highest aspiration of the common people" in its preamble and specifically enshrines the right to an adequate standard of living in Article 25(1). However, as the UDHR was not legally binding, an international human rights bill was discussed and led to the adoption of the 1966 Covenants. For present purposes, the ICESCR of 1966 defined in Article 11(1) the right to an adequate standard of living as follows: "The States Parties to the present Covenant recognise the right of everyone to an adequate standard of living for himself and his family, including adequate food, clothing and housing, and to the continuous improvement of living conditions. The States Parties will take appropriate steps to ensure the realisation of this right, recognising to this effect the essential importance of international cooperation based on free consent".

With this wording, the drafters of the ICESCR put the right to an adequate standard of living at the heart of the universal system of protection of human rights. As observed by Hohmann and Goldblatt, both the projects of improving standards of living and guaranteeing human rights through international covenants lie at the heart of the post-war UN project (2021). Indeed, the UN Charter acknowledges such endeavour as its preamble recognises the need to "promote social progress and better standards of life in larger freedom". More importantly, pursuant to Article 55 of the Charter, "higher standards of living, full employment, and conditions of economic and social progress and development" are considered as necessary conditions "of stability and well-being (…) necessary for peaceful and friendly relations among nations". As underlined by Aguiar, the idea that decent standards of living are a matter of human rights and contribute to guaranteeing freedom was also echoed by several prominent scholars at the time, such as Wells and Lauterpacht (2018: 46).

Such connection between improved standards of living and human rights was then reflected in the wide number of treaties and policy documents that refer to the right to an adequate standard of living or to one or several of its components. Explicit references to this right as such are indeed found in the Convention on the Elimination of All Forms of Discrimination against Women (Art. 14(2)), the International Convention on the Protection of the Rights of all Migrant Workers and Members of their Families (Art. 43(1)), the Convention on the Rights of the Child (Art. 27), and the Convention on the Rights of Persons with Disabilities (Art. 28(1)). Furthermore, the idea of improving or raising people's standards of living is present in the American Declaration on the Rights and Duties of Man, the European Social Charter, the African Charter, the Phnom Penh Statement on the adoption of the Association of Southeast Asian Nations (ASEAN) Human Rights Declaration, and the Arab Charter on Human Rights (Hohmann & Goldblatt, 2021). Likewise, even though regional texts do not explicitly refer to the right to an adequate standard of living in the body of the treaties, they also make reference to the rights to food, to housing, or to other connected rights such as the rights to work or to social security. This wide recognition of the right to an adequate standard of living at universal and regional levels, therefore, demonstrates the importance given to this right in the aftermath of World War II.

Scope and Normative Content

Despite its importance in the post-war context, the content of the right to an adequate standard of living as an autonomous right has been insufficiently explored by international monitoring bodies and the scholarship. Indeed, the ICESCR does not provide indications as to what constitutes an 'adequate standard of living', and the Committee on Economic Social and Cultural Rights (CESCR) has not issued any General Comment that specifically addresses the scope and corresponding obligations of this right. Against this lack of authoritative interpretation, the scholarship has attempted to define it.

Regarding the nature of this right, Henry Shue echoes Roosevelt's famous quote that "necessitous men are not free men" and considers that subsistence rights should be interpreted as basic rights, just like physical security (1993, 2020). As for its content, Eide considers that it entails the ability of everyone "to enjoy their basic needs under conditions of dignity" (2014: 196). As a minimum, it requires that "everyone shall enjoy the necessary subsistence rights: adequate food and nutrition rights, clothing, housing and the necessary conditions of care when such are required" (Eide & Barth Eide, 1999: 523). In the same vein, Shue understands the right to an adequate standard of living as "subsistence rights" that guarantee "unpolluted air, unpolluted water, adequate food, adequate clothing, adequate shelter, and minimal preventive public health care" (1993/2020). The underpinning would be "to have available for consumption what is needed for a decent chance at a reasonably healthy and active life or more or less normal length, barring tragic interventions" (Shue, 1993, 2020). Likewise, Ssenyonjo underlines the idea that no one should "resort to degrading means, such as begging, sex work, child or bonded labour, as the only way to realise access to basic subsistence needs" (2016).

In addition, following the wording of Art. 11(1) ICESCR, the right to an adequate standard of living is normally interpreted as comprising essentially social goods, namely the rights to food, water, and adequate housing. Even though UN monitoring bodies have not provided much detail about the right to an adequate standard of living as such, the same does not hold true regarding its components. Indeed, the CESCR has issued general comments on the three components that are generally understood to make up the right to an adequate standard of living, namely General Comment 12 on the right to adequate food, General Comment 15 on the right to water, and Generals Comments 4 and 7 on the right to adequate housing. Through these General Comments, the CESCR has provided an authenticated interpretation concerning the realisation of these rights and the corresponding obligations falling on states under international law.

According to the CESCR, the right to adequate food is at the core of the right to an adequate standard of living. It should not be understood "in a narrow or restrictive sense which equates it with a minimum package of calories, proteins, and other specific nutrients" and should be realised progressively (General Comment 12, para 6). Additionally, the right to food comprises an absolute minimum standard, the right to be free from hunger and malnutrition under Art. 11(2). Therefore, the right to food is violated in case a state fails to demonstrate that all efforts have been made to satisfy this minimum standard with the resources at its disposal. As for the right to adequate housing, it goes beyond "merely having a roof over one's head" and should be understood as "the right to live somewhere in security, peace, and dignity" (General Comment 4, para. 7). Lastly, even though the right to water is not expressly included in Art. 11(1), it is implied as falling under the scope of the right to an adequate standard of living as it is "one of the most fundamental conditions for survival" and is inextricably related to the other components of the right to an adequate standard of living (General Comment 15, para. 3). It should not be interpreted "by mere reference to volumetric quantities and technologies", but also "as a social and cultural good, and not primarily as an economic good" (General Comment 15). It entitles "everyone to sufficient, safe, acceptable, physically accessible and affordable water for personal and domestic uses" (General Comment, para. 2).

In addition, although it is often "forgotten" by UN bodies and scholars alike (Lott, 2021; Hohmann, 2021), Article 11.1 ICESCR explicitly refers to the right to "the continuous improvement of living conditions" as an integral part of the right to an adequate standard of living (General Comment 4, para. 11). As discussed in the third section of this chapter, this right may be at odds with the limits imposed by finite resources of a finite planet, depending on how it is interpreted. In particular, it raises the question as to whether anyone may enjoy an unlimited right to the improvement of their living conditions, regardless of their socioeconomic status and of the available resources. This dimension of the right to an adequate standard of living, therefore, needs to be explored and debated in a context of climate change and scarcity of fossil fuel resources, all the more as the CESCR has not issued any specific General Comment on this issue.

State Obligations

In terms of obligations, all the definitions of the right to an adequate standard of living point to the idea of ensuring a minimum level of living conditions in line with human dignity, although the use of criteria such as 'adequate' or 'reasonable' still leaves room for interpretation on these minimum thresholds. In this respect, the CESCR held in it is General Comment 12 that the term 'adequate' "is to a large extent determined by prevailing social, economic, cultural, climatic, ecological and other conditions" (CESCR, 1999). The right to an adequate standard of living and its components, therefore, entails the idea of relative standards that must be adapted to the society at stake. Important variations will indeed be found among states in the fulfilment of this right, depending on their level of social and economic development and on the different components of the right to an adequate standard of living. As Eide observed, there are often differences regarding the role that is expected from the state between the different components (2014: 198): for example, interventionist policies are generally expected in the field of health with the provision of social security schemes, while the role of the state in the fulfilment of the right to housing is often more limited. Finally, the right to be free from hunger or the right to water also includes absolute minimum standards that must be guaranteed under any circumstances.

Under international human rights law, states have the obligation to respect, protect, and fulfil the human rights of any person found under their jurisdiction. It should nonetheless be noted that in the case of the right to an adequate standard of living states' obligations are subsidiary to individuals' efforts (Eide, 2014: 197). It is only in case of default on individuals' ability to secure their standards of living that the state—and, at last resort, the international community—should intervene. In this context, states must first and foremost respect individuals' freedom to take the necessary action to fulfil their means of subsistence; then, in a similar fashion to civil and political rights, states must protect individuals' freedom of choice as to the use of resources necessary to fulfil their means of subsistence; and finally, if necessary, states must act as providers in order to ensure that the right to an adequate standard of living is guaranteed for everyone. Usually, the most important obligation of the three in relation to the rights to food, housing, and water requires the state to act as a protector, not as a provider (Eide, 2014: 198). Consequently, the fulfilment of the right to an adequate standard of living will often require a combination of enabling strategies towards the private sector coupled with more interventionist policies.

Furthermore, art. 2(1) ICESCR acknowledges the constraints due to the limited available resources and recognises the principle of progressive realisation of economic, social, and cultural rights, including the right to an adequate standard of living. As Eide underlined, these rights were adopted as standards of achievement (2014: 213). Nonetheless, this principle should not be interpreted so as to deny the legally binding character of economic, social, and cultural rights and use them simply as moral principles orienting public policies. More concretely, according to the CESCR, it should allow states to reach a balance between "the difficulties for any country in ensuring full realisation of economic, social and cultural rights" (General Comment 3), on the one hand, and the obligations stemming from these rights on the other. In particular, states must indeed take steps towards the realisation of the right "as expeditiously and effectively as possible towards that goal" and must respect the principle of non-discrimination at all times. Both obligations are of immediate effect.

Finally, the principle of non-regression enshrined in Art. 2 ICESCR applies. It prohibits states from adopting "deliberately retrogressive measures" in connection with economic, social, and cultural rights. It should nonetheless be noted that this principle is not absolute, as some limitations may be accepted if they are properly justified (General Comment 3, para. 9). Furthermore, as Skogly (2021) reported, a Statement of the CESCR of 2016 seems to restrict this principle to the most vulnerable groups, in order to ensure that retrogressive measures do not negatively impact "disadvantaged and marginalised groups" and "mitigate inequalities" (UN Doc E/C.12/2016/1, para. 4).

An Adequate Standard of Living at the End of the Fossil Fuel Era

Standard of Living and Fossil Fuel Dependence

The concept of a 'standard of living' and its relation to wellbeing have significantly evolved over time. In the tradition of classical economics (David Ricardo and Karl Marx, among others), this concept originally indicated an "irreducible minimum income, a subsistence level of living", associated with the cost of the reproduction of the workforce (Latouche, 1992). The concept of standard of living was still defined in this sense as late as 1934 in the Encyclopedia of the Social Sciences, but its meaning moved from that "irreducible minimum", in the sense of avoiding extreme poverty, to a "desired level of consumption" in the post-war period. Therefore, since that time, in both popular and scientific domains, 'standard of living' generally refers to material wellbeing and is regarded as a measurable concept, similar to per capita gross national product (Latouche, 1992).

Behind this evolution of the concept to a desired level of consumption stands the fundamental role of fossil fuels, which allowed for the unprecedented rise of the standards of living in the post-war period. In fact, the role of these forms of energy in powering modern societies was even increased in the period after the *trente glorieuses* (1945–1975): according to the International Energy Association (IEA) between 1971 and 2019, the world total energy supply increased 2.6 times and in the last two decades of that period, around 80% of all final energy consumed in the world came from fossil fuels (IEA, 2021). This increasing consumption has deeply transformed human societies, based on a "fossil subsidy", that made standards of living considerably higher compared to previous civilisations based on renewable flows of energy (Hagens, 2020). Thus, the widespread availability of cheap fossil fuels in the second half of the twentieth century led to the development of industrialised economies and the growth of global trade. Indeed, the liberation of an important part of human labour, previously dedicated to agriculture and industrial production, opened space to more professions in education and care, in turn, funded by tax income based on activities that require a lot of energy. As a result, the entire infrastructure of human activity is today based on the availability of cheap fossil fuels, so that our current society can be described as a "fossil fuel society" (Rifkin, 2019).

The dark side of this story is twofold: on the one hand, the use of fossil fuels has deteriorated the natural world to the point of altering the same biophysical processes that sustain life on Earth, it is the main cause behind the current climate change and it has an important role in the surpassing of other planetary boundaries (Rockström et al., 2009). In line with the latest evidence, humanity is surpassing at least five out of nine planetary boundaries: climate change; the alteration of the nitrogen and phosphorus cycles; deforestation and changes in land use; the loss of biological diversity; and the release of novel entities in the environment (Persson et al., 2022). Most importantly, in line with this research, surpassing just one of them may irreversibly alter the different balances on which fundamental processes depend to guarantee life in the biosphere and may produce an abrupt change (Rockström et al., 2009). Therefore, more than analysing the impacts that surpassing planetary boundaries may have on humans' standard of living, it would be more appropriate to frame the discourse around the existential threats deriving from the altering of the biophysical planetary balance. To put it simply, these challenges, which derive from the constant growth in material consumption, affect the capacity to satisfy human "existential necessities", such as water and food (Smil, 2022: 8).

On the other hand, the massive use of fossil fuels has also made us deeply dependent on their consumption. At present, despite efforts to decarbonise economies in the last decades, fossil fuels still account for more than 80% of globally consumed primary energy. Moreover, despite dominant narratives about an ongoing energy transition that will achieve net zero emissions in the near future, there are growing evidence that the current transition will not be able to shift our dependence on fossil fuels in key sectors (Khanna et al., 2017; Smil, 2010, 2022). For instance, there are many high heat industrial processes using fossil fuels that have no current alternative using low-carbon technology (Khanna et al., 2017). In fact, as Vaclav Smil famously argues, the "pillars of our civilisation"—namely the production of steel, ammonia, concrete, and plastic—are directly dependent on these fuels and their production on its current scale will be impossible to maintain without them (Smil, 2010). According to this author, a complete decarbonisation of the electric grid is very challenging, and decarbonisation of some basic thermal processes, such as cement (construction) or ammonia (fertilisers for food) production, is progressing very slowly. It is therefore unlikely that these industries will eliminate their dependence on fossil fuels by 2050 (Smil, 2022).

In relation to that, an often overlooked element within green growth narratives is the fact that the expected transition towards a completely renewable energy society bumps into geophysical constraints, such as material limits and the reduction in the productivity of the energy system. According to a relevant study by Capellán-Pérez et al. (2019), a transition achieving a 100% renewable electric system globally by 2060 could cause a substantial re-materialisation of the economy, exacerbating the risk of availability in the future for some key minerals, such as copper and lithium All the mining activities are, up to date, carried out heavily relying on machineries powered by fossil fuels. Furthermore, as these authors foresee, this kind of transition will decrease the productivity of the energy system to levels below the thresholds required to sustain industrial complex societies. This last element points at another blind spot in the energy transition narratives: the fact that a transition from fossil fuels to renewables may power a human society, but not current industrial societies.

Hence, taking into account how dependent our modern societies are on fossil energy, a reduction in the availability of fuels such as crude oil and natural gas will have a significant impact on current standards of living in several ways. Keeping in mind the above-mentioned geophysical constraints, here are just some examples of how the current standard of living might be impacted: first, fossil fuels are a primary source of energy for heating, cooling and powering homes, businesses, and factories; a decrease in the availability of fossil fuels could lead to energy shortages, higher energy costs, and reduced access to energy, which could have a significant impact on daily life, particularly for those who rely heavily on energy for heating and cooling buildings and producing material goods. Second, fossil fuels are also a primary source of energy for transportation, powering cars, buses, tractors, ships, and aeroplanes; again, a decrease in the availability of fossil fuels could lead to significant disruptions to transportation systems, potentially leading to reduced mobility and increased costs for transportation of goods and people, and limiting the access to goods, work, and income. Third, fossil fuels play a critical role in the entire food supply chain, from production to processing and distribution and, very importantly, for the production of inorganic fertilisers; consequently, a reduction in the availability of fossil fuels could have significant impacts on each stage of the supply chain, posing risks to food security and availability.

The Challenge of Satisfying Basic Needs Within the Planetary Boundaries

Given the aforementioned dependency on fossil fuels and its impact on the current ecological overshoot, satisfying basic needs while staying within the planetary boundaries is probably one of the biggest challenges of our time. On the one hand, energy consumption plays a central role in overcoming these boundaries and, thus, global energy consumption should be significantly reduced in order to avoid the ecological collapse. However, on the other hand, and in line with relevant experts in the field, an increase in energy consumption may be crucial to satisfy basic needs in at least the poorest countries in the Global South. Again, Vaclav Smil establishes that in order to approach a "dignified standard of living" around 3.1 billion people will need at least to double (but preferably triple) their per capita energy use, which is deemed necessary to "boost their food production and build essential urban, industrial and transportation infrastructures" (Smil, 2022: 5).

At the same time, there is abundant evidence affirming that it is still possible to significantly reduce human energy consumption at a global level while maintaining a minimum standard of living worldwide, if defined as covering basic needs. For instance, Millward-Hopkins et al. (2020) propose a model to estimate a practical minimum threshold for the final energy consumption required to guarantee "decent material living" to the entire global population. Their main conclusion is that global final energy consumption in 2050 may be reduced to the levels of the 1960s. Of course, achieving this would require drastic changes, particularly radical demand-side changes to reduce consumption to levels of sufficiency.

In a similar vein, O'Neill et al. (2018) argue that it is possible to achieve "a good life for all within planetary boundaries" while significantly reducing the energy and material input globally. According to them, since poorer countries in the Global South will need to increase their energy and material consumption to reach the minimum threshold of a decent standard of living, wealthier nations could (and should) substantially reduce the amount of carbon emitted or materials consumed in order to "free up ecological space". Moreover, they argue that this could be done with no loss of significant wellbeing in richer countries.

Hence, it is important to clarify that it is also possible to significantly reduce global greenhouse gases emissions while covering basic needs. In fact, authors such as Rao and Baer (2012) have addressed the complex issue of establishing a minimum level of emissions, linked to the satisfaction of minimum standards to guarantee a decent standard of living. To do that, they take into account not only individual access to basic goods or services—such as adequate nutrition and housing—but also include other useful services such as refrigeration or mobile phone, as well as the provision by the state of basic infrastructures. Furthermore, they also consider the different energy mixes in the Global North and South, as the dependency on fossil fuels for guaranteeing basic needs is even stronger in poorer nations than in the richer Global North, much more advanced in terms of electrification.

This view is consistent with Henry Shue's distinction between "subsistence emissions" and "luxury emissions" (1993) as he addressed the issue of what a fair allocation of the costs of preventing global warming would be. According to this view, energy transition policies should focus on cutting first "luxury emissions", particularly associated with overconsumption in the Global North. This perspective, grounded on moral philosophy, also coincides with the empirical approach of O'Neill et al. (2018). Indeed, the latter provides evidence that supports that it is possible for richer countries to maintain wellbeing while reducing their carbon footprint, thereby liberating ecological space in form of emissions aimed at covering basic needs in poorer countries.

Adapting the Right to an Adequate Standard of Living to a Finite Biophysical Reality

The prospects of guaranteeing the right to an adequate standard of living are not optimistic in the long run, as human society faces different overlapping crises, such as fossil fuels depletion and climate change. This holds true both from the point of view of guaranteeing the relative standard established by that right (which implies the continuous improvement of living conditions), and from the perspective of the absolute standards, such as the right to be free from hunger or the right to a minimum supply of water that will cover basic uses for consumption and sanitation.

Therefore, there is a strong and urgent need to reflect upon the meaning of this right at the end of the fossil fuel era and find legal and policy tools to guarantee the satisfaction of basic subsistence rights, both at the international and national levels, especially in comparison with an unnecessary material consumption. As it is explored in the next section, *vis-à-vis* those challenges, the premise of this chapter is that the human right to an adequate standard of living provides relevant legal and political tools to design fairer policies in a context of reduced availability of resources, climate change, and increased hardship, but only insofar as this right and the human expectations associated are "resituated" or "readjusted" within the planetary boundaries and include a much-needed reflection on the rights of future generations.

Resituating Human Rights Within Environmental Limits

In general, the consideration of the environment shared both by human rights institutions and mainstream scholarship is anthropocentric, focusing only on environmental impacts on individual human wellbeing (Cançado Trindade, 1992; Shelton, 1991). Thus, including environmental considerations into the traditional human rights framework—such as maintaining habitable conditions on our planet for future generations and putting limits to the development of human societies—calls into question some of the theoretical premises of human rights law.

On the one hand, the consideration of the needs of future generations in human rights law presents significant challenges, as this legal body has traditionally protected the rights of present generations only. Nevertheless, the rights of future generations have been debated in the scholarship for some decades now (Brown Weiss, 1989) and have made their way into positive law recently in some countries, as the ground-breaking Neubauer et al. case of 2021 of the German Constitutional Court has shown. In the concrete case of the human right to an adequate standard of living, the degree of degradation of the natural world unveils the contradictions between the right of present generations to improve their standard of living and the rights of future generations, which need to be addressed (Skogly, 2021). These contradictions emerge when considering the relative standards that imply the world "adequate"; conversely, when addressing absolute standards, fundamental rights such as the right to be free from hunger should be at the forefront and should be prioritised over indeterminate future needs (García-Juanatey, 2018). This line of argument may also apply to other subsistence rights: the satisfaction of the basic needs of human beings currently alive should be met in all circumstances, but, ideally, when satisfying these needs, environmental impacts should be kept within planetary boundaries.

On the other hand, in regard to the concept of planetary boundaries, human rights law may be more receptive, as the notion of limits (such as maintaining public order) is intrinsic to that legal body. Nevertheless, human rights were recognised in a context of perceived abundance of energy sources, particularly fossil fuels, and a very low level of salience of environmental problems. Both the UDHR and the 1966 Covenants were proclaimed prior to the publication of the 1972 *Limits to Growth* report, which intended, with limited success but surprising precision, to raise awareness about the biophysical limits of a process of development based on infinite growth. Arguably, as the conceptualisation of limits and the need to balance different rights lies within the legal foundations of human rights law, the general blindness of that time towards the notion of environmental limits may explain why those were not included.

Conversely, nowadays there have been interesting attempts to link both concepts, human rights/basic needs and planetary boundaries, though the most popular proposals did not emerge from the legal domain. In the field of economic theory, the most well-known framework is probably Kate Raworth's model of "doughnut economics" (2017). Building on the ecological economics conceptualisation of an embedded economy, it proposes a new economic approach based on the objective of guaranteeing human rights and freedoms within planetary boundaries, as conceptualised by Rockström et al. (2009). Expanding on this approach, O'Neill et al. (2018) argue that human rights should be considered as minimum thresholds for the protection of human dignity on a planet with biophysical limits.

Of course, in a world of eight billion people, fulfilling the human right to an adequate standard of living within those boundaries is an increasingly complicated human endeavour. More precisely, when considering the North-South divide, there are clear contradictions between maintaining a high standard of living in richer countries and the need to reduce energy consumption and cut emissions. Hence, the material/energy consumption of the Global North—that according to O'Neill et al. (2018) goes well beyond the satisfaction of basic needs—is currently threatening the possibilities of meeting basic existential needs in the Global South.

Implications for an Adequate Standard of Living

In the present context, characterised by an ecological overshoot, it is urgent to reframe socio-economic rights, as they include rights that have material implications and are energy intensive, such as the right to food, water, and housing.

Firstly, theoretically speaking, this reconceptualisation would entail reframing the right to an 'adequate' standard of living to a 'decent' standard of living. This would imply going back to the dominant consideration of this concept in the period previous to World War II, when it was aimed at covering an "irreducible minimum", in the sense of avoiding extreme poverty. This reconceptualisation may be built on the tradition of scholarly work that, instead of using the standard of living language, focuses on the definition of a set of "subsistence rights" or a "human right to subsistence" (Eide & Barth Eide, 1999; Mancilla, 2019). This also seems to be in line with the recent practice of the CESCR, which has leaned towards "a greater focus on minimal standards" rather than on "human flourishing" in response to poverty (Hohmann, 2021; Graham, 2021).

Consequently, in terms of state obligations, the priority of the human right to an adequate standard of living should be to guarantee minimum core obligations, i.e. basic needs globally and within planetary boundaries, rather than focusing on the improvement of national standards regardless of their impact on other states and future generations. On the one hand, the realisation of the right to an adequate standard of living at the national level should focus primarily on minimum thresholds, such as the right to be free from hunger or to a minimum access to water; at the international level, the subsidiary obligations of the international community to satisfy minimum access to these rights in other countries should be reinforced. On the other hand, once basic needs are satisfied at the national level, the relative concept of 'adequacy' should be closely tied to environmental limits; in this sense, rather than referring to "the maximum of its available resources" in the realisation of this right, the relative standard should be to achieve it "within ecological limits", building upon the considerations made in General Comment 12 on the Right to Adequate Food (CESCR, 1999). To do that, empirical work aimed at identifying the global maximum use of resources should also be taken into account, and policies aimed to satisfy relative standards such as adequate food and housing should be constrained by the policy options that imply less emissions and fewer environmental impacts.

Secondly, the right to a "continuous improvement of living conditions" should be deeply reconsidered in light of the end of the fossil fuel era. Characterised by the post-war context, it seems to refer to an infinite notion of social progress. This notion is still appealing in today's human rights movements, as it motivates human rights activism and scholarship

(Hohmann & Goldblatt, 2021). Nonetheless, it has mostly revolved around the use of resources needed to guarantee certain standards of living, as the different composite rights of the right to an adequate standard of living have, to some extent, forced a materialistic realisation of this right. Such an understanding of the right to continuous improvement of living conditions may then be at odds with the energy transition imposed both by the depletion of natural resources and the climate emergency. As Skogly (2021) underlines, in a world with finite resources, it seems necessary to reflect upon this notion in order to ensure that the right to an adequate standard of living is well implemented.

To do so, different pathways are worth mentioning. One idea may be that the right to continuous improvement of living conditions should probably focus on vulnerable groups, while the wealthiest parts of the population would not be considered. A second idea to explore could be imposing limits on the use of natural resources when realising the right to continuous improvement of living conditions of present generations, in order to secure that same right for future generations (Skogly, 2021). However, these limits should apply only to the aspects of the improvement of *material* living conditions, which are also generally more energy intensive. Therefore, a promising way to reorient this right is to consider non-material and low-energy aspects of wellbeing, which could be expanded to truly improve these conditions. Thus, a third pathway worth exploring consists in rethinking the concepts of 'living conditions' and 'standards of living'. In this respect, some scholars consider that there should be a shift from exclusively social goods, which are still necessary in many parts of the world and for vulnerable groups, to fewer tangible goods (Hohmann & Goldblatt, 2021) that could apply to all, such as culture, nature, art, or social interactions (Skogly, 2021).

Finally, *vis-à-vis* those considerations, it is crucial to recognise that both access to energy services and to electricity are not basic human rights, but rather a means to support essential human needs. As Löfquist convincingly argues: "the rights language should focus on the needs of humans and not be extended too far to include everything that could be beneficial for humans, as such an extension might diminish the attention on the actual aim of human rights: that all humans should have a good enough life" (2020: 711). Similarly, Margaretha Wewerinke-Singh (2022: 18) clarifies that access to energy should be considered more as a "prerequisite to the realisation of human rights". Therefore, formulations such as a 'human right to energy' or a 'human right to electricity' are not adequate in the present context. Conversely, access to energy should be viewed as a secondary entitlement, as it is often required to safeguard fundamental rights such as access to adequate housing, healthcare, education, and even the right to life.

Conclusions

In regard to the multiple and multifaceted legal challenges stemming from the end of the fossil fuel era, this chapter explores the role that the right to an adequate standard of living may have in guaranteeing basic needs throughout this difficult time. Although the right to an adequate standard of living has been recognised in different international instruments, it has not been defined by the ICESCR nor the CESCR. Considering the lack of indications as to what constitutes an 'adequate standard of living', this work delved into the scholarship that has attempted to define it. Building mainly on the definitions suggested by Eide and Shue, the main aim of this right may be ensuring that every human being can enjoy their basic needs under conditions of dignity, which implies enjoying subsistence rights, such as adequate food and nutrition, clothing, housing, and the necessary conditions of care, as well as clean air and water. In terms of obligations, all definitions of this right point to the idea of ensuring a minimum level of living conditions in line with human dignity and include absolute standards, such as the right to be free from hunger or the right to water, which must be guaranteed under any circumstances. However, the use of criteria such as 'adequate' or 'reasonable' still leaves room for interpretation on these minimum thresholds.

In this respect, the Anthropocene challenges the notion of 'adequate standard' as well as the right to a "continuous improvement of living conditions", which need to be reframed in light of the planetary boundaries and society's fossil fuel dependence. Global north societies have been structuring themselves upon high energy-consuming processes, thus also the same subsistence rights have come to depend on access to fossil fuels. The permanent quest to increase the standard of living, especially for a part of the population in the Global North, and its intrinsic link with growing energy and material consumption—thus with resource depletion and emissions—is clearly causing a widespread reduction in quality of life and in the adequacy of living conditions in many parts of the planet. Hence, the scarcity of resources and the need to stay within planetary boundaries impose a necessary and unavoidable reduction of energy and material consumption.

This chapter concludes with a proposal to adapt the right to an adequate standard of living to these challenges. In spite of the discussed limitations of the human rights framework to integrate the needs of future generations as well as the concept of planetary limits, this work is based on the idea that human rights still provide useful tools to navigate the challenges ahead. Nevertheless, human rights were formulated and developed in a context of perceived abundance of energy sources, particularly fossil fuels, and a very low level of salience of environmental problems. Thus, there is an urgent need to reframe human rights, particularly its social and economic branches, to take into account the need to guarantee rights within planetary boundaries and energy scarcity. This would imply a redefinition and recalibration of the right to an adequate standard of living, as it was defined by the international community in the post-war period, orienting this right towards the satisfaction of basic needs of present generations, without overlooking other species and the rights of future generations.

These considerations also raise questions that may open future avenues of research. Some examples are the following: what should the role of international human rights law at the end of the fossil fuel era be: just establishing minimum core obligations or also contributing to human flourishing in other ways? In relation to the previous point, regarding the definition of standard of living, would this standard go beyond the satisfaction of basic material needs? If that would be the case, what non-material aspects should be included? Could access to clean air, vibrant communities, nature, culture, and art forms that are not energy intensive be considered also part of the standard of living guaranteed by this right? To what extent would these non-material aspects be invoked? These are some examples of the legal challenges that lie ahead and may find an answer in the coming years. Nonetheless, at present, a more pressing issue would be for this right to work in the direction of helping humanity out of the vicious circle whereby, by constantly seeking to improve the standard of living of (some) people through intensive use of fossil fuels, it is heading towards a drastic reduction of the habitability of our planet for all.

References

Aguiar Ribeiro Do Nascimento, G. (2018). The Long Road to the International Recognition of Economic and Social Rights: The Right to an Adequate Standard of Living. *The Age of Human Rights Journal, 11*. https://doi.org/10.17561/tahrj.n11.3

Armstrong McKay, D. I., Staal, A., Abrams, J. F., Winkelmann, R., Sakschewski, B., Loriani, S., & Lenton, T. M. (2022). Exceeding 1.5 C global warming could trigger multiple climate tipping points. *Science, 377*(6611), eabn7950. https://doi.org/10.1126/science.abn7950

Brown Weiss, E. (1989). In Fairness to Future Generations: International Law, Common Patrimony, and Intergenerational Equity. United Nations University.

Cançado Trindade, A. (1992). Human rights, sustainable development and the environment. San Josè de Costa Rica/Brasilia, IIDH/BID.

Capellán-Pérez, I., De Castro, C., & González, L. J. M. (2019). Dynamic Energy Return on Energy Investment (EROI) and material requirements in scenarios of global transition to renewable energies. *Energy Strategy Reviews, 26*, 100399. https://doi.org/10.1016/j.esr.2019.100399

CESCR. (1999). General Comment 12 on the Right to Adequate Food. Doc.E/C.12/1999/5, 12 May 1999.

Eide, A. (2014). Adequate Standard of Living. In *International Human Rights Law* (2nd ed., pp. 195–215). Oxford University Press.

Eide, A., & Barth Eide, W. (1999). Article 25. In *The Universal Declaration of Human Rights: A Common Standard of Achievement* (pp. 523–550). Martinus Nijhoff Publishers.

García-Juanatey, A. (2018). Reconciling human rights and the environment: A proposal to integrate the right to food with sustainable development in the 2030 development agenda. *Journal of Sustainable Development Law and Policy (The), 9*(1), 1–33. https://doi.org/10.4314/jsdlp.v9i1.2

Graham, L. D. (2021). The Right to Continuous Improvement of Living Conditions as a Response to Poverty. In *The Right to the Continuous Improvement of Living Conditions. Responding to Complex Global Challenges* (pp. 65–85). Bloomsbury Publishing. Retrieved from https://www.perlego.com/book/2976344/the-right-to-the-continuous-improvement-of-living-conditions-responding-to-complex-global-challenges-pdf

Guterres, A. (2020). *Secretary-General's Remarks to the UN Human Rights Council: "The Highest Aspiration: A Call to Action for Human Rights", Statement made the 24 February 2020 in Geneva, Switzerland.* Retrieved from https://www.un.org/sg/en/content/sg/statement/2020-02-24/secretary-generals-remarks-the-un-human-rights-council-%E2%80%9Cthe-highest-aspiration-call-action-for-human-rights-delivered-scroll-down-for-all-english

Hagens, N. J. (2020). Economics for the future–Beyond the superorganism. *Ecological Economics, 169*, 106520. https://doi.org/10.1016/j.ecolecon.2019.106520

Hohmann, J. (2021). Sources for a Nascent Interpretation of the Right to Continuous Improvement of Living Conditions. The Travaux Préparatoires and the Work of the CESCR. In *The Right to the Continuous Improvement of Living Conditions. Responding to Complex Global Challenges* (pp. 19–40). Bloomsbury Publishing. Retrieved from https://www.perlego.com/book/2976344/the-right-to-the-continuous-improvement-of-living-conditions-responding-to-complex-global-challenges-pdf

Hohmann, J., & Goldblatt, B. (2021). Situating the Right to Continuous Improvement of Living Conditions and Considering its Interpretations and Applications. In *The Right to the Continuous Improvement of Living Conditions. Responding to Complex Global Challenges* (pp. 1–18). Bloomsbury Publishing. Retrieved from https://www.perlego.com/book/2976344/the-right-to-the-continuous-improvement-of-living-conditions-responding-to-complex-global-challenges-pdf

IEA. (2021). *World Energy Balances: Overview, IEA, Paris.* License: CC BY 4.0. Retrieved from https://www.iea.org/reports/world-energy-balances-overview

IPCC. (2023). Summary for Policymakers. In (Core Writing Team) H. Lee and J. Romero (eds.), *Climate Change 2023: Synthesis Report. Contribution of Working Groups I, II and III to the Sixth Assessment Report of the Intergovernmental Panel on Climate Change*, (pp. 1–34). IPCC, Geneva, Switzerland, https://doi.org/10.59327/IPCC/AR6-9789291691647.001

Khanna, N., Fridley, D., Zhou, N., Karali, N., Zhang, J., & Feng, W. (2017). *China's trajectories beyond efficiency: CO2 implications of maximizing electrification and renewable resources through 2050.* Lawrence Berkeley National Laboratory. Retrieved from https://escholarship.org/uc/item/1ph8w8m2

Latouche, S. (1992). Standard of Living. In W. Sachs (Ed.), *The Development Dictionary. A Guide to Knowledge as Power* (pp. 173–189). Zed Books.

Löfquist, L. (2020). Is there a universal human right to electricity? *The International Journal of Human Rights, 24*(6), 711–723. https://doi.org/10.1080/13642987.2019.1671355

Lott, N. (2021). Understanding Forgotten Rights. In *The Right to the Continuous Improvement of Living Conditions. Responding to Complex Global Challenges* (pp. 131–146). Bloomsbury Publishing. Retrieved from https://www.perlego.com/book/2976344/the-right-to-the-continuous-improvement-of-living-conditions-responding-to-complex-global-challenges-pdf

Mancilla, A. (2019). The human right to subsistence. *Philosophy Compass, 14*(9), 1–10. https://doi.org/10.1111/phc3.12618

Masnadi, M. S., & Brandt, A. R. (2017). Climate impacts of oil extraction increase significantly with oilfield age. *Nature Climate Change, 7*(8), 551–556. https://doi.org/10.1038/nclimate3347

Millward-Hopkins, J., Steinberger, J. K., Rao, N. D., & Oswald, Y. (2020). Providing decent living with minimum energy: A global scenario. *Global Environmental Change, 65*, 102168. https://doi.org/10.1016/j.gloenvcha.2020.102168

Mohr, S. H., Wang, J., Ellem, G., Ward, J., & Giurco, D. (2015). Projection of world fossil fuels by country. *Fuel, 141*, 120–135. https://doi.org/10.1016/j.fuel.2014.10.030

O'Neill, D. W., Fanning, A. L., Lamb, W. F., & Steinberger, J. K. (2018). A good life for all within planetary boundaries. *Nature Sustainability, 1*(2), 88–95. https://doi.org/10.1038/s41893-018-0021-4

Persson, L., Carney Almroth, B. M., Collins, C. D., Cornell, S., de Wit, C. A., Diamond, M. L., et al. (2022). Outside the safe operating space of the planetary boundary for novel entities. *Environmental Science & Technology, 56*(3), 1510–1521. https://doi.org/10.1021/acs.est.1c04158

Rao, N., & Baer, P. (2012). "Decent living" emissions: a conceptual framework. *Sustainability, 4*(4), 656–681.

Rifkin, J. (2019). *The Green New Deal: Why the Fossil Fuel Civilization Will Collapse by 2028, and the Bold Economic Plan to Save Life on Earth*. St. Martin's Press.

Rockström, J., Steffen, W., Noone, K., et al. (2009). A safe operating space for humanity. *Nature, 461*, 472–475. https://doi.org/10.1038/461472a

Shelton, D. (1991). Human rights, environmental rights, and the right to environment. *Stanford Journal of International Law, 28*, 103.

Shue, H. (1993). Subsistence emissions and luxury emissions. *Law and Policy, 15*(1), 39–59. https://doi.org/10.1111/j.1467-9930.1993.tb00093.x

Shue, H. (2020). *Basic Rights. Subsistence, Affluence, and US Foreign Policy* (40th Anniversary ed.). Princeton University Press.

Skogly, S. (2021). The Right to Continuous Improvement of Living Conditions and Human Rights of Future Generations—A Circle Impossible to Square? In *The Right to the Continuous Improvement of Living Conditions. Responding to Complex Global Challenges* (pp. 147–164). Bloomsbury Publishing. Retrieved from https://www.perlego.com/book/2976344/the-right-to-the-continuous-improvement-of-living-conditions-responding-to-complex-global-challenges-pdf

Smil, V. (2010). *Energy transitions: History, requirements, prospects*. ABC-CLIO.

Smil, V. (2022). *How the World Really Works: A Scientist's Guide to Our Past, Present and Future*. Penguin UK.

Ssenyonjo, M. (2016). *Economic, Social and Cultural Rights in International Law* (2nd ed.). Hart Publishing.

Wewerinke-Singh, M. (2022). A human rights approach to energy: Realizing the rights of billions within ecological limits. *Review of European, Comparative & International Environmental Law, 31*(1), 16–26. https://doi.org/10.1111/reel.12412

CHAPTER 4

Energy Communities: Why (Sometimes) the Commons Need the State

Clara Esteve-Jordà and Marcos de Armenteras Cabot

Introduction

The current scenario of declining fossil fuel availability and the need to move towards renewable energy sources require a radical reconceptualisation of the hegemonic form of energy production. The new energy system required by the decarbonisation transition must produce energy in an environmentally and socially sustainable way. Currently, natural gas, coal, and crude oil still constitute the primary sources of energy in the European Union (EU). In 2021, fossil fuels accounted for 70% of the EU's gross available energy (Eurostat, 2023a). On a positive note, wind and solar power generated a record 30% of the EU's electricity in the first half of 2024, compared to 27% from fossil fuels (Graham & Fulghum, 2024).

C. Esteve-Jordà (✉)
Institute of Environmental Science and Technology, Autonomous University of Barcelona, Barcelona, Spain
e-mail: clara.esteve@uab.cat

M. de Armenteras Cabot
University of the Balearic Islands, Balearic Islands, Spain

D. Iglesias Márquez et al. (eds.), *Legal Challenges at the End of the Fossil Fuel Era*, Global Issues,
https://doi.org/10.1007/978-3-031-61766-9_4

The energy transition is not only about the substitution of energy sources but also about a socio-cultural change in energy consumption patterns of historic magnitude (Gough, 2015, 2017). This process relies on two preconditions: (1) the electrification of the energy system (Aalto et al., 2021) and (2) a reduction in energy consumption that is compatible with preserving the satisfaction of human needs—which also implies improving energy efficiency. The latter is based on the need to reduce energy consumption by rethinking what and how we produce and consume in all aspects of our lives to achieve a decent standard of living for all with less energy availability (Rao & Min, 2018; Millward-Hopkins et al., 2020; Millward-Hopkins, 2022). For example, through the prioritisation of public transport over private transport, the consumption of local products, the reuse of materials, and the proliferation of community systems of objects and services, among others. The first precondition implies the electrification of the energy system since the energy produced by renewable devices can only be converted into electricity. Final energy consumption in the form of electricity in the EU currently accounts for only 22.8% of total energy consumption (Eurostat, 2023b). For this reason, this analysis does not look at the energy market as a whole but at the electricity market in particular.

The EU has sought to create a competitive market by liberalising the electricity sector, unbundling electricity generation, distribution, and transmission to move away from state monopolies (Sokołowski, 2016). These liberalising reforms have been accompanied by mechanisms to monitor anti-competitive activities. As Fabra and Fabra Utray (2009: 19–20) point out, the characteristics of electricity as a market object (the impossibility of storing electricity, the existence of capacity limits in production and transmission, the volatility of demand and its low or zero price elasticity, and the frequent interaction between a small number of companies protected by barriers to entry, etc.) prevent the electricity market from being a truly competitive market and, consequently, from benefiting the final consumer. This has raised concerns in EU member states about the concentration of power in the hands of a few companies and the lack of competition and consumer choice, which is not only a market problem but also a democratic one.

Market accumulation in the electricity sector can lead to political and environmental difficulties. First, electricity plays a central role in the economic and social life of any state. The concentration of the entire sector in a few private companies implies the accumulation of considerable power in the private sector. Even in a regulated sector, the economic power and political influence that these companies achieve creates glaring inequalities in the relationship between them and consumers and citizens. Secondly, the commodification of access to energy for domestic use generates social externalities, such as energy poverty, and environmental externalities, such as air and water pollution, habitat destruction, and the depletion of natural resources.

The two major challenges to be addressed are (1) the ecological transition in response to the climate crisis and (2) the democratisation of the electricity system to address the first challenge. As has been advanced, this chapter assumes that the climate crisis must be addressed in two parallel directions: on the one hand, through a social and cultural transformation in which consumption and production patterns are changed on a global scale to enable a radical reduction in energy consumption, especially in those countries that are at the centre of the capitalist economy; on the other hand, through the technological substitution of fossil fuels by renewable energy production systems (photovoltaic and wind).[1] This second element is key to the present chapter: in the process of democratisation of the electricity system, the role of the citizens, the state, and private companies must be questioned.

In this sense, the potential of citizens' initiatives in this sector, as agents of production and consumption, will have to be considered. Certainly, in this transition, the state is not likely to cease to be a key player. On the contrary, it will play a crucial role. Political control and the strengthening of its institutional structure will be necessary to ensure a clear commitment to its governance role and a social response to citizens' demands. However, the leadership cannot be exclusively state led. The climate emergency is pressing, and it will not wait for all states to become self-sufficient in managing the energy transition. Civil society, through smaller structures, can take the initiative to drive the energy transition. Citizen initiatives hold the potential to serve as a transformative force in energy policy.

[1] Commission Delegated Regulation (EU) 2022/1214 of 9 March 2022 amending Delegated Regulation (EU) 2021/2139 as regards economic activities in certain energy sectors and Delegated Regulation (EU) 2021/2178 as regards specific public disclosures for those economic activities includes nuclear energy and fossil gas in its taxonomy for tackling climate mitigation, as some member states' plans include nuclear energy alongside renewables among the energy sources to be used to meet climate targets. However, here it is considered, as in the Delegated Regulation (Recital 10), that the central role should be played by renewable energies.

They can empower citizens to collectively disengage from the electricity oligopoly, thereby reducing its influence and bolstering resilience against eventual energy challenges. Furthermore, examining energy production and distribution through the lens of the commons may unveil the democratic potential of alternative approaches to organising and producing energy, ultimately leading to transformative eco-social outcomes. These initiatives, framed as community practices for accessing energy, align with the concept of 'common goods'. Common goods represent a distinct third way of conceiving ownership, separate from both the public and the private, characterised as communal.

Fortunately, in recent years, there has been an increase in initiatives and movements supporting fundamental changes in the energy supply system that represent the new energy transition paradigm. Community practices, conflicts, innovations, etc., on the part of trade unions, cooperatives, neighbourhood organisations, athenaeums, etc., can provide alternatives to the capitalist market, based on proposals that involve the weaving of strong public-community alliances. These are what Santiago (2023: 140) calls laboratories of emancipatory socio-cultural innovation.

In this sense, the purpose of this chapter is not to explain the role of energy communities, for which the literature is quite fertile and robust (Campos, 2020; Debizet et al., 2022; Loebbe et al., 2022). Instead, it aims to explore how the state must play an active role in driving energy communities forward to ensure a just, fair, and inclusive energy transition. The chapter is structured in three parts. First, it presents an analysis of energy within the theoretical framework of the commons. Second, it examines energy communities through the prism of democratic governance theory, outlining their benefits and possible limiting barriers. Third, it discusses the role of the state in removing these barriers. Finally, some conclusions are drawn.

Energy beyond the Private/Public Dichotomy: The Commons

The academic literature addressing the conceptualisation of energy democracy acknowledges the challenge of defining a concept that remains dynamic and evolving (Szulecki, 2017; van Veelen & van der Horst, 2018). Nevertheless, a common theme in this literature is the association of energy democracy with the aspiration to establish an alternative energy production and generation paradigm distinct from the prevailing one. This aspiration encompasses citizen initiatives aimed at forming energy communities, which collectively oversee energy production and generation to achieve independence from the electricity oligopoly and prevent market dominance. Importantly, this shift away from market dynamics is not initiated by the state or driven primarily by a public agenda; instead, it arises from the collaborative efforts of individual agents who, by deviating from market-oriented approaches, advocate for community collaboration to address a fundamental need.

Private, Public, and Common

The prevailing approach throughout the twentieth century was based on the ideological tension that distinguished the private sphere, associated with private relationships between individuals and their exchanges in the market, from the public sphere, associated with the state, as the sole mechanisms for social distribution and exchange. This dualism would be translated in terms of private property and public or state property. However, in 1990, in the light of Hardin's critique of the 'tragedy of the commons', Elinor Ostrom shed light on communal practices throughout history and explained the existence of successful cases where community management of certain resources overcame the public/private dichotomy without succumbing to the 'tragedy' prophesied by Hardin. According to Caffentzis and Federici (2014: 100) 'the common/s appears as a possible "third" space besides and equal to the state and the market'.

The notion of the commons has been developed within the disjunction between public and private property, on the assumption that it is neither 'public' nor 'private', but a new category independent of both.[2] The commons would then represent a form of property independent of private and public property regimes. One of the prominent proponents of this position, Ugo Mattei (2011: 34), argues that in the Western constitutional system, state sovereignty and private property 'colonise the imagination' since all political programmes (left-wing and right-wing) advocate for one or the other without departing from this dualism. For this reason, he argues that both options derive from the same economic logic, 'which

[2] However, this distinction does not always incorporate a third category, as the 'commons' are sometimes examined not so much in terms of ownership but rather from the perspective of their governance or management.

marginalises the common, erases ecological perspectives, and subdues general intelligence, resulting in a single mode of thinking: the relentless logic of capital accumulation'.

This perspective, also defended by other authors (Negri & Hardt, 2009: viii–ix; Caffentzis & Federici, 2014), suggests that the defence of the primacy of either the state or the private sector responds to the same logic, which, according to Mattei, is the capitalist system. In this sense, the idea underlying these arguments is that the state is co-opted by private interests,[3] thereby serving the preferences of specific private sectors over the common good. According to this view, there would be no democratic alternative within the state-market dualism, with the only option being the commons.

In her research, Ostrom (1999) found shared practices among the success stories that led to the effective governance of these assets. These practices relate to access to resources, participants' rights, deliberative mechanisms, and conflict resolution mechanisms among participants. Ostrom's conclusions on the governance of the common-pool resources can be linked to Mattei's thesis: since the state has been co-opted by the market, the commons are the only possible alternative to the state-market dualism. If so, then it is not so much the role of the state that should be of concern, but rather the study of best practices to ensure the success of energy communities: what are the deliberative processes, what are land titles like, how is access regulated, and so on.

However, the division between public and common, while applicable to certain goods, cannot be universally applied to all types of goods or collective initiatives. The effectiveness of this configuration depends on the political and economic circumstances of the moment (Lloredo Alix, 2020: 225). Conceptualising common property as entirely distinct from private and public property may hinder the development of grassroots initiatives, which, by their nature, require a symbiotic relationship with the state. The

[3] In the current situation, this dualism is indeed of great interest. There is no doubt that the oligarchic powers seek to co-opt the state to promote their interests rather than the common good. This has been evident in the EU policymaking through phenomena such as the 'revolving doors' (Silva, 2019), where influential EU officials (politicians and civil servants) have seamlessly become lobbyists. This phenomenon has also been observed in the energy sector and, particularly, in the electricity market. Corporate Europe Observatory has created *RevolvingDoorWatch*, a database of EU officials and politicians who have moved from public office to the industry and lobbies. Its reports include analyses of the revolving door phenomenon in different EU member states, where the energy sector is at the forefront.

analysis of the specific situation in the given moment can help to understand how to articulate and identify the necessary forces to constitute the commons. In the case of energy communities, as with other initiatives, large initial investments, highly specialised technical capacity, and broad agreement among participants are required, so state involvement is crucial. This goes beyond funding and regulation; it may include empowering citizens to create them.

Energy Production and Distribution through the Prism of the Commons

The energy transition cannot and must not be side-lined by questions of justice. The most vulnerable communities have contributed the least to the depletion of natural resources and have benefited the least from the use of fossil fuels. Such a transition cannot simply consist of replacing coal-fired power plants with for-profit solar energy farms (Dawson, 2022), again run by oligopolies. Indeed, there are numerous cases of massive for-profit renewable energy projects being developed in ecologically rich and sparsely populated areas where opposition is less likely (Franquesa, 2023). These populations receive few benefits from the energy produced and have little control over it (Giotitsas et al., 2022: 2). Instead, they suffer from the impact of these investments on their natural environment, which also leads to many social conflicts. Often, local governments protect the interests of large companies because they benefit financially, making citizen protests futile (Yenneti et al., 2016: 96).

Now that the transition to an energy system in which coal, oil, and fossil gas[4] are no longer the protagonists, it is time to think about a model in which all communities can use and produce clean energy. It seems plausible, in this sense, to frame the study of this new model within the theoretical framework of the commons. That is, to consider the possibility of communities operating outside government control and private ownership, which attribute a common value to energy and make sustainable use of it, through management rules (Bollier & Helfrich, 2019). In other words, commons are social practices of creating and/or managing a resource through the institutions created and managed by a community of

[4] As mentioned above, the EU has included fossil gas in its taxonomy to address climate neutrality. However, fossil gas does not figure prominently in the regulation, as the criteria for fossil gas projects to attract investment are not easy to meet (Annex I, Section 4.29).

producers or users (Ostrom, 1990). Thus, the commons could provide a general framework for constructing a holistic and sustainable alternative to the current socio-economic configuration that permeates virtually all facets of human activity, i.e. profit-maximising market relations. Thus, practices of communal ownership and local democracy seem not only appropriate but desirable.

In this sense, solar and wind energy flows are natural resources from which no one can be excluded, as they cannot be enclosed. In this case they would be 'commons' in the economic sense of the term: they are neither exclusive nor rivalrous. However, it is the process of transforming these energy flows into electricity that may place citizens in a situation of vulnerability *vis-à-vis* those who have the technical and material capacity to carry out this process (i.e. the electric oligopoly), since electricity production and distribution are confinable and excludable. But through distributed and decentralised energy systems with appropriate institutional mechanisms, multiple users may benefit from energy while optimising the use of natural resources. Such consumers, now 'prosumers' (Butenko, 2016), can collectively co-produce and create common value over the supply of energy, thus making this resource a common good. Additionally, energy communities can play an important role in the social acceptability of renewable energy development by encouraging citizen participation and co-responsibility in projects (Otamendi Irizar et al., 2022: 2). The active participation of the prosumers in the production, storage, and management of clean energy strengthens cooperation and cohesion, stimulates financial responsibility—i.e. energy savings—and promotes the proper maintenance of the installations.

This conception of energy production and distribution as a common good is reinforced by some research, indicating that the distribution of governance responsibilities through decentralisation of power from the national to the regional and local levels contributes significantly to the capacity and willingness of nations to achieve renewable energy deployment (de Laurentis, 2020: 3). It is therefore important to decentralise power and think about organised economies on a small scale (Giotitsas et al., 2020: 3; Giotitsas et al., 2022: 3). Among other reasons, this can increase sustainability, democratic density, and citizens' commitment to the energy transition and, consequently, strengthen social cohesion (Santiago, 2023: 124). Distributed energy systems can thus be likened, to some extent, to common property regimes in which self-governing entities generate, integrate, store, and respond to the demand for a resource, in this case renewable energy. Electricity, in this sense, should move from being a private commercial commodity supplied on a public grid to a co-produced common.

Producing Electricity Together: What Are Energy Communities?

Energy communities are groupings of individuals associated with the pursuit of objectives related to energy supply and production. These can be legal entities owned by individuals, small companies, or local authorities. They are usually located in the vicinity of renewable energy projects, and their primary purpose is to provide environmental, economic, or social benefits, rather than speculative gains (Gruber et al., 2021). The ownership and/or management of the energy is communal, so we can speak of common ownership schemes (Pappalardo, 2022). Community members are beneficiaries of the community renewable energy system (REScoop, 2021). Energy communities are systems of distributed ownership of energy generation sources, i.e. localised, and controllable microgrids. The users of the energy community (EC) can therefore be owners and managers of the project, i.e. enjoy not only the ownership but also the functionality that the system provides.

Such an organisational structure allows citizens to collectively produce, consume, store, share, and sell renewable energy, involving a high level of citizen participation. Membership of the community must be free and voluntary (through an opt-in and opt-out system). Citizens have democratic control over energy investments by becoming co-owners of renewable energy installations, usually through the principle of 'one member, one vote' (Caramizaru & Uihlein, 2020: 21). Although the forms of participation of each member can be decided by the community itself, it is important to regulate measures that prohibit a single member from having too high a percentage of votes (e.g. a maximum of 51% of the votes), to avoid interference by traditional actors. It must be borne in mind that the aim of the energy community is not the individual's profit motive, and therefore, it must be carefully monitored, and safeguards must be put in place to ensure that there is no individual profit within the energy community.

The term 'energy community' (EC) has been conceptually introduced in the regulatory field of the European Union, with at least two alternative formal descriptions in two separate Directives included in the Clean Energy Package: 'renewable energy community (REC)' and 'citizen energy community (CEC)'. The main framework for the RECs was introduced by the Renewable Energy Directive (RED II) 2018/2001, on the promotion of the use of energy from renewable sources. Art. 2. (16) defines it as a legal entity (a) which, in accordance with the applicable national law, is based on open and voluntary participation, is autonomous, and is effectively controlled by shareholders or members that are located in the proximity of the renewable energy projects that are owned and developed by that legal entity; (b) the shareholders or members of which are natural persons, SMEs, or local authorities, including municipalities; (c) the primary purpose of which is to provide environmental, economic, or social community benefits for its shareholders or members or for the local areas where it operates, rather than financial profits.

The concept of CEC was defined in Art. 2 (11) of the Internal Electricity Market Directive (IEM) 2019/944 as a legal entity that (a) is based on voluntary and open participation and is effectively controlled by members or shareholders that are natural persons, local authorities, including municipalities, or small enterprises; (b) has for its primary purpose to provide environmental, economic, or social community benefits to its members or shareholders or to the local areas where it operates rather than to generate financial profits; and (c) may engage in generation, including from renewable sources, distribution, supply, consumption, aggregation, energy storage, energy efficiency services or charging services for electric vehicles, or provide other energy services to its members or shareholders.

The main differences between the two definitions—REC and CEC—relate to the rules on membership, admissible generation technologies, the geographic scope, and the allowed activities. CECs focus exclusively on electricity use and can cover more activities, namely generation, distribution and supply, consumption, aggregation, storage, energy efficiency services, charging services for electric vehicles, or other energy services. Meanwhile, RECs only include renewable energy technologies and can only cover production, consumption, and selling of energy, although they can cover any energy sector, such as transport or air conditioning (heating and cooling). Shareholders or members of RECs must be located in the proximity of the projects owned and developed by the REC, while there is no geographic limitation for CECs (EUEA, 2021).

The benefits of energy communities are significant. Firstly, this model promotes a distributed grid system, based on completely different organisational principles than the centralised and hierarchical system of electricity supply (Wolsink, 2020: 3). The level of complexity of the structure can be reduced compared to the current market structure. On a technical level, a distributed energy system can reduce grid losses and increase system efficiency and reliability of supply. In addition, the characteristics and needs of each community foster the capacity for innovation of locally adapted systems. Secondly, it encourages the participation and empowerment not only of citizens but also of local authorities and small and medium-sized enterprises in renewable energy projects. This can generate more local acceptance of renewable energies and increase the decarbonisation process of the economy, i.e. move towards the energy transition at a fairly appropriate pace (Wolsink, 2020).

Furthermore, this system has a positive and direct economic impact on consumers, as energy costs are minimised. This can lead to a reduction in energy poverty, as cheaper service tariffs are applied. In addition to households, this system can bring benefits to any other type of end-user, such as farms, factories, schools, shops, offices, public bodies, etc. The characteristics of the community users determine the management and administration possibilities, their social capital, etc., so that each community can adapt to its own needs. In addition, energy communities can generate local employment and prevent the outflow of financial resources from the region (Kunze & Becker, 2014: 63). Be that as it may, it is important that energy communities are designed as a non-profit system, in which energy speculation is eliminated.

In any case, the advantages are not purely economic but can also bring environmental and social benefits, as surpluses can be reinvested in environmental or social improvements in the surroundings, such as energy advice services, sports clubs, environmental conservation programmes, etc. (CAG, 2021: 15). Likewise, the more renewables are installed in urban, peri-urban, degraded, and/or anthropised areas—such as roofs, gardens, facades, fields, basements, attics, garages, and industrial estates—the less space will eventually be needed on agricultural, forest, or natural land. On a social level, this organisational form tends to be more equitable and fairer than traditional energy systems. Energy becomes more affordable under this system, and not scarce, as energy is actually a fairly abundant commodity. And not only is it socialised, but it also allows for the inclusiveness of hitherto vulnerable consumers. This model can also contribute to creating a sense of community, of trust, thus increasing local social cohesion (Caramizaru & Uihlein, 2020: 21).

Nonetheless, it is essential to consider potential barriers that could limit the establishment of energy communities. Firstly, regulatory and legal obstacles may arise due to the absence of clear definitions in current legislation, beyond those outlined in the EU Directives mentioned earlier. Additionally, bureaucratic and administrative challenges must be considered. Secondly, despite the increasing prominence of energy communities, there remains a relative lack of knowledge regarding technology, regulations, subsidies, and related aspects, particularly in certain countries and regions where such practices are unfamiliar (Palm, 2021: 39). Thirdly, the absence of access to financial support, such as subsidies or grants, can be a significant obstacle, particularly during the initial installation phase (Lazdins et al., 2021: 2). This hurdle can be especially challenging for groups with limited disposable income. In addition, it is necessary to consider the different barriers that may be encountered between urban and rural areas (Streimikiene et al., 2021).

Why Am I Not Part of an Energy Community If They Are So Good?

As mentioned above, energy saving, community development, emancipation from electricity monopolies, reduction of the power held by electric companies, the democratisation of energy production and distribution, among other factors, present compelling reasons to urgently consider co-creating an energy community. There may be various reasons for not considering doing so: lack of financial resources, unfamiliarity with the administrative process, lack of a sense of community, lack of free time to collaborate with others in setting up such a project, or lack of trust in the ethics of (the few) people who drive them (Proka (REScoop.eu), 2023). These factors may prevent the widespread adoption of renewable energy initiatives 'from below'.

According to Romero-Rubio and de Andrés Díaz (2015: 405), there were six key factors that contributed to the development of energy communities in Germany: (1) strong political support and stable regulations; (2) financial incentives and preferential access to the grid to facilitate collective ownership models; (3) availability of credit capital with preferential terms and a sufficient number of people with sufficient financial capacity to invest; (4) relatively high sensitivity to environmental issues; (5) a tradition of local energy activism; and (6) the 'Bioenergievillages' (*Bioenergiedorfs*) initiative launched by the German government in 2005 and aimed to promote the creation of rural sustainable energy communities focused on bioenergy. In addition, other research (BECoop et al.,

2023: 10) has concluded that the barriers resulting from poor regulation and complex administrative procedures can be important constraints for the development of energy communities.

To ensure a just energy transition, all segments of society should have equal opportunities to actively participate in energy communities. There are social elements and potential problems in organisational and production practices around the governance of energy community regimes. What is more, to be a truly socially inclusive alternative that ensures equity in resource allocation, emphasis must be placed on the inclusion of vulnerable consumers (Reis et al., 2020). Issues such as age, illness, gender, income level, and cultural specificities need to be considered and addressed in the context of the energy commons. In gender terms, for instance, the impact can be positive, as a cooperative approach with stronger support networks and training and capacity building enables more women to participate in the energy system. However, policy makers might need to take more direct action and adopt regulations and incentives to make the energy community accessible to more groups in society (Lazoroska et al., 2021: 10). To move forward in terms of collective resilience, the inclusive approach can only be tackled with sufficient awareness and cooperation between very different actors, in different sectors and at different levels.

Most community energy projects are developed from a bottom-up local initiative, with an important role for citizens committed to the preservation of the environment and the local social fabric (IDAE, 2019: 26). Be that as it may, and despite these basic premises for the governance of energy communities, part of the challenge of a just energy transition lies in the institutional integration of these changes into a resilient socio-legal sphere. Indeed, this will require new rules and safeguards to avoid old government monopolies or new private monopolies (Heldeweg & Saintier, 2020: 7). The role of national and international governmental actors in strengthening this new governance will remain important but will have to operate on new foundations. The organisational principles of this public intervention are discussed below.

The Dependency between Public and Common: The Role of the State in the Development of the Energy Communities

It has already been suggested that even in a theoretical framework where energy can be conceived as a common good, it is difficult to imagine a scenario where the state is completely absent—at least for the time being—because the good governance of the production and consumption of electricity produced by energy communities is very often and largely dependent on the state. Admittedly, a basic element for the soundness of a fair energy system is the relationship between the users of the resource and the providers of public infrastructure which should not involve the plundering of vulnerable territories to obtain the materials needed for panels and batteries.

Indeed, energy communities are not isolated entities but integrated into the public energy system and therefore governed not only by internal rules but integrated within a specific institutional framework (Anderies et al., 2004). An energy supply that relies on interconnected decentralised systems requires a transition from centralised hegemonic models to polycentric, hybrid, flexible, and adaptive systems that facilitate the deployment and development of decentralised energy systems in smart microgrids with great variety, flexibility, and resilience (Gui et al., 2017). In this sense, the state has a central and irreplaceable role in the ecological transition. The state should not only tolerate new organisational forms but assist them with political and government support. In this scenario, then, the hitherto central actors no longer exercise hierarchical power but facilitate the establishment of collective networks of decentralised energy systems.

Some of the public values involved in government regulation (national, regional, local) are safety, fairness, inclusiveness, and environmental protection. The state and other levels of government should therefore institutionalise processes that create optimal conditions for self-governance in energy communities. Considering the six factors that Romero-Rubio and de Andrés Díaz (2015: 405) found in the German case, it is relevant to highlight at least four types of support instruments: (1) the clear definition of a legal framework favourable to the administrative and technical legalisation of local initiatives; (2) support in the early stages of the formation of an energy community; (3) support for project financing through direct subsidies or tax incentives; and (4) the dynamisation of a critical social mass and energy culture (IDAE, 2019: 26). These measures or actions require an active role of the public administration in many aspects of the public sphere.

Favourable Legal Frameworks

Legal frameworks must be aligned with environmental and energy transition objectives, which encourage the generation and consumption of renewable energy sources and promote energy efficiency. According to a COME-RES report (2023), regulatory and administrative aspects are the most challenging for energy communities. Enabling a favourable legal framework consists of developing a clear and appropriate regulatory framework to define energy communities as legal entities and to support their development. It has already been seen that the EU, for example, has already defined them—'renewable energy communities (REC)' and 'citizens energy communities (CEC)'. However, as of 2023, many states are still in the process of transposing this EU legislation (Bankwatch Network and REScoop, 2022). The regulatory framework must consider the legal entities that may become local energy communities, such as those that arise from associations, cooperatives, industrial park managers, residents' associations, port areas, etc. (MITECO, 2020). Some flexibility is therefore important. Regulation must be broad enough to cater for different types of energy communities, of varying legal nature and size. One size does not fit all, and the rules must be adaptable to various community models.

The Directive (EU), 2019/944 obliges member states to provide a favourable legal framework for energy communities that ensures various aspects. These include, among others, that participation in the community is open to all interested individuals and entities and is voluntary, so its members can join or leave it. Furthermore, members of energy communities should not lose their rights and obligations as domestic energy customers or owners of energy assets because of their participation. Legal provisions should ensure that individuals and entities retain control over their assets and are not adversely affected by participation in the energy community. In addition, a simple and clear process is required for prosumers to sell excess energy produced to the grid and share their self-produced electricity among members, neighbourhoods, and properties, as well as if the energy community is unable to self-supply.

Moreover, such regulatory frameworks might require cooperation between energy communities and distribution system operators. This cooperation should be based on fair compensation mechanisms set by regulators to ensure that distribution system operators have an incentive to cooperate with energy communities. Likewise, energy communities must be subject to fair, proportionate, and transparent procedures and charges, including registration and licensing, as well as clear, fair, and non-discriminatory grid access tariffs. This includes ensuring that they contribute adequately and in a balanced way to the overall sharing of system costs, and that they are not unfairly disadvantaged compared to traditional energy suppliers. Members of energy communities must also enjoy the same consumer protection rights as traditional energy customers. Legal frameworks must protect against abuse and ensure that members are informed and empowered.

This step also includes the generation of legal frameworks that facilitate decision-making processes on land use, infrastructure, and system design, and developing protocols and rules for resolving conflicts between stakeholders. In other words, stakeholder and community engagement in projects should be encouraged. This might require new organisational principles, i.e. adapting the rules of use and supply to local circumstances (who is entitled to what, who provides what, etc.). This also includes structural changes, such as adapting the current energy grid structure, and changes in urban and land-use planning regulations. To this end, long-term stability must be ensured. Laws must provide stability and predictability for energy communities, as uncertainty can deter investment and participation. Clear, long-term policies and regulations are essential.

Initial Support

Municipal support for energy communities is crucial for their successful establishment and growth. Particularly in the early stages, local authorities can play a key role by providing various forms of material, administrative, and capacity building support. This includes the provision of local resources for the establishment of the community, such as municipal land, buildings, roofs, and services to stimulate and facilitate collective participation processes, legal advice, etc. It also means carrying out feasibility studies, drawing up the financing plan, etc., and adopting and adapting standards of use and supply to local conditions. Municipalities can also assist with grid connection and infrastructure development.

It is also likely that bureaucracy will need to be simplified and clarified, and local authorities will need to be empowered and trained to provide adequate advice to energy communities. Municipalities should provide legal and regulatory guidance, i.e. clear and accessible information on local regulations and permits related to the establishment and operation of energy communities. This includes guidance on zoning, land use, permitting, and building code compliance. Streamlining permitting and licensing procedures for energy community projects can significantly reduce administrative burdens and costs. Municipalities can create a specific contact point or one-stop shop for energy community applications.

Technological support and the creation of information centres with expert assistance that promote local participation and development will be essential (Sperling, 2017: 886). Municipalities, together with the support of civil society organisations, can offer technical expertise to assist energy communities in project planning, site selection, and selection of renewable energy technologies. As noted, this type of assistance can be particularly valuable for communities in the early stages of development. Local authorities should also provide relevant data, such as energy consumption patterns, renewable energy potential, and grid infrastructure information. Access to this data can help energy communities make informed decisions about their projects.

It is also crucial to adapt the initial administrative and licensing procedure to the obligations and risks associated with energy communities. Mechanisms should be established to monitor and report on the progress and impact of energy community projects. This information can be used to evaluate the success of support programmes and make necessary improvements. This may require one-stop-shops—which may be the same as for applications—to provide technical assistance to citizens, community energy initiatives, local authorities, and SMEs, i.e. a specific and specialised support service, a single point of contact between the administration and potential energy communities to simplify these processes (Bankwatch Network and REScoop, 2022).

Moreover, administrative support must be ensured to guide and advise citizens in setting up energy communities so that no one is left behind. This requires support at the local level to promote energy communities that include the participation of people in vulnerable situations. Municipalities can assist energy communities in choosing the most appropriate legal structure and provide guidance on compliance with relevant legal and fiscal regulations. Indeed, municipalities can become potential initiators, investors, and members of energy communities, acting as facilitators, awareness-raisers, and creators of networks and best practices.

Municipalities can also help build trust and legitimacy for community energy initiatives (BECoop, et al., 2023). They can advocate for the interests of communities at higher levels of government and facilitate networking opportunities to share best practices and lessons learned.

As an example, the Barcelona Provincial Council (Spain) has created the Office for the Promotion of Energy Communities (OICE), a service that will accompany local authorities in the province throughout the process of creating new communities. The project plans to provide advice to around a hundred educational centres in the province and a catalogue of workshops, resources, technical seminars, and training courses (Diputació de Barcelona, 2023). In any case, it is important that the collaboration with energy communities is conceived as a long-term, ongoing partnership. Municipalities must be prepared to adapt their support according to changing community needs and evolving energy system conditions.

Financial Incentives

On the third point, the need for inclusive access to the distributed energy system has already been stressed. It is important to avoid high transaction costs that make energy communities impossible (e.g. due to upfront investment costs). Financial incentives can take the form of subsidies for initial investment (vouchers, loans, and rebate schemes), reduced or zero rates or exemptions for renewable energy system equipment, property tax incentives at the community project site, incentives for environmental projects, personal or business loans, and low-interest loan and grant schemes (e.g. from cooperative banks). Measures to support social groups adversely affected by the change should also be considered. Once the energy communities are in place, it is also important to consider the possibility of creating support structures for the model, such as local and national governance associations and federations of energy communities, to ensure some control over the individual energy community. As well as feeding off each other—through good practices, initiatives, inspiration—a certain cooperative grassroots movement can be encouraged.

The initial investment required to set up an energy community requires support from the state, which may take the form of direct investment, assistance in accessing private credit, or tax deductions. Even though commons are seen as collective mechanisms for social transformation, they still require a boost from the state. We cannot expect individuals with significant financial resources to enter these sectors where the barriers to entry are high. Other examples, such as housing cooperatives, community gardens, or non-profit cooperative markets, also require state support due to the market dominance of these sectors. This support enables the collectives initiating these activities to secure the economic resources needed to develop the projects. These initiatives can be spearheaded by groups of individual citizens, trade unions, or neighbourhood associations seeking to establish an alternative model of communal living, distinct from the self-interest and competition that prevails within the capitalist system. Several examples of community energy financing initiatives can be found in different European countries, for instance, Goiener and Som Energia (Spain), E'nostra (Italy), Ecopower (Belgium), Énergie Partagée Coopérative (France), Electra Energy (Greece), and Bürgerwerke (Germany) (SCCALE 20 30 50, 2023).

Empowering Citizens: Beyond the Neutrality of the State

It must be borne in mind that even if the level of citizens' welfare can provide sufficient purchasing power and capital to cover investments, this may not be enough. Potential prosumer collectives must not only possess the financial capacity to invest but also be convinced that the changes can be positive for them and for the environment. For a genuine energy transition to occur, the state must make efforts to convince or assemble a critical mass of people and create an 'energy culture'. In this sense, the promotion of participation and advisory tools, as well as education, training, and awareness-raising programmes, are crucial for fostering this social conviction.

Pushing citizens through public mechanisms to become 'environmental citizens' (Dobson, 2007) is one of the keys to the development of this type of initiative. Tax incentives and financial support to motivate citizens to become more 'environmental' do not always have optimal results. It is common to think that taxes that incorporate the externalities generated by an activity, or that seek to discourage a particular activity, can be the key to any environmental policy that seeks to change citizens' behaviour. However, this perspective is not always sufficient. Acceptance of the rules is a reason to act, but this acceptance is the key to maintaining this type of regulation over time. It is therefore necessary not only to implement public policies that help to direct citizens' behaviour towards a given common good but also to make citizens want to be part of the change towards that common good.

Dobson (2007: 280), in his exploration of 'environmental citizens', suggests that behaviour driven solely by self-interest, often associated with regulations designed to induce behavioural change, does not always protect the common interest. Instead, the 'environmental citizen' should be committed to achieving the common good. This shift implies that the citizen is no longer merely an individual agent existing within a neutral system within a framework of justice but, rather, an active agent actively pursuing the common good. The next question is: where does this 'environmental citizen' come from?

The education system has a crucial role to play in fostering 'green' behaviours and environmental awareness. However, it is not enough to have knowledge about environmental issues and collective responsibility in our relationship with our biophysical base. Environmental problems also require collective participation in decision-making and the empowerment of citizens to engage in cooperative initiatives for the common good. The promotion of energy communities should not rely solely on the good intentions of a group of individuals; the state must also be involved to encourage citizens to be part of this necessary change.

In this context, state support should include promotion through institutional advertising and effective communication. This argument might seem to contradict the principle of political neutrality, where the state should remain neutral so that citizens can choose between remaining individual, rational, and free agents, joining an energy community, or continuing to contract with an oligopolistic private company. Nevertheless, interpreting such neutrality in this context would be a misinterpretation, since the freedom of individual agents may be influenced by private powers—i.e. electricity companies—that can influence the decisions and choices of citizens.

Therefore, the establishment of the commons relies not only on the voluntary actions of a group but also on a concept of the common good that may in fact depend on or be influenced by the actions of the state. The insistence on state neutrality, where the state refrains from promoting a particular use or consumption of goods through its public influence mechanisms, creates a vacuum that private powers can exploit. State neutrality, when understood as the absence of action in certain aspects of social life, can inadvertently lead to the co-option of public discourse by those with significant social power. The state has a duty to intervene and encourage forms of community participation, including through the media. Consequently, media advertising and the promotion of participatory and consultative tools, as well as education, training, and awareness-raising programmes, play a key role in fostering this societal engagement. In addition, strengthening the capacity of neighbourhood associations, trade unions, NGOs, and other organisations would facilitate the promotion of public awareness and involvement in these initiatives.

Conclusions

The urgency of the transition away from fossil fuels underscores the need for new institutional and community structures that facilitate a fair, rapid, and sustainable energy paradigm shift. The direction in which such transformations should go is towards a distributed and decentralised system, which implies a redefinition of energy legislation, governance frameworks, fiscal systems, land-use planning, and property regimes. In light of this need, this chapter has highlighted the transformative potential of studying energy under the theoretical framework of the commons and underlined the benefits and possible limitations of energy communities on the road to energy transition.

In all likelihood, the energy transition will require a large critical mass of citizens convinced of the need for change towards a new energy system. This will require empowering local communities and individuals to make decisions about their energy sources, consumption patterns, and the development of renewable energy projects. The direct involvement of civil society in energy management not only promotes self-sufficiency and community resilience but is also a potential catalyst for more equitable and democratic societies.

Nevertheless, it has been stressed that the promotion of bottom-up governance depends on robust support from national and local energy policies and government processes. Indeed, the role of the state in establishing energy communities is of great importance. In particular, the state must ensure fair and sustainable processes through supportive legal frameworks, initial assistance such as training and infrastructure, and the provision of financial incentives. In addition, any incentives must be accompanied by energy efficiency and energy saving policies. It will be important to document such processes so that these models can be replicated and improved, and to ensure that this potential symbiosis between the state and civil society is smooth, effective, and peaceful.

Acknowledgments Part of this work has been supported by the Agència de Gestió d'Ajuts Universitaris i de Recerca (AGAUR) and the EU (NextGenerationEU) under Investigo grant (200076ID12).

References

Aalto, P., Haukkala, T., Kilpeläinen, S., & Kojo, M. (2021). Chapter 1—Introduction: Electrification and the Energy Transition. In *Electrification. Accelerating the Energy Transition* (pp. 3–24). Academic Press, Elsevier. https://doi.org/10.1016/B978-0-12-822143-3.00006-8

Anderies, J. M., Janssen, M. A., & Ostrom, E. (2004). A Framework to Analyze the Robustness of Social-ecological Systems from an Institutional Perspective. *Ecology and Society, 9*(1), 18. Retrieved from http://www.jstor.org/stable/26267655

Bankwatch Network and REScoop. (2022). *Energy Communities: A Brief Explainer for Managing Authorities in Central and Eastern Europe*. Briefing. Retrieved August 22, 2023, from https://www.rescoop.eu/toolbox/energy-communities-a-brief-explainer-for-managing-authorities-in-central-and-eastern-europe

BECoop, CEES, PowerPoor, eCREW, W4RES, UP-STAIRS, COME RES, EC2, NRG2PEERS, & Sun4All. (2023). *Are Renewable Energy Communities a Vehicle to Mitigate the Energy Crisis and Lift People Out of Energy Poverty?* March 2023. Collaborative briefing. Retrieved September 7, 2023, from https://ieecp.org/wp-content/uploads/2023/04/Collaborative-briefing-energy-communities-and-the-energy-crisis-2023.pdf

Bollier, D., & Helfrich, S. (2019). *Free, Fair, and Alive: The Insurgent Power of the Commons*. New Society Publishers.

Butenko, A. (2016). *User-Centered Innovation and Regulatory Framework: Energy Prosumers' Market Access in EU Regulation (May 30, 2016)*. TILEC Discussion Paper No. 2016-015. Retrieved from https://ssrn.com/abstract=2797545

Caffentzis, G., & Federici, S. (2014). Commons Against and Beyond Capitalism. *Community Development Journal, 49*, 92–105. https://doi.org/10.1093/cdj/bsu006

CAG Consultants. (2021). *Devon Community Energy: Socio Economic Impact Assessment*. Final Report. Devon County Council. Retrieved July 25, 2023, from https://devoncc.sharepoint.com/sites

Campos, I. (2020). Regulatory Challenges and Opportunities for Collective Renewable Energy Prosumers in the EU. *Energy Policy, 138*, 1–12. https://doi.org/10.1016/j.enpol.2019.111212

Caramizaru, A. & Uihlein, A. (2020). *Energy Communities: An Overview of Energy and Social Innovation*. EUR 30083 EN, Publications Office of the European Union, Luxembourg, 2020, ISBN 978-92-76-10713-2. https://doi.org/10.2760/180576, JRC119433.

Dawson, A. (2022). *People's Power. Reclaiming the Energy Commons*. OR Books, LLC.

de Laurentis, C. (2020). Mediating the Form and Direction of Regional Sustainable Development: The Role of the State in Renewable Energy Deployment in Selected Regions. *European Urban and Regional Studies, 27*(3), 1–15. https://doi.org/10.1177/0969776420904989

Debizet, G., Pappalardo, M., & Wurtz, F. (2022). *Local Energy Communities. Emergence, Places, Organizations, Decision Tools.* Routledge.

Diputació de Barcelona. (2023). *1,3 M€ per a la nova Oficina d'Impuls a les Comunitats Energètiques.* 9 October. Retrieved October 30, 2023, from https://www.diba.cat/mediambient/-/13m-nova-oficina-d-impuls-a-comunitats-energetiques

Directive (EU) 2018/2001 of the European Parliament and of the Council of 11 December 2018 on the promotion of the use of energy from renewable sources.

Directive (EU) 2019/944 of the European Parliament and of the Council of 5 June 2019 on common rules for the internal market for electricity and amending Directive 2012/27/EU.

Dobson, A. (2007). Environmental Citizenship: Towards Sustainable Development. *Sustainable Development, 15*, 276–228. https://doi.org/10.1002/sd.344

European-Ukrainian Energy Agency (EUEA). (2021). *What is the Difference: "Renewable Energy Communities" and "Citizen Energy Communities"?* 30.03.2021. Retrieved July 25, 2023, from https://euea-energyagency.org/en/news/market-news/what-is-the-difference-renewable-energy-communities-and-citizen-energy-communities/

Eurostat. (2023a). *Fossil Fuels Stabilised at 70% of Energy Use in 2021.* 30.01.2023. Retrieved August 21, 2023, from https://ec.europa.eu/eurostat/web/products-eurostat-news/w/ddn-20230130-1#:~:text=The%20EU%20continues%20to%20largely,same%20level%20as%20in%202020

Eurostat. (2023b). *Shedding Light on Energy.* European Union. Retrieved August 21, 2023, from https://ec.europa.eu/eurostat/web/interactive-publications/energy-2023

Fabra, N., & Fabra Utray, J. (2009). Competencia y poder de mercado en los mercados eléctricos. *Cuadernos Económicos de ICE, 79*, 17–43. https://doi.org/10.32796/cice.2010.79.5982

Franquesa, J. (2023). *Molinos y gigantes.* Errata Naturae.

Giotitsas, C., Nardelli, P. H. J., Kostakis, V., & Narayanan, A. (2020). From Private to Public Governance: The Case for Reconfiguring Energy Systems as a Commons. *Energy Research & Social Science, 70*, 1–7. https://doi.org/10.1016/j.erss.2020.101737

Giotitsas, C., Nardelli, P. H. J., Williamson, S., Roos, A., Pournaras, E., & Kostakis, V. (2022). Energy Governance as a Commons: Engineering Alternative Socio-technical Configurations. *Energy Research & Social Science, 84*, 1–6. https://doi.org/10.1016/j.erss.2021.102354

Gough, I. (2015). Climate Change and Sustainable Welfare: The Centrality of Human Needs. *Cambridge Journal of Economics, 39*(5), 1191–1214. https://doi.org/10.1093/cje/bev039

Gough, I. (2017). Recomposing Consumption: Defining Necessities for Sustainable and Equitable Well-Being. *Philosophical Transactions of the Royal Society A, 375*(2095), 1–18.

Graham, E., & Fulghum, N. (2024). *Wind and solar overtake EU fossil fuels in the first half of 2024.* EMBER. Retrieved August 3, 2024, from https://ember-climate.org/insights/research/eu-wind-and-solar-overtake-fossil-fuels/#supporting-material

Gruber, L., Bachhiesl, U., & Wogrin, S. (2021). The Current State of Research on Energy Communities. *Elektrotechnik und Informationstechnik, 138*, 515–524. https://doi.org/10.1007/s00502-021-00943-9

Gui, E. M., Diesendorf, M., & MacGill, I. (2017). Distributed Energy Infrastructure Paradigm: Community Microgrids in a New Institutional Economics Context. *Renewable and Sustainable Energy Reviews, 72*, 1355–1365. https://doi.org/10.1016/j.rser.2016.10.047

Heldeweg, M. A., & Saintier, S. (2020). Renewable Energy Communities as 'Socio-legal Institutions': A Normative Frame for Energy Decentralization? *Renewable and Sustainable Energy Reviews, 119*, 1–14. https://doi.org/10.1016/j.rser.2019.109518

IDAE. (2019). Guía para el Desarrollo de Instrumentos de Fomento de Comunidades Energéticas Locales. Madrid, España: Instituto para la Diversificación y Ahorro de la Energía (IDAE). Documento de trabajo.

Kunze, C., & Becker, S. (2014). *Energy Democracy in Europe: A Survey and Outlook.* Rosa Luxemburg Foundation.

Lazdins, R., Mutule, A., & Zalostiba, D. (2021). PV Energy Communities—Challenges and Barriers from a Consumer Perspective: A Literature Review. *Energies, 14*(16), 4873. https://doi.org/10.3390/en14164873

Lazoroska, D., Palm, J., & Bergek, A. (2021). Perceptions of Participation and the Role of Gender for the Engagement in Solar Energy Communities in Sweden. *Energy, Sustainability and Society, 11*(35), 1–12. https://doi.org/10.1186/s13705-021-00312-6

Lloredo Alix, L. (2020). Bienes comunes. *EUNOMÍA. Revista en Cultura de la Legalidad, 19*, 214–236. https://doi.org/10.20318/eunomia.2020.5709

Loebbe, S., Sioshansi, F., & Robinson, D. (Eds.). (2022). *Energy Communities. Customer-Centered, Market-Driven, Welfare-Enhancing?* Academic Press-Elsevier.

Mattei, U. (2011). *Beni comuni. Un manifesto.* Laterza.

Millward-Hopkins, J. (2022). Inequality can double the energy required to secure universal decent living. *Nature Communications, 13*, 5028. https://doi.org/10.1038/s41467-022-32729-8

Millward-Hopkins, J., et al. (2020). Providing decent living with minimum energy: A global scenario. *Global Environmental Change, 65*(2020), 102168. https://doi.org/10.1016/j.gloenvcha.2020.102168

Ministerio para la Transición Ecológica y el Reto Demográfico (MITECO). (2020). *Consulta pública previa.* Comunidades Energéticas Locales. Retrieved September 7, 2023, from https://energia.gob.es/en-us/participacion/paginas/detalleparticipacionpublica.aspx?k=358#

Negri, A., & Hardt, M. (2009). *Commonwealth.* Harvard University Press.

Ostrom, E. (1990). *Governing the Commons: The Evolution of Institutions for Collective Action.* Cambridge University Press.

Ostrom, E. (1999). *Design Principles and Threats to Sustainable Organizations That Manage Commons.* W99-6, Workshop in Political Theory and Policy Analysis, Indiana University, 17 February. Retrieved July 6, 2023, from https://hdl.handle.net/10535/5465

Otamendi Irizar, I., Grijalba, O., Arias, A., Pennese, C., & Hernéndez, R. (2022). How Can Local Energy Communities Promote Sustainable Development in European Cities? *Energy Research & Social Science, 84*(102363), 1–15. https://doi.org/10.1016/j.erss.2021.102363

Palm, J. (2021). *Energy Communities in Different National Settings—Barriers, Enablers and Best Practices, Deliverable D3.3.* New Clean Energy Communities in a Changing European Energy System (NEWCOMERS). https://doi.org/10.13140/RG.2.2.23798.32327

Pappalardo, M. (2022). Energy Communities and Commons. Rethinking Collective Action Through Inhabited Spaces. In G. Debizet, M. Pappalardo, & F. Wurtz (Eds.), *Local Energy Communities. Emergence, Places, Organizations, Decision Tools* (1st ed.). Routledge.

Proka, A. (2023). *Barriers and Opportunities for the Development of Energy Communities with Municipal Involvement. Results from LIFE LOOP Survey (D2.3).* LIFE LOOP. Retrieved October 30, 2023, from https://energy-cities.eu/wp-content/uploads/2023/04/D2.3-LIFE-LOOP-BO-report_final.pdf

Rao, N. D., & Min, J. (2018). Decent living standards: Material prerequisites for human wellbeing. *Social Indicators Research, 138*, 225–244. https://doi.org/10.1007/s11205-017-1650-0

Reis, I., Gonçalves, I., Lopes, M. A. R., & Henggeler Antunes, C. (2020). A study of the inclusion of vulnerable consumers in energy communities with peer-to-peer exchanges. In *2020 International Conference on Smart Energy Systems and Technologies (SEST), Istanbul, Turkey* (pp. 1–6). https://doi.org/10.1109/SEST48500.2020.9203312

REScoop. (2021). *Q&A: What are 'Citizen' and 'Renewable' Energy Communities?* RESCoop.eu. Retrieved October 30, 2023, from https://www.rescoop.eu/uploads/rescoop/downloads/QA-What-are-citizens-energy-communities-renewable-energy-communities-in-the-CEP.pdf

Romero-Rubio, C., & de Andrés Díaz, J. D. (2015). Sustainable energy communities: A study contrasting Spain and Germany. *Energy Policy, Elsevier, 85*, 397–409. https://doi.org/10.1016/j.enpol.2015.06.012

Santiago, E. (2023). *Contra el mito del colapso ecológico*. Arpa Editores.

SCCALE 20 30 50. (2023). *Financing Guide for Energy Communities*. Retrieved October 30, 2023, from https://www.sccale203050.eu/wp-content/uploads/2023/02/SCCALE203050_financingguide_energycommunities.pdf

Silva, M. (2019). The European Union's Revolving Door Problem. In D. Dialer & M. Richter (Eds.), *Lobbying in the European Union*. Springer. https://doi.org/10.1007/978-3-319-98800-9_20

Sokołowski, M. (2016). *Regulation in the European Electricity Sector*. Routledge.

Sperling, K. (2017). How Does a Pioneer Community Energy Project Succeed in Practice? The Case of the Samsø Renewable Energy Island. *Renewable and Sustainable Energy Reviews, 71*, 884–897. https://doi.org/10.1016/j.rser.2016.12.116

Streimikiene, D., Baležentis, T., Volkov, A., Morkūnas, M., Žičkienė, A., & Streimikis, J. (2021). Barriers and Drivers of Renewable Energy Penetration in Rural Areas. *Energies, 14*(20), 6452. https://doi.org/10.3390/en14206452

Szulecki, K. (2017). Conceptualizing Energy Democracy. *Environmental Politics, 27*, 21–41. https://doi.org/10.1080/09644016.2017.1387294

van Veelen, B., & van der Horst, D. (2018). What is Energy Democracy? Connecting Social Science Energy Research and Political Theory. *Energy Research & Social Science, 46*, 19–28. https://doi.org/10.1016/j.erss.2018.06.010

Wolsink, M. (2020). Distributed Energy Systems as Common Goods: Socio-political Acceptance of Renewables in Intelligent Microgrids. *Renewable and Sustainable Energy Reviews, 127*, 1–14. https://doi.org/10.1016/j.rser.2020.109841

Yenneti, K., Day, R., & Golubchikov, O. (2016). Spatial Justice and the Land Politics of Renewables: Dispossessing Vulnerable Communities Through Solar Energy Mega-projects. *Geoforum, 76*, 90–99. https://doi.org/10.1016/j.geoforum.2016.09.004

CHAPTER 5

Linking Energy Poverty, Environmental Justice and Forced Displacement: Controversies and Ways Forward

Francesca Rosignoli

Introduction

Over the last decade, the number of forcibly displaced people has steadily increased, recording a total number of 89.3 million displaced people worldwide in 2021 (UNHCR, 2022b). According to the Internal Displacement Monitoring Centre (IDMC), the majority of them are internally displaced people (IDPs), which amounted to 59.1 million at the end of 2021 (IDMC, 2022). More specifically, data shows that the current trend is marked by a significant impact of disasters triggering more than 60% of the internal displacements due to weather-extreme events such as storms and floods (IDMC, 2022). Although environmental factors are significantly impacting the trend of current displacement, people who find

F. Rosignoli (✉)
Rovira i Virgili University, Tarragona, Spain
e-mail: francesca.rosignoli@urv.cat

D. Iglesias Márquez et al. (eds.), *Legal Challenges at the End of the Fossil Fuel Era*, Global Issues,
https://doi.org/10.1007/978-3-031-61766-9_5

themselves in vulnerable situations in the context of climate change and environmental degradation receive limited (or any) legal protection.

At the time of writing, existing legal instruments to protect IDPs are the following:[1]

- The 1998 Guiding Principles on Internal Displacement
- The 2009 African Union Convention for the Protection and Assistance of Internally Displaced Persons in Africa (known as Kampala Convention)
- The 2013 Peninsula Principles on Climate Displacement within States (hereinafter the Peninsula Principles).

However, all these legal instruments have shown significant shortcomings when it comes to internally displaced people in the context of climate change and disasters (Felipe Pèrez, 2018; Rosignoli, 2022). Indeed, although the 1998 Guiding Principles on Internal Displacement refers to 'environmentally displaced people' by including disasters within the causes of displacement, this is a non-legally binding instrument, thus involving no obligations to states to implement these rules.

As enshrined in Article 2,

> Internally displaced persons are persons or groups of persons who have been forced or obliged to flee or to leave their homes or places of habitual residence, in particular as a result of or in order to avoid the effects of armed conflict, situations of generalised violence, violations of human rights or natural or human-made disasters, and who have not crossed an internationally recognised state border. (Guiding Principles on Internal Displacement, 1998)

The only existing legally binding instrument to protect IDPs is the 2009 African Union Convention for the Protection and Assistance of

[1] Apart from a few exceptions, it is worth mentioning that International Law prohibits any form of forced displacement. Further, although there is no specific right to protection against forced displacement, it is implicit in several human rights, such as the right to protection against interference with privacy, family, home or correspondence (Art. 12 UDHR), the right to freedom of movement and residence within the borders of each state (Art. 13 UDHR), and the right to a standard of living adequate for the health and well-being of himself and of his family, including food, clothing, housing and medical care and necessary social services (Art. 25 UDHR).

Internally Displaced Persons in Africa. The so-called Kampala Convention explicitly mentions and protects 'those who have been internally displaced due to natural or human-made disasters, including climate change.'[2] However, this instrument is very limited in scope as those people are protected only in the African States that ratified the Convention.[3]

Finally, the 2013 Peninsula Principles represent the first formal policy addressing internally climate-displaced peoples.[4] Although they might have the potential to pave the way for adopting a global legally binding Convention such as the Kampala Convention, the Principles are a soft-law instrument with limited scope that does not trigger any obligations to states.

As a result, given the limited efficacy of existing legal instruments and the increasing number and vulnerability of IDPs in the context of climate change and environmental degradation, which policies can be advanced to tackle this emerging policymaking area? This chapter seeks to answer this question by proceeding as follows. First, it provides a survey of existing literature to explain the reasons behind the weaknesses of International Law in addressing this issue. Second, it analyses the energy poverty-environmental justice-displacement nexus through the concept of energy vulnerability. Finally, it argues the urgent need to design policies to address this emerging policymaking area and provides for a proposal based on the strategic role of energy communities to strengthen the collective capabilities of those unable (trapped populations) or unwilling to leave (immobility).

[2] African Union Convention for the Protection and Assistance of Internally Displaced Persons in Africa (adopted 22 October 2009, entered into force 6 December 2012), Article 5, para 4.

[3] At the time of writing, the number of states that ratified the Kampala Convention amounts to 33 out of 55.

[4] See Principle 2 of the Peninsula Principles on Climate Displacement within states which defines climate displacement as 'the movement of people within a state due to the effects of climate change, including sudden and slow-onset environmental events and processes, occurring either alone or in combination with other factors.'

Climate Change and Disaster Displacement: A Critical Examination of the Literature

As mentioned above, International Law has shown many shortcomings when it comes to protecting internally displaced people in the context of climate change and displacement. In particular, it is worth mentioning the lack of a legally binding treaty addressing this emerging category of internally displaced people.

At first glance, the reasons behind these limitations in International Law might rely on the empirical challenges in isolating 'environmentally' displaced people as an identifiable and measurable group of people (Mayer, 2016, 2019). The lack of a common working definition and, therefore, of a legal definition included in a legally binding treaty might make it difficult to provide reliable data concerning the number of 'environmentally' displaced people (Gemenne, 2011; Jakobeit & Methmann, 2012). In addition, the multi-causal nature of displacement might further complicate these calculations, since people can be displaced as a result of various factors such as wars, armed conflicts, disasters and environmental degradation (Foresight., 2011). As pointed out by the literature (Brzoska & Fröhlich, 2016; Burrows & Kinney, 2016; Lee, 2020; Reuveny, 2007), the impacts of climate change and/or disasters are increasingly triggering armed conflicts and violence due to the scarcity of resources so that it might be hard to disentangle those factors and provide separate data.[5] A telling example is the Sahel region, where environmental degradation and climate change-related challenges coexist with (and more often trigger) indiscriminate violence perpetrated by armed actors, thus resulting in displacement and humanitarian crises in the whole region (UNHCR, 2022a).

On closer inspection, however, many progresses have been made to overcome the critical issues discussed above over the last decades. As for the empirical challenges in disentangling displaced by conflicts and climate-related impacts, significant improvements have been made by the Internal Displacement Monitoring Centre (IDMC). Born in 1998 as part of the Norwegian Refugee Council (NRC), IDMC provides verified estimates of the number of internally displaced by conflict and disasters. More specifically, it distinguishes between internal displacements (i.e. each new

[5] A critical stand against this nexus has been raised by some authors opposing that armed conflicts rather depend on several political and social variables. See, among others, Salyean (2008).

forced movement of a person within her home country during the year) and internally displaced people, IDPs (i.e. the total number of people who live in internal displacement at the end of the year).

Not surprisingly, the total number of IDPs is usually higher than the number of internal displacements because the first data also includes the number of people displaced in previous years who still live in internal displacement (the so-called stocks), while the second data estimates the number of internal displacements, including repeated displacements, recorded during the year (the so-called flows). As for the legal challenges related to the multi-causal nature of displacement, most recent initiatives advanced by the United Nations High Commissioner for Refugees (UNHCR) are paving the way for future recognition of those displaced in the context of climate change and disasters. Although reluctant to use the term 'climate refugees' and still prefer to use the expression 'persons displaced in the context of disasters and climate change,' the UNHCR has recognised that people may have a valid claim for refugee status if the adverse effects of climate change interact with armed conflict and violence. Based on its recent contributions (UNHCR, 2018, 2020), the UNHCR has recognised the importance of climate change as a driver of displacement and the urgent need to protect people displaced in the context of disasters and environmental degradation. Thus, rather than a limitation, such an interplay between environmental degradation and conflicts might be well seen as a promising pathway for protecting climate refugees in the long term.

Notwithstanding all the progress made over the years, the protection gap surrounding the figure of IDPs in the context of climate change and disasters is still a reality. This misrecognition implies that such IDPs more often end up in camps where they experience human rights violations (e.g. gender-based violence), protracted detention between 10 and 26 years (Ferris, 2018) and further environmental degradation. Indeed, not only IDPs in camps have limited or no access to healthcare, education, jobs and basic services but are extremely exposed to energy poverty and vulnerability (see next section of this chapter). Indeed, energy needs in camps are usually overlooked when responding to a displacement crisis for the following reasons (Mastor et al., 2018):

- The lack of dedicated funding and expertise for providing long-term energy solutions.
- Scarce (if any) long-term investments in sustainable and efficient energy services.

- The absence of energy-related data collection for estimating IDPs' energy needs and costs.

Ultimately, organisations providing assistance and first relief to IDPs in the context of disasters do not consider that most of the people in camps will remain there between 10 and 26 years, thus requiring long-term energy solutions able to take their energy needs into account, while avoiding overstress already scarce resources. Given that climate change and climate change-induced displacement are permanent phenomena and that the legal gap is preventing to offer IDPs adequate protection, which policies can be advanced to tackle this permanent and structural phenomenon? The next section will stress the importance of energy policies to cope with this issue by exploring the energy poverty-environmental justice-displacement nexus.

The Energy Poverty-Environmental Justice-Displacement Nexus

Most of the literature has focused on the environmental injustices deriving from the disproportionate impacts of CO_2 emissions on the poorest and less responsible developing countries (Bettini, 2013; Bettini et al., 2017; Caney, 2014; Klepp & Herbeck, 2016; Mastaler, 2018; Park & Pellow, 2019; Turhan & Armiero, 2019). The main argument is that although most emissions are emitted by the energy systems of developed countries, the hardest consequences of climate change affect developing countries, especially through forced displacement and South-South climate migration.

Extractivism and megaprojects by multinationals—sometimes promoted by states and financial organisations—are also producing a large-scale displacement of those living in areas rich in natural resources and commodities (mostly indigenous people) and land dispossession (Vindal Ødegaard & Rivera Andía, 2019). The 'energy' question is key here and it is more often framed through the lens of climate justice with a focus on forced displacement and migration in the context of white, colonial, patriarchal remaking of environmental injustice (Whyte, 2017). By contrast, a rather neglected aspect in the analysis of climate change and disaster displacement is the environmental justice issues raised by the extreme energy poverty and vulnerability experienced by those people who end up in IDP camps for more than a decade.[6] In other words, the environmental injustice suffered by the 'trapped populations' (Foresight., 2011), i.e. those unable to leave, has been rather overlooked by the literature so far.[7]

The term 'energy poverty' is conceived here as the insufficient access to essential energy services like appropriate heating, cooling and lighting, and the energy requirements to power *forcibly displaced people camps* (adapted from DellaValle & Czako, 2022: 2).

In this regard, severe forms of 'energy' deprivation of IDPs in camps are:

- The lack of or limited access to alternative, efficient and sustainable energy services, with firewood and charcoal being the most used fuels, especially for cooking (Mastor et al., 2018; UNHCR, 2019).
- The lack of or limited access to (clean) electricity as diesel generators are still preferred above, for example, solar mini-grids, particularly in rural settlements in sub-Saharan Africa (Baldi et al., 2022).
- The lack of or limited access to heating, which relies almost exclusively on the warmth of the cooking fire or more often remains an unmet or unfulfilled energy need (Food and Agriculture Organization of the United Nations et al., 2019; Lahn & Grafham, 2015).

While major investments would be needed to meet these energy needs, improve technical expertise and raise awareness, projects going in this direction are relatively rare[8] as the energy needs of IDPs are poorly understood and still (erroneously) perceived as a matter of temporary relief.

[6] For a comprehensive overview of definitional controversies around energy poverty, vulnerability and justice, see Bouzarovski (2018).

[7] Among the few exceptions, see Zickgraf (2019).

[8] Among the existing projects and initiatives, see the Energy Monitoring Framework (https://eis.unhcr.org/home), the UNHCR Global Strategy for Sustainable Energy (UNHCR, 2019), the Moving Energy Initiative (https://www.chathamhouse.org/about-us/our-departments/environment-and-society-programme/moving-energy-initiative-sustainable) and the projects Humanitarian Engineering and Energy for Displacement, HEED (http://heed-refugee.coventry.ac.uk/), Renewable Energy for Refugees (https://data.humdata.org/organization/practicalaction) and Energy Solutions for Displacement Settings (https://energypedia.info/wiki/Energy_Solutions_for_Displacement_Settings).

All this prevents those people from meeting their basic human needs and full control of their lives, thus resulting in a form of energy vulnerability and, ultimately, injustice. In this regard, energy vulnerability[9] is conceived here as a form of unequal and unfair form of deprivation that derives from environmental injustices 'by design' and 'by default.'[10] For the purpose of this chapter, environmental injustices 'by design' are defined as those originating from the disproportionate sharing of environmental hazards depending upon CO_2 emissions. More specifically, this type of environmental injustice is generated by multinationals extracting fossil fuels from the ground and states that more often let them pollute in the name of economic growth and national interest. Crucially, these subjects would have the power but little interest in changing energy production and consumption patterns at a global scale. In a capitalistic system, the role of states is far from being neutral in confronting environmental injustice 'by design.' Apart from a few exceptions, such as Pulido and Pellow (2017), most environmental justice scholars focus on this type of injustice and claim mitigation policies advanced by states to force multinationals and companies to reduce CO_2 emissions. In other words, these scholars tackle environmental injustices embedded in the system (i.e. injustices embedded into the design) but identify the states (i.e. the system) as suitable subjects to redress them. However, as outlined by Pulido and Pellow (2017), rarely those who created the problem will be able to solve it.

Environmental injustices 'by default,' instead, are those injustices affecting people's everyday life, such as the lack of access to environmental services and amenities, which ultimately result in capabilities deprivation of individuals and communities. The focus here is more on redressing environmental injustice affecting a specific community for a specific purpose: advancing adaptation policies able to activate protection mechanisms from below. Scholars engaged with this topic tackle injustices that take

[9] Cf. the broader definition of energy vulnerability by Bouzarovski: 'energy vulnerability thinking is probabilistic: it highlights the factors that affect the likelihood of becoming poor' (Bouzarovski, 2018: 19). Among the factors that describe the likelihood of experiencing a socially and materially inadequate level of energy services, the author includes the affordability–access factor, the nature and structure of the built environment of the home, social practices and energy needs, the access to infrastructures, and injustices (Bouzarovski, 2018: 19–21).

[10] The distinction between environmental injustice 'by design' and 'by default' is part of my proposal.

place on the ground at a local scale (i.e. injustices by default). In the context of this research, environmental injustices 'by default' materialise in the misrecognition of IDPs' energy needs, deprivation of their capabilities and pressure on already scarce environmental resources.

This latter aspect might be crucial given the specific conditions of IDP camps in the context of disasters and climate change-related impacts. In the absence of sustainable alternatives, most camps rely on the extensive use of firewood for cooking and heating (Lahn & Grafham, 2015). In turn, it results in massive deforestation that ultimately affects the environment of a state already affected by environmental degradation. More specifically, deforestation is a significant problem in developing countries as they host 86% of forcibly displaced people worldwide. Data shows that while 64,700 acres of forest are burned by forcibly displaced people annually, developing countries already suffer from environmental degradation and deforestation (ibid, 2015). Therefore, such unsustainable production and energy consumption patterns exceed the environment's carrying capacity, increase resource depletion and may fuel violent conflicts over scarce resources. In turn, when environmental degradation interacts with armed conflict and violence, people may be compelled to move internally or even leave their country and seek international protection. It follows that energy policies are urgent to alleviate the energy vulnerability and poverty of IDPs in camps, to tackle environmental injustices 'by default' and, accordingly, to prevent IDPs' anticipatory leaves.

Strengthening Collective Capabilities Through Energy Communities: Towards a Community-Based Approach

As mentioned above, the mainstream literature has focused on environmental injustices 'by design,' i.e. unfair impacts of CO_2 emissions mainly produced by developed countries on least responsible developing countries. This contribution, instead, focuses on environmental injustices 'by default.' As said, those include the misrecognition of IDPs' energy needs, deprivation of their capabilities and pressure on already scarce environmental resources.

Compared to classic energy justice studies typically concerned with unequal distribution of resources/hazards, procedures affecting meaningful involvement of all people in the decision-making process and misrecognition of different socio-cultural identities (Jenkins et al., 2016; McCauley et al., 2013; Sovacool et al., 2016; Walker & Day, 2012), I am adding here the capability dimension.[11] So far, such capability dimension has been explicitly explored in the energy sector by several authors (Day et al., 2016; Melin et al., 2021; Sovacool & Dworkin, 2016; Wood & Roelich, 2019).[12] More specifically, these authors pointed out the multidimensional capability deprivation derived from the lack of access to energy services. In particular, Day, Walker and Simcock provided a useful definition of energy poverty through the lens of the capability approach as 'an inability to realise essential capabilities as a direct or indirect result of insufficient access to affordable, reliable and safe energy services, and taking into account available reasonable alternative means of realising these capabilities' (Day et al., 2016: 260). While these reflections emphasise the access to energy services among the necessary conditions for individuals and communities to flourish and live a life they have reason to value (Sen, 1999), there has been scant application of a capabilities lens in the context of forced displacement correlated with climate change and disasters.[13]

Yet, specific conditions of IDP camps in the context of disasters and climate change can be well seen as a pivotal case of (and to address) environmental injustices 'by default.' Above all, most camps are usually located in the so-called climate hotspots: regions 'for which potential climate change impacts on the environment or different activity sectors can be particularly pronounced' (Giorgi, 2006: 1). Indeed, as a result of the impacts of climate change (e.g. extreme weather events, slow and rapid onset events), climate hotspots are increasingly becoming hubs of vulnerability where the affected populations more often end up in camps as IDPs because they are either unable to leave (trapped populations) or unwilling

[11] As pointed out by Schlosberg (2007), Environmental Justice encompasses four different dimensions of justice: distribution, recognition, participation and capabilities (Schlosberg, 2007).

[12] The first attempt to shed light on the operationalisation of the capability approach in the field of energy justice was made at the Conference 'Energy Justice and the Capability Approach – interdisciplinary perspectives' held in 2018 in Malmö (Sweden). For an overview of recent contributions on energy justice and the capability approach, see (Melin et al., 2021).

[13] Among the few exceptions, see (Rosignoli, 2022; Van der Land, 2018).

to do so (immobility).[14] A challenging question is then the following: how can we design just energy policies able to alleviate the energy vulnerability and poverty of IDPs 'trapped' in camps?

A community-based approach is to be adopted for turning such hubs of vulnerability (i.e. camps) into hubs of resistance and resilience (i.e. energy communities). The main goal of this proposal is to strengthen IDPs' collective capabilities[15] to cope with energy poverty and vulnerability. As said, those people are about to stay 'trapped' in camps for many years. As a result, equipping camps with long-term energy solutions able to take IDPs' energy needs into account and avoid overstress already scarce resources is definitely urgent.

To this end, this proposal suggests turning camps, currently designed for providing temporary shelter and assistance, into energy communities driving for innovative and durable energy solutions. In this context, energy communities are conceived as emerging subjects, scale, place, network, process and identities (Walker, 2011) able to construct 'alternative forms of organisation and practices that might be scaled up, replicated and multiplied horizontally' (Schlosberg, 2020: 312). In other words, camps should be transformed into innovation hubs and conceived as pilot cases to be scaled up in different contexts. The focus here is not just on replacing unsustainable patterns of production and consumption of energy but rather on creating energy communities willing to be part of a broader transformative project.

In this respect, this proposal slightly differs from one of the most recent initiatives advanced by the UNHCR: the 2019–2025 UNHCR Global Strategy for Sustainable Energy (UNHCR, 2019). First, such proposal differs in scope as it targets IDPs in the context of climate change and disasters. By contrast, the UNHCR strategy mostly applies to refugees and may be extended to IDPs only in some cases. Second, my proposal has the

[14] For an overview on trapped populations and immobility, see (Foresight., 2011; Zickgraf, 2019). A key example of trapped populations can be found in Somalia, where pastoralists are unable to move because of armed conflicts. These latter hinder pastoralists from relocating in the face of drought and limit access to humanitarian organisations assisting affected populations (Foresight., 2011). Cases of immobility mainly driven by religion can be observed in Pacific Small Island Developing States (PSIDS). In Tuvalu, some people do not want to migrate as they still maintain faith in God's protection of their villages. Similarly, the elderly based in Fiji do not want to leave ancestral burial grounds (Cf Zickgraf, 2019).

[15] Collective capabilities are conceived here as capabilities exercised by collective subjects acting 'in order to secure a capability for the members of that group' (Robeyns, 2017: 116).

ambitious goal of turning IDP camps into energy communities potentially promoting energy citizenship, i.e. converting passive beneficiaries into active citizens participating in climate actions (Cf. DellaValle & Czako, 2022). IDPs, in fact, are citizens from the same home country displaced in camps. For this reason, it might be easier to involve them in climate actions while preventing them to cross the border to look for a better place to meet their energy needs and a life they value. In this view, this proposal differs from the UNHCR strategy, which is more focused on meeting refugee households' energy needs than creating a community. Indeed, this UNHCR strategy and initiatives like cash-based interventions (UNHCR, 2016) confirm a distributional rather than a capability approach so that IDPs are not (or scarcely) involved as active agents of a transformative agenda, being more considered as indirect beneficiaries than direct participants. More specifically, the UNHCR's overall strategy is based on the 2018 Global Compact on Refugees' principles of burden- and responsibility-sharing to ease the burden on host countries and to benefit refugees and host community members. Accordingly, the main goals are more likely to be greening infrastructures, promoting sustainable management of natural resources and patterns of energy production and consumption, and delivering clean energy to improve the quality of human settlements.[16]

The community-based approach proposed here, instead, suggests taking energy communities as a unit of analysis and conceiving them as a collective subject. I defined a collective subject as 'any group acting to make a right working in the concrete life of both the group as a collective subject and its members, improving their living conditions' (Rosignoli, 2018: 820). In this context, the overall improvement of IDPs' living conditions will be achieved through coordinated actions with national and international stakeholders for strengthening their collective capabilities (Rosignoli, 2018). I categorised collective capabilities into two types: resistant and resilient capabilities (Rosignoli, 2018). Resistant capabilities refer to the group's collective capacity to resist the impacts of climate change while undertaking measures/actions aimed at preventing the territory from being uninhabitable (Rosignoli, 2018, 2022). Such actions shall imply the active participation of energy communities and may include, but are not limited to, building sea walls, dams, seawater desalination

[16] See Global Compact on Refugees, paras 78 and 79.

plants, artificial coral reefs and raising the ground level. For example, energy communities may participate by producing and sharing clean energy to green their camps' infrastructures and to support these projects.

Resilient capabilities have to do with the capacity of any group to react constructively and cope with environmental losses and damages in the aftermath of a disaster (Rosignoli, 2018, 2022). In this regard, a significant engagement of energy communities is needed to bolster adaptive strategies for those who decide to stay/are unable to leave. In particular, energy communities might be key actors in activating civil protection/Red Cross mechanisms to strengthen cooperation between international, national and non-state actors to improve response to disasters and better support most vulnerable groups' needs.

Such climate actions aim not only to provide long-term energy solutions to improve the living conditions of IDPs but also to meet their energy needs and strengthen their collective capabilities through active participation. Those actions also aim to avoid overstressing already scarce resources by advancing sustainable production and consumption patterns within a broader environmental justice agenda. This means expanding the network and enhancing the coordination with various actors, including non-state actors and the subnational level, which are becoming ever more relevant actors in the global climate governance established by the Paris Agreement (Rosignoli, 2022, Chapter 5).

The aim is to re-integrate IDPs in the context of disasters and environmental degradation within their home country's society, thus preventing them from opting for anticipatory leaves in the lack of an alternative, concrete and equal chance to rebuild their lives in the aftermath of disasters. By preventing forced moves, such a community-based approach ultimately protects the basic capability of communities to flourish: the 'ability to continue and reproduce the traditions, practices, cosmologies, and the relationships with nature that tie native peoples to their ancestral lands' (Schlosberg & Carruthers, 2010: 12).[17]

To conclude, such climate actions might potentially redress, or at least reduce, environmental injustices 'by default' for the following reasons:

[17] In particular, see indigenous environmental justice claims typically focused on the importance of attachment to place, cultural survival and social reproduction (Di Chiro, 2008) for a community to function, flourish and ensure its members the capability to have the life they have reason to value (Holland, 2020).

- They raise awareness of IDPs' energy needs.
- They increase IDP's direct participation and responsibility for fighting against climate change and for environmental justice.
- They enhance collective actions involving various actors at different levels of government able to reduce the pressure on scarce resources and create alternative, sustainable and more accessible infrastructures.
- They prevent anticipatory leaves and environmental injustices-related claims.

Conclusion

This chapter has provided a community-based approach to fill the protection gap that surrounds the figure of Internally Displaced Persons (IDPs) in the context of disasters and climate change. More specifically, it has shed light on justice-oriented energy policies that might meet IDPs' energy needs, strengthen their collective capabilities and avoid pressure on already scarce environmental resources. Given that existing legal instruments are insufficient to cope with the increasing number and vulnerability of IDPs, this chapter has emphasised the need for energy policies able to redress environmental injustices suffered by those unwilling/unable to leave (trapped populations) that end up in camps in protracted detention.

Unlikely the rare existing strategies mostly considering IDPs as passive beneficiaries of Intergovernmental Organizations (IGOs') agendas, the community-based approach employed in this chapter suggests adopting a capability rather than a sole distributional approach to justice so that IDPs are meaningfully involved and conceived as active participants of a global transformative agenda. To this end, I proposed to turn camps into energy communities. These latter have been conceived as collective emerging subjects, scale, place, network, process and identities able to ignite climate collective actions for constructing alternative practices to be scaled up and replicated in different contexts. Such actions might have advantages that go beyond the sole acknowledgement of IDPs' energy needs.

In particular, these actions might also include the creation of networks and the significant involvement of various actors at different levels of government able to reduce the pressure on scarce resources and build sustainable and more accessible infrastructures. In this regard, the engagement of non-state actors and the subnational level might be crucial to resist and cope with climate change within the global climate governance established by the Paris Agreement. Finally, energy communities' collective actions might strengthen their collective capabilities to resist and adapt to the impacts of climate change, thus preserving their social reproduction and cultural survival.

In the long term, these actions might serve to re-integrate IDPs into their own societies and communities while preventing anticipatory leaves from becoming their last-resort option. In this view, this community-based approach ultimately seeks to dismantle the energy poverty-environmental injustice-internal displacement nexus by providing a way forward for the energy vulnerability of people trapped in camps and climate hot spots. As argued in this chapter, the link between energy poverty, environmental justice and internal displacement can be found in the concept of energy vulnerability conceived here as an unequal and unfair form of deprivation stemming from environmental injustices 'by design' and 'by default.' With this distinction, I have proposed a categorisation of environmental injustices that distinguishes between the injustices embedded into the capitalistic system at a global scale (by design) and those affecting people's everyday life at a local level (by default). While the environmental justice mainstream has mainly focused on environmental injustices 'by design' and the urgent need for mitigation policies to phase out fossil fuels' extractivism, this contribution has tried to complement this analysis by emphasising the importance of environmental injustices 'by default' in the specific case of IDP camps. This case has shown that focusing on environmental injustices 'by default' might be crucial for coping with capability deprivation and advancing adaptation policies to be scaled up, replicated and multiplied horizontally. The ultimate goal is, indeed, activating protection mechanisms from below confronting energy vulnerability. Camps transformed into energy communities might turn passive beneficiaries into active citizens able to represent IDPs' needs in the specific context of disasters and climate change while undertaking self-help initiatives and or claiming to join initiatives involving international, national and non-state actors to improve response to disasters and better support most vulnerable groups' needs.

A meaningful engagement of various actors, including affected communities in the first place, to alleviate IDPs' energy vulnerability might help dismantle the vicious circle linking energy poverty, environmental injustice and internal displacement. The literature exploring this link is still scant, especially when it comes to environmental injustices 'by default.' As a result, future research on environmentally just energy policies able to tackle environmental injustices experienced by IDPs, people unwilling to leave their countries or unable to do so (i.e. trapped populations) in the aftermath of environmental disruptions is needed and strongly encouraged.

References

Baldi, D., Moner-Girona, M., Fumagalli, E., & Fahl, F. (2022). Planning Sustainable Electricity Solutions for Refugee Settlements in Sub-Saharan Africa. *Nature Energy, 7*(4), 369–379. https://doi.org/10.1038/s41560-022-01006-9

Bettini, G. (2013). Climate Barbarians at the Gate? A Critique of Apocalyptic Narratives on "Climate Refugees". *Geoforum, 45*, 63–72. https://doi.org/10.1016/j.geoforum.2012.09.009

Bettini, G., Nash, S. L., & Gioli, G. (2017). One Step Forward, Two Steps Back? The Fading Contours of (In)justice in Competing Discourses on Climate Migration. *Geographical Journal, 183*(4), 348–358. https://doi.org/10.1111/geoj.12192

Bouzarovski, S. (2018). Understanding Energy Poverty, Vulnerability and Justice. In *Energy Poverty* (pp. 9–39). Palgrave Macmillan. https://doi.org/10.1007/978-3-319-69299-9_2

Brzoska, M., & Fröhlich, C. (2016). Climate Change, Migration and Violent Conflict: Vulnerabilities, Pathways and Adaptation Strategies. *Migration and Development, 5*(2), 190–210. https://doi.org/10.1080/21632324.2015.1022973

Burrows, K., & Kinney, P. L. (2016). Exploring the Climate Change, Migration and Conflict Nexus. *International Journal of Environmental Research and Public Health, 13*(443), 1–17. https://doi.org/10.3390/ijerph13040443

Caney, S. (2014). Two Kinds of Climate Justice: Avoiding Harm and Sharing Burdens. *Journal of Political Philosophy, 22*(2), 125–149. https://doi.org/10.1111/jopp.12030

Day, R., Walker, G., & Simcock, N. (2016). Conceptualising Energy Use and Energy Poverty Using a Capabilities Framework. *Energy Policy, 93*, 255–264. https://doi.org/10.1016/j.enpol.2016.03.019

DellaValle, N., & Czako, V. (2022). Empowering Energy Citizenship Among the Energy Poor. *Energy Research and Social Science, 89*, 1–18. https://doi.org/10.1016/j.erss.2022.102654

Di Chiro, G. (2008). Living Environmentalisms: Coalition Politics, Social Reproduction, and Environmental Justice. *Environmental Politics, 17*(2), 276–298. https://doi.org/10.1080/09644010801936230

Felipe Pèrez, B. (2018). Beyond the Shortcomings of International Law: A Proposal for the Legal Protection of Climate Migrants. In *Climate Refugees: Beyond the Legal Impasse?* (pp. 214–229). Routledge. https://doi.org/10.4324/9781315109619

Ferris, E. (2018). *When Refugee Displacement Drags on, is Self-Reliance the Answer?* Retrieved December 9, 2022, from https://www.brookings.edu/blog/order-from-chaos/2018/06/19/when-refugee-displacement-drags-on-is-self-reliance-the-answer/

Food and Agriculture Organization of the United Nations, Deutsche Gesellschaft für Internationale Zusammenarbeit GmbH, Global Alliance for Clean Cookstoves, The International Organization for Migration, The Moving Energy Initiative, Practical Action, et al. (2019). *The Global Plan of Action for Sustainable Energy Solutions in Situations of Displacement: Framework for Action*. Retrieved from https://www.unitar.org/ptp/sustainable-energy

Foresight. (2011). *Final Project Report*. The Government Office for Science.

Gemenne, F. (2011). Why the Numbers Don't Add Up: A Review of Estimates and Predictions of People Displaced by Environmental Changes. *Global Environmental Change, 21*(Suppl. 1), S41–S49. https://doi.org/10.1016/j.gloenvcha.2011.09.005

Giorgi, F. (2006). Climate Change Hot-Spots. *Geophysical Research Letters, 33*(8), 1–4. https://doi.org/10.1029/2006GL025734

Holland, B. (2020). Capabilities, Well-being, and Environmental Justice. In B. Coolsaet (Ed.), *Environmental Justice* (pp. 64–77). Routledge. https://doi.org/10.4324/9780429029585-7

IDMC. (2022). *Global Report on Internal Displacement 2022*.

Jakobeit, C., & Methmann, C. (2012). 'Climate Refugees' as Dawning Catastrophe? A Critique of the Dominant Quest for Numbers. In J. Scheffran, M. Brzoska, H. G. Brauch, P. M. Link, & J. Schilling (Eds.), *Climate Change, Human Security and Violent Conflict: Challenges for Societal Stability* (pp. 301–314). Springer Berlin Heidelberg. https://doi.org/10.1007/978-3-642-28626-1_16

Jenkins, K., McCauley, D., Heffron, R., Stephan, H., & Rehner, R. (2016). Energy Justice: A Conceptual Review. *Energy Research and Social Science, 11*, 174–182. https://doi.org/10.1016/j.erss.2015.10.004

Klepp, S., & Herbeck, J. (2016). The Politics of Environmental Migration and Climate Justice in the Pacific Region. *Journal of Human Rights and the Environment, 7*(1), 54–73. https://doi.org/10.4337/jhre.2016.01.03

Lahn, G., & Grafham, O. (2015). *Heat, Light and Power for Refugees: Saving Lives, Reducing Costs. Chatham House Report for the Moving Energy Initiative*. Retrieved from https://www.chathamhouse.org/publication/heat-light-and-power-refugees-saving-lives-reducing-costs

Lee, J. R. (2020). *Environmental Conflict and Cooperation: Premise, Purpose, Persuasion, and Promise*. Routledge.

Mastaler, J. S. (2018). Social Justice and Environmental Displacement. *Environmental Justice, 12*(1), 17–22. https://doi.org/10.1089/env.2018.0029

Mastor, R., Dworkin, M., Landa, M., & Duff, E. (2018). Energy Justice and Climate-Refugees. *Energy Law Journal, 39*(1), 139–171.

Mayer, B. (2016). *The Concept of Climate Migration: Advocacy and its Prospects. The Concept of Climate Migration: Advocacy and its Prospects*. https://doi.org/10.4337/9781786431738

Mayer, B. (2019). Definitions and Concepts. In R. McLeman & F. Gemenne (Eds.), *Routledge Handbook of Environmental Displacement and Migration* (pp. 323–328). https://doi.org/10.4324/9781315638843-25
McCauley, D., Heffron, R. J., Stephan, H., & Jenkins, K. (2013). Advancing Energy Justice: The Triumvirate of Tenets. *International Energy Law Review, 32*(3), 107–110.
Melin, A., Day, R., & Jenkins, K. E. H. (2021). Energy Justice and the Capability Approach—Introduction to the Special Issue. *Journal of Human Development and Capabilities, 22*(2), 185–196. https://doi.org/10.1080/19452829.2021.1909546
Park, L. S.-H., & Pellow, D. (2019). Forum 4: The Environmental Privilege of Borders in the Anthropocene. *Mobilities, 14*(3), 395–400. https://doi.org/10.1080/17450101.2019.1601397
Pulido, L. (2017). Conversations in Environmental Justice: An Interview with David Pellow. *Capitalism, Nature, Socialism, 28*(2), 43–53. https://doi.org/10.1080/10455752.2016.1273963
Reuveny, R. (2007). Climate Change-Induced Migration and Violent Conflict. *Political Geography, 26*(6), 656–673. https://doi.org/10.1016/j.polgeo.2007.05.001
Robeyns, I. (2017). *Wellbeing, Freedom and Social Justice—The Capability Approach Re-Examined.* Open Book Publishers.
Rosignoli, F. (2018). Categorizing Collective Capabilities. *Partecipazione e Conflitto, 11*(3), 813–837. https://doi.org/10.1285/i20356609v11i3p813
Rosignoli, F. (2022). *Environmental Justice for Climate Refugees.* Routledge. https://doi.org/10.4324/9781003102632
Salyean, I. (2008). From Climate Change to Conflict? No Consensus Yet. *Journal of Peace Research, 45*(3), 315–326.
Schlosberg, D. (2007). *Defining Environmental Justice: Theories, Movements, and Nature.* Oxford Univ. Press. https://doi.org/10.1093/acprof:oso/9780199286294.001.0001
Schlosberg, D. (2020). Sustainable Materialism and Environmental Justice. In B. Coolsaet (Ed.), *Environmental Justice* (pp. 303–315). Routledge. https://doi.org/10.4324/9780429029585-27
Schlosberg, D., & Carruthers, D. (2010). Indigenous Struggles, Environmental Justice, and Community Capabilities. *Global Environmental Politics, 10*(4), 12–35. https://doi.org/10.1162/GLEP_a_00029
Sen, A. (1999). *Development as Freedom.* Oxford University Press. https://doi.org/10.1215/0961754X-9-2-350
Sovacool, B. K., & Dworkin, M. H. (2016). *Global Energy Justice: Problems, Principles, and Practices.* Cambridge University Press. https://doi.org/10.1080/1523908x.2015.1055857
Sovacool, B. K., Heffron, R. J., McCauley, D., & Goldthau, A. (2016). Energy Decisions Reframed as Justice and Ethical Concerns. *Nature Energy, 1*, 1–6. https://doi.org/10.1038/nenergy.2016.24

Turhan, E., & Armiero, M. (2019). Of (Not) Being Neighbors: Cities, Citizens and Climate Change in an Age of Migrations. *Mobilities, 14*(3), 363–374. https://doi.org/10.1080/17450101.2019.1600913

UN High Commissioner for Refugees (UNHCR). (2018). In *Harm's Way: International Protection in the Context of Nexus Dynamics Between Conflict or Violence and Disaster or Climate Change*, December 2018. Available at: https://www.refworld.org/docid/5c2f54fe4.html [Accessed 9 February 2023].

UN High Commissioner for Refugees (UNHCR). (2020). *Legal considerations regarding claims for international protection made in the context of the adverse effects of climate change and disasters*, 1 October 2020. Available at: https://www.refworld.org/docid/5f75f2734.html [Accessed 9 February 2023].

UNHCR. (2016). *UNHCR Strategy for the Institutionalisation of Cash-Based Interventions 2016–2020.*

UNHCR. (2019). *Global Strategy. For Sustainable Energy.*

UNHCR. (2022a). *Climate Change and Disaster Displacement.* Retrieved April 7, 2023, from https://www.unhcr.org/climate-change-and-disasters.html

UNHCR. (2022b). *Global Trends Forced Displacement in 2021.* Retrieved from https://www.unhcr.org/62a9d1494/global-trends-report-2021

Van der Land, V. (2018). *Migration and Environmental Change in the West African Sahel: Why Capabilities and Aspirations Matter. Migration and Environmental Change in the West African Sahel: Why Capabilities and Aspirations Matter.* Routledge. https://doi.org/10.4324/9781315440163

Vindal Ødegaard, C., & Rivera Andía, J. J. (2019). *Indigenous Life Projects and Extractivism. Indigenous Life Projects and Extractivism.* Palgrave Macmillan. https://doi.org/10.1007/978-3-319-93435-8

Walker, G. (2011). The Role for "Community" in Carbon Governance. *Wiley Interdisciplinary Reviews: Climate Change, 2*(5), 777–782. https://doi.org/10.1002/wcc.137

Walker, G., & Day, R. (2012). Fuel Poverty as Injustice: Integrating Distribution, Recognition and Procedure in the Struggle for Affordable Warmth. *Energy Policy, 49*, 69–75. https://doi.org/10.1016/j.enpol.2012.01.044

Whyte, K. (2017). Is it Colonial déjà vu?: Indigenous Peoples and Climate Injustice. In *Humanities for the Environment: Integrating Knowledge, Forging New Constellations of Practice.* Routledge. https://doi.org/10.4324/9781315642659

Wood, N., & Roelich, K. (2019). Tensions, Capabilities, and Justice in Climate Change Mitigation of Fossil Fuels. *Energy Research and Social Science, 52*, 114–122. https://doi.org/10.1016/j.erss.2019.02.014

Zickgraf, C. (2019). Keeping People in Place: Political Factors of (Im)mobility and Climate Change. *Social Sciences, 8*(8), 1–17. https://doi.org/10.3390/socsci8080228

CHAPTER 6

Theorising a Human Rights-Based Approach to Energy Transition and Its Justiciability in International and Domestic Jurisprudence

Simona Fanni

Introduction

The international community has widely acknowledged the importance of promoting the energy transition, recognising that decarbonising our energy system is crucial to pursue the goals set by the Paris Agreement.[1] This view has been incorporated into many prominent initiatives and

[1] The energy transition is an indispensable process. In fact, as the United Nation's 'Theme Report on Energy Transition. Towards the achievement of SDG 7 and net-zero emissions' (p. IV) has clarified, "An energy revolution based on renewables and energy efficiency is urgently needed not just to accelerate economic progress and development, but also to slash emissions that are rapidly warming our planet. The energy sector today, dominated by fossil fuels, accounts for 73% of human-caused greenhouse gas emissions. Global CO_2 emissions must be halved by 2030 to avoid an increasing frequency and severity of dangerous and unprecedented weather extremes, including heatwaves, devastating floods and droughts, risks to food and water security, population displacement, and loss of lives and livelihoods".

S. Fanni (✉)
University of Seville, Seville, Spain

D. Iglesias Márquez et al. (eds.), *Legal Challenges at the End of the Fossil Fuel Era*, Global Issues,
https://doi.org/10.1007/978-3-031-61766-9_6

documents at global and regional level.[2] In its Report 'Climate Change 2022: Mitigation of Climate Change', the International Panel on Climate Change (IPCC) has stated that achieving the goal of limiting global warming consistently with the Paris Agreement is dependent on the implementation of energy transition.[3] As was to be expected, discussing the energy transition was one of the priorities of the 27th Conference of the Parties (COP27) hosted in Egypt, which took place in November 2022, at a particularly dramatic time for the international community, in which the climate crisis and the energy crisis—along with other complex challenges such as the global food insecurity—were being exacerbated by the impact of the conflict in Ukraine and the associated international sanctions imposed on Russia. Nevertheless, several scholars have voiced criticism for the fact that "a reform of States' GHG emissions' responsibilities was definitely not prioritized" (Sharati & Tillet, 2022), and that no reference was made to an accountability mechanism that would encompass both production-based accountability and consumption-based accountability.

The European Union (EU) has developed an ambitious approach to energy transition: according to the European Green Deal and the European Climate Law, the EU aims to be climate-neutral by 2050. Moreover, the EU 2030 Framework for Climate and Energy Policies sets some "more ambitious targets [...] [such as] a reduction in greenhouse gas emissions of 40% [by 2030 relative to 1990 levels], [...] [and] an increase of renewable sources in EU energy to 32% [for the amount of renewable energy in the EU's energy consumption by 2030]" (Renou-Maissant et al., 2022: 526). However, although EU Member States face common energy and environmental challenges, their respective "energy systems and their performances vary greatly from one country to another" (Renou-Maissant et al., 2022: 526). In fact, such structural characteristics as "national energy endowments, demographics, geography, climate, past national

[2] Besides the Report 'Climate Change 2022: Mitigation of Climate' of the IPCC and the COP27, Resolution n. 3/2021 of the Inter-American Court of Human Rights on 'Climate emergency: scope of the Inter-American Human Rights Obligations' as well as the 'European Green Deal' and the 'Clean energy for all Europeans package' are significant examples. Interestingly, this perspective has also been recently adopted and stated by the Committee on Economic, Social and Cultural Rights in its General Comment n. 26 (2022), of 24 January 2023, on 'Land and Economic, Social and Cultural Rights', E/C.12/GC/26, at para. 56.

[3] International Panel on Climate Change (IPCC), Report 'Climate Change 2022: Mitigation of Climate Change', see especially pp. 55 and 628 ff. This view is widely restated in the Report, and Chapter 6 is particularly significant in this regard.

energy policies and historical development" can have a significant impact. For instance, various "countries belong[ing] to the former Eastern Bloc [...] still suffer[...] from an entrenched specialization in heavy industries under the impulse of central planners" (Renou-Maissant et al., 2022: 537), while wide reliance on nuclear energy is a significant factor with regard to France (Renou-Maissant et al., 2022: 528).

Despite the complexity of the international climate agenda and of the challenges that it poses, climate litigation has been achieving interesting results, more widely tackled below, especially through the promotion of intergenerational and intragenerational equity and extraterritoriality, at both national and international level. For the purpose of this chapter, it deserves to be stressed that climate litigation has a significant potential for the promotion of energy transition litigation through a human rights-based approach, in particular through the definition of specific States' obligations and by paving the way to the affirmation of the business sector's accountability. Moreover, intergenerational and intragenerational equity may help the justiciability of the responsible use of resources, and extraterritoriality may play an important role to fill the gap of accountability of both States and corporations with respect to the exploitation of such mineral resources—as cobalt—that are crucial for renewable energy infrastructures (as solar arrays, wind turbines and batteries).

Starting from the national landscape, an interesting wave of climate litigation has been spreading before domestic courts all over the world, "asking state and corporate actors to reduce greenhouse gas emissions and redress the harms associated with the impacts of climate change" (Savaresi & Setzer, 2022: 7), and seeking innovative ways to close the accountability gap. This trend is global, although, in practice, some differences can be observed, since "the trends in Global South climate litigation reflect the priorities of the jurisdiction in which the action is commenced [...], [and] while a number of landmark cases of strategic climate litigation in the Global North are targeted at driving governmental ambition on climate change, litigants from the Global South are more likely to use litigation to compel governments to enforce existing policies for mitigation and adaptation, attempting to overcome implementation constraints" (Setzer & Benjamin, 2019: 3).

However, several common features characterise the approach of climate litigation globally and make it innovative in various respects: firstly, the adoption of a proactive interpretive approach that combines human rights law and international environmental law, with courts often citing in their legal reasoning instruments such as the United Nations Framework Convention on Climate Change (UNFCCC) and the Paris Agreement; secondly, the reference to the narrative of sustainability and future generations.

The beginning of a new climate litigation trend can also be observed in the international human rights landscape. This phenomenon has not yet achieved the same proportions as the domestic climate litigation wave; however, climate cases are multiplying before the European Court of Human Rights (ECtHR or 'Strasbourg Court')—although none of them has been decided yet. Again, the Committee on the Rights of the Child (CRC) has tackled important questions relating to young generations and climate litigation, which, to some extent, from the perspective of climate-vulnerable inhabitants of low-lying islands as well as of the cultural identity of indigenous peoples and their spiritual connection with their land, were tackled by the Human Rights Committee (HRC) in the recent, important case *Daniel Billy* et al *v. Australia* (the *Torres Strait Islanders* case). Last but not least, the Inter-American Court of Human Rights (IACtHR) has provided a broad interpretation of the concept of 'effective control' and has set the stage for a renewed perspective on the extraterritoriality of States' human rights obligations in the environmental field.

On this premise, this chapter aims to explore this interesting case law and outline a critical analysis of its strengths and weaknesses. For this purpose, Section "The Language of Human Rights as a Source of Mitigation Obligations: The Efforts of Domestic Courts" focuses on domestic courts, in order to analyse how human rights law and constitutional rights have been used as a source of States' mitigation obligations; Section "Domestic Courts and Some Interesting Views on the Complex Issues of the *Locus Standi* of Future Generations and Extraterritoriality" delves into the efforts of national courts to tackle such complex issues as the *locus standi* of future generations and, to some extent, the extraterritorial scope of mitigation obligations; subsequently, Section "A Rising Climate Litigation Wave Before Judicial and Quasi-Judicial International Human Rights Bodies" deals with the case law of judicial and quasi-judicial international human rights bodies. In light of the analysis carried out in the previous sections, Section "From the Climate Litigation Wave to the Justiciability of the Energy Transition: Some Proposals" assesses the current and the potential role of climate change litigation to pave the way to the justiciability of the energy transition, by theorising a human rights-based approach to this complex issue, from a multi-level viewpoint. Finally, some conclusions are formulated.

The Language of Human Rights as a Source of Mitigation Obligations: The Efforts of Domestic Courts

To elucidate how domestic courts have developed their approaches, the starting point is analysing how they have articulated the language of international human rights law and of constitutional rights in their views on climate change, as a standard for assessing the States' efforts to comply with their mitigation obligations.

Reliance on human rights (see Voigt, 2022) is not uncommon with respect to climate change: international human rights bodies, such as the Committee on Economic, Social and Cultural Rights (CESCR) and the Committee on the Elimination of Discrimination against Women (CEDAW)—that provide interpretive guidance respectively on the scope of the rights enshrined in the International Covenant on Economic, Social and Cultural Rights and the Convention on the Elimination of All Forms of Discrimination against Women—have elucidated the inherent interconnection between human rights and States' obligation to mitigate climate change (Mayer, 2021a: 410). Indeed, as recalled by Mayer (2021a: 410), the CESCR has "suggested that '[i]n order to act consistently with their human rights obligations',[4] States parties should revise the nationally determined contributions (NDCs) to global mitigation action that they have communicated under the Paris Agreement' (Mayer, 2021a: 410) and the CEDAW has clarified that States have an obligation 'to effectively mitigate ... climate change in order to reduce the increased disaster risk'".[5]

As highlighted by Mayer (2021a: 412), many scholars have welcomed the idea that human rights treaties are the source of mitigation obligations; among them, Michael Burger and Jessica Wentz have recognised that there is a "growing consensus that a mitigation obligation does exist under international human rights law" (Burger & Wentz, 2015: 205). Occasionally, some criticism has been raised against this view (for instance, by such scholars as Fanny Thornton and Lavanya Rajamani, whose views are recalled later), but, interestingly, the climate litigation case law under

[4] Mayer (2021a, 410) refers to the Committee on Economic, Social and Cultural Rights (CESCR), Statement: Climate Change and the International Covenant on Economic, Social and Cultural Rights, para. 6, UN Doc. E/C.12/2018/1 (31 October 2018).

[5] Committee on the Elimination of Discrimination Against Women (CEDAW), General Recommendation No. 37 on the Gender-Related Dimensions of Disaster Risk Reduction in the Context of Climate Change, para. 14, UN Doc. CEDAW/C/GC/37 (13 March 2018).

consideration has also provided some responses to these objections, as discussed later in this chapter.

One of the most significant examples of how domestic courts have embraced the idea that an interconnection exists between human rights and international environmental law, especially environmental treaties and, prominently, the Paris Agreement, is the recent judgement of the Brazilian Supreme Court in *PSB* et al *v. Brazil*.[6] The Court declared that environmental treaties are a particular type of human rights treaties and, as such, they enjoy a 'supranational' or 'superlegal' status which, in terms of legal hierarchy, makes them prevail over conflicting domestic laws. The *ratio decidendi* of the ruling conveys the core of the view that the Brazilian Supreme Court has developed, where it affirms that the

> Executive Branch has the constitutional duty to ensure the Climate Fund remain operational, and to annually allocate its resources for the purpose of mitigating climate change. The withholding of such resources is prohibited due to the government's constitutional duty to protect the environment [...], to affirm the fundamental right to a healthy environment, and to uphold its international commitments [...], as well as the constitutional principle of separation of powers. (Barroso, 2022; Neumann Ciuffo, 2022)

Some commentators have questioned whether the Paris Agreement, as an environmental treaty, can be said to belong to the category of human rights treaties, as maintained by the Brazilian Supreme Court; indeed, the Agreement has only one reference to human rights, in its Preamble (Neumann Ciuffo, 2022). What is more, "[w]hen the Paris Agreement was negotiated, much disagreement ensued over the adoption of human rights obligations in the substantive part of the text" (Neumann Ciuffo, 2022, citing the view expressed by Maria Antonia Tigre in an interview). However, some objections may be made to this statement: firstly, according to Article 31 of the Vienna Convention on the Law of the Treaties,

> [t]he fact that this was placed in the Preamble of the Paris Agreement only emphasizes further that these duties form part of the objects and purposes of the treaty, and should be used as part of the interpretation of the Paris Agreement. This is infinitely a more direct treaty device for integrating human rights into the formulation and assessment of any State Party's climate action, rather than previous attempts by the UN Special Rapporteur

[6] *PSB* et al. *v Brazil*, Supreme Court of Brazil, ADPF 760 (2020).

on Human Rights and Environment that focused on drawing an interpretive nexus between human rights obligations and environmental duties of States. (Desierto, 2021: 2)

Secondly, as Desierto (2021) further stressed, the scope of the reference to human rights is remarkable, as "[t]he language used is deliberate, imperative, specific, and comprehensive in covering all of human rights law", by expressly emphasising the States' duty to protect, respect and consider their human rights obligations and several rights (interestingly, the right to health, the right to development and intergenerational equity) when interpreting and implementing the Paris Agreement. From this perspective, the same author (Desierto, 2021: 2) focused on the NDCs submissions of the largest emitters (China, the European Union and the United States) to assess whether and to which extent they have incorporated the human rights language in this framework. In this respect, Desierto stressed that only the European Union is "explicitly integrating the dimensions of human rights and gender equality by States in all their national plans and strategies under the EU Energy Union Governance Regulation" (EU NDC, 12).

In this sense, an illustrative example of how human rights, as well as constitutional rights, can be used for assessing States' compliance with their mitigation obligations, is the judgement delivered by the German Constitutional Court in the *Neubauer*[7] case. In its judgement, the Court "unanimously declared the Federal Climate Protection Act partly unconstitutional because it does not sufficiently protect people against future infringements and limitations of freedom rights in the wake of gradually intensifying climate change" (Kotzé, 2021, 1424; see *Neubauer*, especially paras. 183 ff., 195 ff. and 243), especially insofar as it did not provide specific steps for the achievement of Germany's post-2030 goals to reduce GHG and mitigate climate change. As stressed in scholarship, "the fact that the post-2030 period was left largely undetermined violated fundamental rights" (Bäumler, 2021: 2).

The language of fundamental rights also underlies the legal reasoning of the Dutch Supreme Court in the *Urgenda* case,[8] which represents a

[7] *Neubauer*, et al. *v. Germany*, Bundesverfassungsgericht [BVerfG], 24 March 2021, Case No. BvR 2656/18/1, BvR 78/20/1, BvR 96/20/1, BvR 288/20.

[8] *Staat der Nederlander v. Urgenda*, Hoge Raad 20 December 2019, ECLI:NL:HR:2019:2006.

milestone in the landscape of climate litigation, as it was the first case in which a court ordered a State to reduce its GHG emissions (by at least 25% below 1990 levels before the end of 2020) by relying on a human rights framework and on the European Convention on Human Rights (Bergkamp, 2020; Leijten, 2019). When assessing whether Article 2 and Article 8 of the European Convention on Human Rights—which protect respectively the right to life and the right to private and family life—oblige the State to take measures, the Court found that those provisions give rise to a positive duty for the State "to take appropriate steps if there is a real and immediate risk to persons and the state in question is aware of that risk" (*Urgenda*, para 5.2.2). The positive duty "to take preventive measures to counter the danger, [arises] even if the materialisation of that danger is uncertain [...] consistent with the precautionary principle" (*Urgenda*, para. 5.3.2), and includes adaptation and mitigation measures. Therefore, the Court recognised human rights as a source of mitigation obligations, specified the scope of those obligations also in line with the precautionary principle and, importantly, ruled out that States could have a margin of appreciation on whether to take any measure and limited the States' discretion in choosing the steps to be taken (*Urgenda*, para. 5.3.2).

The landscape explored is eloquent, and some other cases might be recalled to provide further evidence of the width and the significance of climate litigation all over the world. *Shrestha*[9] is a significant example, as the Nepali Supreme Court found that the absence of a climate change law amounted to a breach of the right to a healthy environment, protected under the Constitution (p. 12, para. 5), and that "[t]o address the effects of climate change through adaptation and mitigation and the high risks seen in the ecology of, *inter alia*, higher mountainous areas and to restore, including but not limited to, its ecology, accommodating following topics, a consolidated law related to climate change Act needs to be necessarily enacted" (*Shresta*, p. 13, para. 6). In a similar fashion, in *Leghari*,[10] the Lahore High Court found that the Government's climate inaction amounted to a breach of such fundamental rights as the right to life—which, according to the Court, included the right to a healthy and clean environment and the right to human dignity (*Leghari*, paras. 12 ff.).

[9] *Shrestha v. Office of the Prime Minister* et al., Supreme Court of Nepal, 25 December 2018, Decision no. 10210.

[10] *Ashgar Leghari v. Federation of Pakistan* et al., Case No: W.P. No. 25501/2015, 25 January 2015.

However, beyond these important achievements, courts have also had to tackle two further important issues characterising the climate litigation wave: *locus standi* and extraterritoriality.

Domestic Courts and Some Interesting Views on the Complex Issues of the *Locus Standi* of Future Generations and Extraterritoriality

Locus standi or legal standing is one of the most complex issues that courts have to deal with in the context of the wave of climate change litigation. Interestingly, in this respect, one of the above-mentioned objections to the "growing consensus that a mitigation obligation does exist under international human rights law" is "whether the causal links between a state's mitigation action and the enjoyment of human rights may be 'too intricate and diffuse' to ground a convincing legal case" (Mayer, 2021a, 412, recalling the view of Fanny Thornton). For example, in practice, some legal orders do not permit *actiones populares* (see, for instance, the *Friends of the Irish Environment v. Government of Ireland* case and, internationally, the approach of the ECtHR), whilst they may be an effective tool for the promotion of public interest litigation, including for the purpose of overcoming the obstacles related to the *locus standi* in the environmental field.

The rising tide of climate change litigation in the domestic courts has dealt with the issue of legal standing from various viewpoints, including from the perspective of future generations, which is inherently connected to the question of 'non-identity' (Parfit, 1984). This approach has helped courts to address the long-term and intertemporal impact of climate change, which takes on special importance also for the protection of young generations.

Several solutions have been suggested. In the United States, domestic courts have relied on the doctrine of 'public trust', which reflects the conceptualisation provided by Brown Weiss (1992). This approach was effectively conveyed by the statement of the Supreme Court of Pennsylvania in *Robinson Township v. Commonwealth*,[11] where it affirmed that "[t]he Commonwealth's obligations as trustee to conserve and maintain the public natural resources for the benefit of the people, including

[11] *Robinson Twp. v. Commonwealth*, 83 A.3d 901, 974 (Pa. 2013).

generations yet to come, create a right in the people to enforce the obligations" (Slobodian, 2020: 580 ff.). In *Juliana v. the United States*[12] and in *Urgenda Foundation v. State of the Netherlands*,[13] the courts have not directly addressed the issue, by respectively saying that it was "unnecessary to address standing of future generations plaintiffs because the youth plaintiffs had adequately alleged current harm" (*Juliana*, para. 252; Slobodian, 2020: 577) and that "it is without a doubt plausible that the current generation of Dutch nationals, in particular but not limited to the younger individuals in this group, will have to deal with the adverse effects of climate change in their lifetime" (*Urgenda*, para. 37). Similarly, in the case *Future Generations v. Ministry of the Environment and Others*,[14] the Colombian Constitutional Court did not go beyond recognising that the applicants could file the *acción de tutela*, since it "can be filed as long as it [...] shows the connection between the violation of collective and fundamental or individual rights" (*Future Generations*, pp. 12–13), without making any more specific statement on the particular *locus standi* of future generations.

Interestingly, it is necessary to look back to the 1990s to find a more specific—and, possibly, fascinating—articulation of the legal standing of future generations. The judgement of the Supreme Court of the Philippines in the *Minors Oposa* case[15] did not deal with climate change, as it concerned an environmental case in a broader sense. Specifically, the Court was ruling on a class suit and took the opportunity to affirm that "[b]ecause of the very broadness of the concept of 'class' here involved – membership in this 'class' appears to embrace everyone living in the country whether now or in the future – it appears to me that everyone who may be expected to benefit from the course of action petitioners seek to require public respondents to take, is vested with the necessary *locus standi*" (*Minors Oposa*, p. 16, Separate Concurring Opinion of Judge Feliciano). Interestingly enough, the idea underlying this conceptualisation of the legal standing of the "[p]etitioners minors [...] [as] represent[atives] [of]

[12] *Juliana v. the United States*, 217 F. Supp. 3d 1224, 1260 (D. Or. 2016).

[13] *Netherlands v. Urgenda*, Case No.200.178.245/01, Hague Court of Appeal (2018); *Stichting Urgenda v. Government of the Netherlands* (Ministry of Infrastructure and the Environment), Rechtbank Den Haag, 24 June 2015.

[14] *Luis Armando Tolosa Villabona*, Corte Suprema de Justicia [C.S.J.] [Supreme Court], Sala Civ. 5 de abril de 2018, STC4360-2018, Radicación no. 11001-22-03-000-2018-00319-01 (Colom.).

[15] *Minors Oposa v. DENR*, Supreme Court of the Philippines, 33 I.L.M. 173 (1994).

their generation as well as generations yet unborn" was based on the principle of intergenerational equity (*Minors Oposa*, p. 8).

The concept of intergenerational equity reflects the approach developed by domestic courts to deal with the wave of climate change litigation, which also employs the concepts of sustainability and of the judicious and responsible management not only of natural resources but also of the CO_2 budget and its impact on the freedoms of future generations. Once again, the *Neubauer* case stands out for its approach: the German Constitutional Court held that "one generation must not be allowed to consume large portions of the CO_2 budget while bearing a relatively minor share of the reduction effort, if this would involve leaving subsequent generations with a drastic reduction burden and expose their lives to serious losses of freedom" (*Neubauer*, para. 192). The eloquence of this statement also recalls the principle of distributive environmental justice.

Before delving into the international jurisprudential landscape, one last question deserves special attention, namely extraterritoriality.[16] This issue is still opaque in domestic climate case law. Nevertheless, in the *Neubauer* case, the German Constitutional Court "provided for some benchmarks in Germany's global responsibilities" (Bäumler, 2021). First off, the Court said that "no violation of a duty of protection arising from fundamental

[16] The extraterritoriality of States' human rights obligations has been tackled especially by human rights bodies on various occasions. While it has been usually been linked to State's control over the right-holder, a different understanding of the scope of extraterritoriality has progressively gained ground in the jurisprudence of international human rights bodies (this point is further discussed in this Chapter, at p. 10), due to the affirmation of the concept of 'effective control', intended as State's control over a potential cause of harm to the right-holder. This conception was adopted and supported by the Inter-American Court of Human Rights in its Advisory Opinion No. 23 'The Environment and Human Rights' (an interesting analysis was made by Berkes, 2018; some criticism was expressed by Besson, 2020). However, from a wider viewpoint, it seems interesting to recall the complex debate that scholarship has developed with respect to extraterritoriality under international law. In this regard, it was stressed that extraterritoriality should not be confused with transboundary harm (as tackled in the early days in *Trail Smelter*); however, it should also be recalled that the International Court of Justice, in its Advisory Opinion on the Legality of the Threat or Use of Nuclear Weapons, has said that "the existence of the general obligation of States to ensure that activities within their jurisdiction [...] respect the environment of other States [...] is now part of the corpus of international law" (para. 29) (see Laukkanen & Laukkanen, 2022). The debate is still in progress and, from a human rights perspective, it will be interesting and important to see how the European Court of Human Rights will tackle extraterritoriality of States' obligations in its rising wave of climate litigation. An interesting analysis on extraterritoriality and human rights was developed by Milanovic (2011).

rights is ascertainable vis-à-vis the complainants who live in Bangladesh and Nepal" (*Neubauer*, para. 173). Nevertheless, as Bäumler (2021) has clearly illustrated,

> [w]hile not directly confirming Germany's extraterritorial duty to protect, it still further evaluated the content of such a duty, including by stressing that the duty to protect individuals abroad might be different and, in fact, lower than the duty to protect those living in Germany (para. 181). In arguing that Germany fulfilled its duty to mitigate climate change by partaking in the global effort, by joining the Paris Agreement, but by also pointing towards Art. 9 of the Paris Agreement that obliges developed countries to financially assist developing countries, the Court ultimately sketched out Germany's global obligations towards individuals living abroad. (see also *Neubauer*, paras. 101, 175–178)

The climate litigation wave has succeeded in 'challenging' domestic courts to provide interesting responses to the States' mitigation obligations, through a human rights-based approach, at times—as can be observed—through the mechanism of strategic litigation[17] (Setzer & Higham, 2022, 15). Despite the complexity of the climate agenda and the room for improving the results achieved so far, it may be argued that this case law, taken overall, has provided several responses to the doubts expressed by some scholars (as Fanny Thornton, whose view this chapter expressly recalls, and Alan Boyle) as to the classification of climate mitigation duties as human rights obligations. In this sense, turning again to the analysis made by Mayer (2021a: 412), Lavanya Rajamani's objection that many negotiators "were skeptical of the utility of a human rights approach, given the complex and laden agenda of the climate process, and the limited space for new methodological or conceptual approaches" (Rajamani, 2018) has also been addressed in several ways. Side by side with this fascinating domestic judicial landscape, there is a growing corpus of cases brought before international human rights bodies. The next section focuses on this dimension.

[17] "These are cases where the claimants' motives go beyond the concerns of the individual litigant and aim at advancing climate policies, creating public awareness, or changing the behaviour of government or industry actors" (Setzer & Higham, 2022: 15).

A Rising Climate Litigation Wave Before Judicial and Quasi-Judicial International Human Rights Bodies

The climate litigation wave is rising also before judicial and quasi-judicial international human rights bodies which, similarly to national courts, are called to tackle such issues as human rights and mitigation obligations, the protection of young and future generations, and the extraterritoriality of States' duties.

To date, only two cases have been adjudicated by international bodies, namely *Daniel Billy* et al. *v. Australia*,[18] decided by the HRC, and *Sacchi* et al. *v. Argentina* et al.,[19] on which the CRC has adopted five almost identical decisions of inadmissibility, one for each respondent state. This chapter focuses on the latter case, which was a diagonal claim,[20] and was dismissed on a procedural basis, because the petitioners had not exhausted domestic remedies. Although the authors of the complaint had relied on Article 7(e) of the Optional Protocol to the Convention on the Rights of the Child on a communications procedure (OPCP), which provides that exhaustion of domestic remedies "shall not be the rule where the application of the remedies is [...] unlikely to bring effective relief", the CRC considered that the "no prospect of success test" had not been met in the case (Gradoni & Mantovani, 2022: 11).

Nevertheless, the fact that the CRC considered that the case did not fall within the scope of Article 7(e) of the OPCP did not prevent the Committee from delivering some eloquent findings, firstly by recognising that "climate crisis is a children's rights crisis" and that "by perpetuating life-threatening climate change, the State party is violating the authors' rights to life, health, and the prioritization of the child's best interests, as well as the cultural rights of the authors from indigenous communities" (*Sacchi* et al., para. 3.8; see also paras. 10.2–10.14).

[18] *Daniel Billy* et al. *v. Australia*, Human Rights Committee, CCPR/C/135/D/3624/2019 (22 September 2022).

[19] *Chiara Sacchi*, et al *v Argentina, Brazil, France, Germany and Turkey*, Committee on the Rights of the Child, CRC 104/2019-108/2019 (23 September 2020). The excerpts cited in this chapter are taken from *Chiara Sacchi*, et al *v Argentina*, Committee on the Rights of the Child (CRC/C/88/D/104/2019) (8 October 2021).

[20] Diagonal claims are "transboundary claims or claims brought by individuals or groups against States other than their own" (Feria-Tinta, 2021, 2).

Consistently with children's heightened vulnerability to climate change, the CRC defined their victimhood, which was framed in terms of intertemporality by saying that "as children, the authors are particularly impacted by the effects of climate change, both in terms of the manner in which they experience such effects as well as the potential of climate change to affect them throughout their lifetime, in particular if immediate action is not taken" (*Sacchi* et al., para. 10.13). This implies that since "children are entitled to special safeguards, including appropriate legal protection, States have heightened obligations to protect children from foreseeable harm" (*Sacchi* et al., para. 10.13). It seems interesting to recall that a similar view was expressed later by the HRC in the *Daniel Billy* case, where it affirmed that

> environmental degradation, climate change and unsustainable development constitute some of the most pressing and serious threats to the ability of present and future generations to enjoy the right to life" and that "States parties should take all appropriate measures to address the general conditions in society that may give rise to direct threats to the right to life or prevent individuals from enjoying their right to life with dignity. (*Daniel Billy et al.*, para. 8.3)

In a fashion that seems to echo the words of the late Judge Cançado Trindade, who authoritatively said that "[h]uman solidarity manifests itself not only in a spatial dimension [...] but also in a temporal dimension – that is, among the generations who succeed each other in the time, taking the past, present, and future altogether",[21] the CRC also elucidated the extraterritoriality of States' obligations to mitigate climate change. This point deserves special attention: indeed, providing an interesting example of judicial dialogue, the Committee incorporated into its legal reasoning the concept of the "new extraterritorial jurisdictional link based on control over domestic activities with extraterritorial effect" (Berkes, 2018), which was recognised for the first time by IACtHR in its Advisory Opinion n. 23 'The Environment and Human Rights' (AO n. 23). This conceptualisation hinges on the State of origins' exercise of "effective control over the activities carried out that caused the harm and consequent violation of human rights" (AO n. 23, para. 104(h)).

[21] *Bamaca-Velasquez v. Guatemala*, Judgement, Inter-Am. Ct. H.R. (ser. A) No. 11.129, ¶ 23 (2002), (Separate Opinion of Trindade, J.).

Consistently, the Committee affirmed that "when transboundary harm occurs, children are under the jurisdiction of the State on whose territory the emissions originated for the purposes of article 5 (1) [jurisdiction] of the Optional Protocol if there is a causal link between the acts or omissions of the State in question and the negative impact on the rights of children located outside its territory, when the State of origin exercises effective control over the sources of the emissions in question" (*Sacchi* et al., para. 10.7).

The adoption of the jurisdictional extraterritorial link of effective control—which derives from a broad application of the principle of due diligence (Berkes, 2018; Besson, 2020)—can be particularly beneficial to young applicants' diagonal claims, as they can help to hold States accountable not only in those situations where States have control over the right-bearers but also when they have control over a potential cause of harm to the right-holder with an extraterritorial impact.

The fact that the Committee adopted this approach was not a foregone conclusion, since such broad understanding of jurisdiction is not generally accepted, as some reference to the most significant cases recalled below demonstrates.

Indeed, in some cases, human rights bodies have adopted a similar view when determining jurisdiction. For example, the HRC, in *Sergio Euben Lopez Burgos v. Uruguay* (para 12), affirmed that "it would be unconscionable to so interpret the responsibility under article 2 of the Covenant as to permit a State party to perpetrate violations of the Covenant on the territory of another State, which violations it could not perpetrate on its own territory". Again, in its General Comment no. 36, the HRC made an even more explicit reference to 'effective control' when addressing the concept of jurisdiction and said that States have an obligation to protect "all persons over whose enjoyment of the right to life it exercises power or effective control. This includes persons located outside any territory effectively controlled by the State, whose right to life is nonetheless impacted by its military or other activities in a direct and reasonably foreseeable manner" (para. 63). By contrast, the ECtHR relies on a narrower concept of jurisdiction, given that "[s]ince its judgment in *Banković v Belgium*, it has largely displayed two models of jurisdiction: one based on territorial control, and one based on personal control" (Keller & Heri, 2022:161; Besson, 2021). Recently, the CESCR has articulated the concept of extraterritoriality with regard to States' obligations to respect, to protect and to fulfil in its General Comment no. 26 (2022) on land and economic, social and cultural rights (paras. 40 ff.). In this respect, the CESCR has stressed that the extraterritorial obligation to respect requires States "to take specific measures to prevent their domestic and international policies and actions, such as [...] climate change-mitigation policies, from interfering, directly or indirectly, with the enjoyment of human rights" (para. 41).

That being said, the CRC, consistently with "the principle of common but differentiated responsibility, as reflected in the Paris Agreement", found "that the collective nature of the causation of climate change does not absolve the State party of its individual responsibility that may derive from the harm that the emissions originating within its territory may cause to children, whatever their location" (*Sacchi* et al., para. 10.10). The Committee also affirmed the foreseeability of the "potential harm of the State party's acts or omissions regarding the carbon emissions originating in its territory", since such harm has been known for decades and "[i]n light of existing scientific evidence showing the impact of the cumulative effect of carbon emissions on the enjoyment of human rights, including rights under the Convention" (*Sacchi* et al., para. 10.11).

The decision of the CRC in *Sacchi* et al. may provide a helpful reference for the Strasbourg Court, before which climate cases are multiplying and becoming more heterogeneous. At present, we can only speculate about how the ECtHR will handle climate litigation; nevertheless, the fact that to date the Court has usually fast-tracked these cases is an indication of the importance it ascribes to them. The competent Chamber relinquished jurisdiction in favour of the Grand Chamber in cases *KlimaSeniorinnen v. Switzerland*, *Duarte Agostinho and Oth. v. Portugal and Oth.*, and *Carême v France*; importantly, in *Duarte Agostinho*, the Court, when it communicated the case, asked the parties to also comment on Article 3 of the ECHR. Although it is true that the Court "is master of the characterisation to be given in law to the facts of the case" (*ex multis*: *Guerra v. Italy*, *Radomiljia and Oth. v. Croatia*), consistently with the *jura novit curia* principle, it should also be recalled that, so far, Article 3 of the ECHR has not been successfully invoked in the Court's environmental case law.

As suggested in scholarship (Keller et al., 2022), dealing with the impact of climate change on the young generations—as well as future generations—in terms of ill-treatment may be an interesting way of tackling their specific vulnerability and, thus, the victim status of this category of applicants. One of the most complex questions is whether the severity threshold can be met; in its jurisprudence, the ECtHR has found that this threshold was met, for example, due to the "actual bodily injury or intense physical or mental suffering" (in case *Bouyid*) or to the "permanent state of anxiety and uncertainty" about one's future (in case *E-Masri*), and "a sense or feeling of vulnerability" (in case *Doğanay*) (Heri, 2020). Since the assessment that the Court carries out is "relative and context dependent" (Heri, 2020), one could speculate that, for example, the young age of the applicants could be a relevant factor. Heri (2020) also suggests that a possible consequence of the application of Article 3 of the ECHR may be the recognition of "extensive positive obligations", which might require "adequate legislative and administrative measures to curb carbon emissions, including by corporate actors". This idea is particularly significant when the focus is set on energy transition and on the crucial role that corporations play in this area.

The analysis of the international landscape may be used to provide a response to what is possibly one of the most serious doubts raised about the "growing consensus that a mitigation obligation does exist under international human rights law", namely that "[w]hile climate change has obvious implications for the enjoyment of human rights, it is less obvious whether, and to what extent, such effects can be qualified as human rights violations in a strict legal sense" (Mayer, 2021a: 412).

After the above analysis of the broad domestic and international landscape, the next section will discuss why the approach developed by this case law may be helpful for addressing the energy transition and its justiciability.

From the Climate Litigation Wave to the Justiciability of the Energy Transition: Some Proposals

The language of human rights, intragenerational and intergenerational equity and extraterritoriality may help promote energy transition litigation.

Although no specific decision has expressly dealt with the energy transition so far, some interesting cases have begun to be brought before national courts and international human rights bodies. Judicial dialogue may be an effective way to incorporate the results achieved by the climate case law explored by this chapter into this new energy case law. Again, some reference to the interpretive guidance provided by such bodies as the IACtHR, in its Advisory Opinion no. 23 'The Environment and Human

Rights' and its Resolution no. 3/2021, and the HRC in its General Comment no. 26 (2022) may help.

Firstly, it can be suggested that the recognition of human rights as a source of States' mitigation obligations may offer a suitable legal framework for the definition of States' duties to promote the energy transition and decarbonise their energy systems.

Some cases offer the opportunity to address this issue, with special reference to the climate complaints that were filed before the ECtHR. In this regard, the *Greenpeace Nordic and Oth. v. Norway* case is a particularly interesting example from the perspective of the promotion of fossil fuel phase-out and the States' obligations to mitigate climate change by reducing their emissions. This case was defined as an outlier, as it addresses "a specific type of GHG-intensive project, [...] [r]ather than economy-wide GHG reductions" (Nordlander & Monti, 2022: 2). In particular, the applicants complained that Norway "has failed to take precautionary measures of prevention and protection required of it under ECHR Articles 2 and 8", and that "[by] licensing continuing exploration for oil and gas in new areas of the Arctic (Barents Sea), Respondent aims to bring new fossil fuels to market from 2035 and beyond, even though the best available science shows that the emissions from already proven reserves of fossil fuels exceed the remaining carbon budget, given the 1.5 °C temperature target set in the Paris Agreement, increasing the risk that global warming surpasses the 1.5 °C limit" (paras. 2 and 3 of the Application). Judicial dialogue with judgements such as those in *Urgenda* and *Neubauer* and with the CRC decision in *Sacchi* et al. may be helpful for "integrating human rights into the formulation and assessment of any State Party's climate action" (as argued by Desierto, 2021: 2), consistently with the Paris Agreement. This may be an effective way to address the 'production gap', that is as highlighted by the applicants in the *Greenpeace Nordic* case, the "significant difference between planned fossil fuel extraction and climate goals" (*Greenpeace Nordic* case, Application of Greenpeace Nordic, Natur og Ungdom, and the six young individual Applicants, 6). The statement contained in the 'Individual opinion by Committee Member Duncan Laki Muhumuza', in the decision of the HRC on *Daniel Billy*, provides an interesting example of the approach that is suggested here, insofar as it recognises that

> State Party has not taken any measures to reduce greenhouse gas emissions and cease the promotion of fossil fuel extraction and use, which continue to

> affect the authors and other islanders, endangering their livelihood, resulting in the violation their rights under article 6 of the [International] Covenant [on Civil and Political Rights] (para. 11).

Some further support for this interpretation can be found in soft law, namely in Resolution no. 3/2021 of the IACtHR on 'Climate emergency: scope of the Inter-American Human Rights Obligations', which affirmed that "[t]hose States in the region that have passed legislation on climate change and the energy transition and those that have made a commitment to develop and update their [...] [NDCs] must incorporate a human rights approach into their construction and implementation" (para. 2). Again, some important interpretive support can be sought in General Comment no. 26 (2022), where the HRC stressed that "[m]itigation policies should lead to absolute emissions reductions through the phasing out of fossil fuel production and use" (para. 56).

The applicants in the *Greenpeace Nordic* case have also addressed another important question which deserves specific attention, that is the incorporation of science. In this regard, referring to the "existing scientific evidence showing the impact of the cumulative effect of carbon emissions on the enjoyment of human rights, including rights under the Convention" (*Sacchi* et al., para. 10.11) not only plays an important role for assessing the foreseeability of the "potential harm of the State party's acts or omissions regarding the carbon emissions originating in its territory" (*Sacchi* et al., para. 10.11) but can also heighten the States' obligations to protect, as the CRC has stressed.

From a scientific viewpoint and consistently with the precautionary principle, a narrow interpretation of the States' margin of appreciation could be suggested, as the Hague Court has held in the *Urgenda* case (see the above-mentioned para. 5.3.2 of the judgement). Indeed, as Clark et al. (2020) have said with reference to the ECtHR, States' "'contact with the vital forces of their country' is of limited relevance when it comes to determining the extent of their respective obligations to mitigate the global problem that is climate change [and, since] [...] giving States a broad margin of appreciation in this area would be incompatible with the imperative of achieving that goal [...], the margin of appreciation is to be confined to 'choice of means'". This view may be incorporated through judicial dialogue, since it would help to recognise that States' margin of appreciation is limited to the *quomodo* of the promotion of the energy transition (that may be contemplated by the NDCs, for example).

Judicial dialogue could also be helpful with regard to the results achieved by climate litigation with respect to intergenerational equity and the intertemporal scope of the States' obligations. In this regard, particularly interesting guidance was provided by the German Supreme Court in the above-mentioned paragraph 192 of the *Neubauer* judgement, on the impact of the consumption of large portions of CO_2 by one generation, with a relatively minor share of the reduction effort, on the freedom of future generations. This intertemporality of intergenerational equity, which implies sustainable use of resources and sustainable consumption of the CO_2 budget, may be effective for addressing the stewardship of resources and the promotion of renewable energy systems.

This perspective may also benefit from some reference to the view expressed by the HRC in the *Daniel Billy* case, where it said that

> State party's failure to adopt timely adequate adaptation measures to protect the authors' collective ability to maintain their traditional way of life, to transmit to their children and future generations their culture and traditions and use of land and sea resources discloses a violation of the State party's positive obligation to protect the authors' right to enjoy their minority culture (para. 8.14),

when finding the breach of Article 27 of the International Covenant on Civil and Political Rights.

A similar approach to intergenerational equity may also be helpful for tackling the claims brought by young applicants before the ECtHR, as in the cases of *Duarte Agostinho*, *Greenpeace Nordic* and *Soubeste and Oth. v. Austria and Oth.*, which have articulated the extraterritorial reading of States' human rights obligations and the need for States to "take responsibility for their contributions to 'overseas' emissions" (Clark et al., 2020: 1).

Extraterritoriality is one of the most complex question to tackle; providing an effective solution is crucial, since it is assuming growing importance in the 'new energy litigation', especially that brought before the ECtHR. Indeed, by referring to the concepts of presumptive responsibility and 'indivisible injury', the young applicants in the *Duarte Agostinho* case not only complained about States' domestic emissions but also addressed their 'overseas' emission "entailed by [...] their export of fossil fuels, [...] the import of goods containing 'embodied' carbon and [...] the contributions to emissions abroad of entities domiciled within their respective jurisdictions (e.g. via fossil fuel extraction elsewhere or its financing)" (Clark et al., 2020, 1). In the *Greenpeace Nordic* case, the applicants expressly addressed the emissions from Norway's oil and gas exports.

It is not reasonable to expect the Strasbourg Court to deeply reconsider its conception of the extraterritorial jurisdictional link. However, in the case of *Ilascu v Russia and Moldova*, the ECtHR held that a "State's responsibility may [...] be engaged on account of acts which have sufficiently proximate repercussions on rights guaranteed by the Convention, even if those repercussions occur outside its jurisdiction" (as recalled by Clark et al., 2020: 4). Moreover, Sandvig et al. (2021) pointed out that

> th[e] mechanism for emission accounting [under the Paris Agreement] was not meant to detract from the well-established 'no-harm principle' of international law [...] [, which] prohibits States from knowingly conducting or permitting activities within their jurisdiction or control which cause transboundary harm to the environment in other States or to the properties or persons therein.

The situation is complex at the domestic level too. In the *People v. Arctic Oil*[22] case (from which the *Greenpeace Nordic* case originated), the Supreme Norwegian Court held that "if activities abroad over which the Norwegian authorities have a direct control, or against which they can implement measures, cause harm in Norway, they fall within the scope of the provision", and that "[o]ne example is combustion abroad of oil or gas produced in Norway, when it leads to harm in Norway as well". However, "the duty to take care under Article 112 [which protects the right to a healthy environment] does not grant corresponding individual rights to challenge petroleum-related activities [and does not apply extraterritorially]" (paras. 78–142; 149). Possibly, the view recalled above from the *Neubauer* judgement is more open and could be helpful.

In this framework, international cooperation seems the more viable solution, as provided for by the Paris Agreement and as suggested by Resolution no. 3/2021 of the IACtHR, which highlights that "States have an obligation to cooperate in good faith in order to prevent pollution of the planet, which [...] involves [...] move towards a clean and just

[22] *Greenpeace Nordic Association v Ministry of Petroleum and Energy* (2020) Case no 20-051052SIV-HRET (Norwegian Supreme Court) (*People v Arctic Oil*).

energy transition", consistently with the principle of common but differentiated responsibilities (para. 11).

Subsequently, at para. 12, Resolution no. 3/2021 addresses the business sectors, where it holds that "States must ensure that both public and private entities reduce their GHG emissions" and that the "programs [aimed at achieving this goal] must be constructed with a cross-cutting approach to rights". This is a very important issue and developing this kind of approach may be crucial for promoting energy transition litigation, including for closing the accountability gap in relation to the business sector. In this regard, the need to address corporations' accountability and promote justiciability should be one of the priorities. A human rights-based approach may help and might be successful, and it could be developed consistently with States' obligation to protect and the principle of due diligence.

Pursuing this purpose is so important also because, with respect to the business sector, an interesting case law trend is rising before domestic courts, as several civil actions are beginning to be filed against fossil fuel corporations to oblige them to reduce their emissions. The judgement of the District Court of The Hague in *Milieudefensie v. Shell* was the first decision of this kind to be issued. Similarly to the *Urgenda* ruling, the Court ordered Shell to reduce all carbon dioxide resulting from its global operations—including the emissions from the combustion of oil and gas products by its customers—by 45% by 2030 compared with 2019. The legal core of the Court's reasoning is the duty of care under Dutch tort law, but the judgement also relied on human rights law and the Paris Agreement. Although at the moment Shell has appealed the decision and therefore the situation is still evolving, the judgement of the District Court of the Hague is ground-breaking. Some criticism has been expressed in scholarship (Mayer, 2021b: 3), recalling "the tension that Koskenniemi describes between ascending and descending legal reasoning". Indeed, Mayer has maintained that the "Court clearly has a predilection for descending reasoning [...] [as it] applies to Shell a standard inferred from political objectives and scientific studies, notwithstanding its lack of recognition by oil companies" and that "[a] more convincing judgment would have sought more balance between ascending and descending reasoning—between what we believe should be done and what companies generally do". The criticism expressed by Mayer (2021b) may be a good starting point for improving courts' results when they address corporations' duties in the field of the energy transition.

It seems necessary that States, for example in line with their obligation to protect and the related principle of due diligence, take appropriate action in this respect. This is all the more desirable since the *Milieudefensie* judgement[23] is not an isolated example of proactive litigation. The number of cases is increasing, as shown by the recent lawsuit filed by ClientEarth against Shell's directors for failing to prepare the multinational oil and gas company for net zero and the legal actions underway in Germany against *BMW*[24] and *Wintershall Dea.*[25]

Appropriate consideration of human rights, international environmental law and the business sectors is of crucial importance for ensuring a just energy transition. In this sense, an interesting case has been brought before the ECtHR, namely *Soubeste and Oth. v. Austria and Oth.* In an unprecedented way, the *Soubeste* case combines human rights, climate change and investment law, and focuses on whether the Energy Charter Treaty has inhibited the respondent states from taking measures to deal with climate change, in violation of the applicants' rights protected under Articles 2, 3 and 8 of the ECHR. Despite some concerns were expressed in scholarship, especially about "the possibility for the Court to render a judgment that affects the participation of ECHR parties in other international conventions" (Nordlander & Monti, 2022), the potential of the case is remarkable.

It becomes even more evident when one thinks about the importance of renewable energy investments and of the mining activities that are necessary for the energy transition (as recognised by Resolution of the IACtHR no. 3/2021 at paras. 56 and 55). In fact, "there is a rush for the mineral resources needed to build solar arrays, wind turbines, and batteries" (Sourgens, 2022: 2). The incorporation of the views developed by the climate jurisprudence, with appropriate contextualisation, will be of fundamental importance for preventing exploitation, especially of populations and resources in the Global South.

[23] *Milieudefensie v. Royal Dutch Shell*, District Court of the Hague (DC), 26 May 2021, ECLI:NL:RBDHA:2021:5337 (Milieudefensie), English translation ECLI:NL:RBDHA:2021:5339.

[24] *Deutsche Umwelthilfe (DUH) v. Bayerische Motoren Werke AG (BMW).*

[25] *Barbara Metz* et al. *v. Wintershall Dea AG.*

Conclusions

A rising tide of climate (and energy) litigation is challenging courts and human rights bodies, domestically and internationally, to tackle the climate crisis. The evolving body of case law has led to the recognition of human rights as a source of mitigation obligations for States and to the emergence of an interesting conceptualisation of the intertemporality and the extraterritoriality of States' obligations. In this way, climate litigation has proved to be capable of promoting States' accountability in the complex framework of climate change.

In Michael Burger and Jessica Wentz words, there is a "growing consensus that a mitigation obligation does exist under international human rights law" (Burger & Wentz, 2015: 205), and it has so far been transversally and consistently expressed by both domestic courts and international human rights bodies in their climate case law.

In this sense, it seems significant to recall that one of the most recent relevant national judgements, namely the Brazilian Supreme Court's decision in *PSB* et al *v. Brazil*, has taken a step further by declaring that environmental treaties are a particular type of human rights treaties.

Although this view has received some criticism (Neumann Ciuffo, 2022), it should be kept in mind the role that the reference to human rights in the Preamble of Paris Agreement plays, consistently with the interpretive rules enshrined in Article 31 of the Vienna Convention on the Law of Treaties. As stressed by Desierto (2021), "[the incorporation of the reference to human rights in the Preamble of the Paris Agreement] is infinitely a more direct treaty device for integrating human rights into the formulation and assessment of any State Party's climate action".

Domestic courts' approach analysed in this chapter appears consistent with this perspective, since national judges have relied on human rights as a standard for assessing the adequacy of States' legislation and policies on climate change. Such judgements as those issued in *Urgenda*, *Neubauer*, *Shrestha* and *Leghari* are paradigmatic examples of the role that national jurisprudence has been playing.

The ideas of intragenerational and intergenerational equity as well as of extraterritoriality that both domestic and international case law have been promoting are important too and can make a remarkable difference in the successful and effective protection of human rights and the definition of States' obligation not only in the field of climate change but also with respect to the energy transition. Judicial dialogue and cross-fertilisation, including from a multi-layered viewpoint, can be very useful, as they would help a mutual improvement of the results that domestic case law and international jurisprudence have achieved.

In this sense, the results achieved in the landmark judgement *Minors Oposa* may be helpful for overcoming the difficulties related to the *locus standi* or legal standing. The concept of 'class' that the Supreme Court of the Philippines has adopted may result to be particularly helpful for domestic and international bodies when dealing with young petitioners' applications. Sharing the conceptualisation of the legal standing of the "[p]etitioners minors [...] [as] represent[atives] [of] their generation as well as generations yet unborn" based on the principle of intergenerational equity (*Minors Oposa*, p. 8) could be helpful to overcome the difficulties related to the non-identity and non-existence (at least, in the present) of future generations and would enhance the legal standing of young generations. This approach may be particularly beneficial for those domestic legal orders that do not allow *actiones populares* or that provide restrictive requirements with regard to environmental class actions.

The adoption of the conception of extraterritoriality based on 'effective control' (see Berkes, 2018) may be helpful from various points: first off, it may help to deal with the difficulties related to diagonal claims, since it can help to hold States accountable not only in those situations where States have control over the right-bearers but also when they have control over a potential cause of harm to the right-holder with an extraterritorial impact.

The decision of the CRC in the *Sacchi* et al. case provides a significant example, which is also the result of judicial dialogue and of the thriving reference to soft law as an interpretive instrument. In fact, by referring to the notion of 'effective control' elucidated by the IACtHR in its AO no. 23 'The environment and Human Rights', the CRC has defined States' responsibility and its articulation, and has clarified that, consistently with "the principle of common but differentiated responsibility, as reflected in the Paris Agreement", "the collective nature of the causation of climate change does not absolve the State party of its individual responsibility that may derive from the harm that the emissions originating within its territory may cause to children, whatever their location" (*Sacchi* et al., para. 10.10).

As some commentators have observed (Berkes, 2018; Besson, 2020), the jurisdictional extraterritorial link of effective control derives from a broad application of the principle of due diligence, and this chapter suggests that this may help to address the business sector and to hold it accountable for the impact of its activities on climate change.

This idea may also fit with regard to the energy transition, with special regard to the mining activities that are necessary for this process since, as Resolution of the IACtHR no. 3/2021 (para. 55) has recognised "the transition to a low-carbon future requires the extraction of minerals necessary for the construction of products and infrastructure that allow the operation of the renewable energy matrix". As recalled in the chapter, in fact, "there is a rush for the mineral resources needed to build solar arrays, wind turbines, and batteries" (Sourgens, 2022: 2), and it may become a threat especially for the populations and the resources in the Global South. The rush to cobalt in the Democratic Republic of Congo is a paradigmatic example (as widely set out by Sourgens, 2022).

In order to prevent inequity and exploitation, and to promote distributive justice as well, it is crucial to recognise that "States must comply with all their human rights and environmental obligations in the context of mining activities for energy transition purposes" (Resolution of the IACtHR no. 3/2021, para. 55).

In this sense, the approach that the climate litigation has developed so far may be strategic: intragenerational and intergenerational equity as well as the notion of extraterritoriality based on 'effective control' may pave the way to the sustainable and equitable use of the natural resources (with appropriate contextualisation, the view enshrined in *Neubauer*, para. 192, is relevant in this sense) used for the promotion of the energy transition and the construction of renewable energy facilities. The adoption of a human rights-based approach, consistently with the ideas of intertemporality and extraterritoriality of States' obligations, would provide an effective paradigm for conveying the stewardship of natural resources and the prevention of the risks of exploitation. This would be an effective way to ensure the justiciability of the energy transition.

This approach, however, could not prescind from the adoption of appropriate States' action, which also implies regulatory efforts. In this regard, again, soft law provides important guidance, as Resolution no. 3/2021 of the IACtHR has clarified that "States should focus their efforts on all those strategies that rapidly promote investments in climate-resilient infrastructure, forms of mobility and energy free of polluting emissions, the reduction of the use of fossil fuels" (para. 56), and that "States should consider minimizing all subsidies for fossil fuels, creating taxes on them and redistributing revenues towards clean, renewable, and non-polluting energy systems, such as wind or solar, putting respect for human rights at the center for the development of these energy sources" (para. 57).

The idea that "[m]itigation policies should lead to absolute emissions reductions through the phasing out of fossil fuel production and use" (General Comment no. 26 (2022), para. 56) and that States have specific human rights obligations in this sense, was explicitly stated by the recent CESCR in its General Comment no. 26 (2022).

The adoption of adequate States' action and measures could also be an effective way to provide an appropriate legal framework and guidance for courts in those cases where civil actions are filed against fossil fuel corporations to oblige them to reduce their emissions. This could be a way to reconcile courts' current approach—as that observed in the *Milieudefensie* judgement—with the criticism expressed by Mayer (2021b), for the purpose of seeking "more balance between ascending and descending reasoning – between what we believe should be done and what companies generally do" (Mayer, 2021b, 3).

Since this jurisprudential trend is growing—as the recent lawsuit filed by ClientEarth against Shell's directors for failing to prepare the multinational oil and gas company for net zero and the legal actions underway in Germany against *BMW* and *Wintershall Dea* demonstrate—it is crucial that States provide appropriate and prompt responses. In this respect, the fact that "States must ensure that both public and private entities reduce their GHG emissions" and that the "programs [aimed at achieving this goal] must be constructed with a cross-cutting approach to rights" was expressly held by Resolution no. 3/2021 of the IACtHR (para.12).

Conclusively, it could be questioned what the future holds for the energy transition and, especially, for energy transition litigation. The latest steps taken by human rights bodies are promising, with reference to the decision adopted by the HRC in *Daniel Billy* and the view expressed by the CESCR in its General Comment no. 26 (2022). However, to have some further answers, it will be interesting to observe the approach that the ECtHR will decide to take in relation to the rising wave of climate litigation that, as this chapter has stressed before, has a significant potential also in term of energy transition litigation and States' obligations related to mitigation and the phasing-out of fossil fuels. *Greenpeace Nordic*, *Duarte Agostinho* and *Soubeste* offer a great opportunity to the Strasbourg Court to promote a thriving human rights-based case law on issues related to the energy transition.

However, to succeed to achieve this goal, the ECtHR will need to address some issues and challenges, as overcoming the narrower concept of jurisdiction it has so far adopted, given that, as some commentators have stressed, "[s]ince its judgment in *Banković v Belgium*, it has largely displayed two models of jurisdiction: one based on territorial control, and one based on personal control" (Keller & Heri, 2022: 161; Besson, 2021). In this sense, judicial dialogue with the IACtHR, the CRC and the HRC may be crucial for the Strasbourg Judges, and reference to the soft law sources considered in this chapter (i.e. AO n. 23 'The Environment and Human Rights' and Resolution no. 3/2021 of the IACtHR, and General Comment no. 26 (2022) of the CESCR) may be a very helpful interpretive support for elucidating States' human rights and mitigation obligations in the field of the energy transition.

The fact that a rising wave of energy litigation has started to spread at the domestic level is promising too, despite the challenges that national courts will have to tackle as well, especially when they are called to hold the business sector accountable. By now, it is not possible to predict how this landscape will evolve specifically, but it can be said the time seems ripe for the development of a thriving energy transition litigation wave.

References

Barroso, L. J. (2022, December 5). Facing Climate Change in the Brazilian Supreme Court: The Right to a Healthy Environment as a Human Right. *I·CONnect*. Retrieved December 31, 2022, from http://www.iconnectblog.com/2022/12/facing-climate-change-in-the-brazilian-supreme-court-the-right-to-a-healthy-environment-as-a-human-right/

Bäumler, J. (2021, June 8). Sustainable Development made Justiciable: The German Constitutional Court's Climate Ruling on Intra- and Inter-generational Equity. *EJIL:Talk!*. Retrieved December 31, 2022, from https://www.ejiltalk.org/sustainable-development-made-justiciable-the-german-constitutional-courts-climate-ruling-on-intra-and-inter-generational-equity/

Bergkamp, L. (2020). The Dutch Supreme Court's Climate Judgment: Its Consequences and Implications for Business—Revolution Through Litigation. *European Energy and Environmental Law Review, 29*(3), 89–97. Retrieved December 31, 2022, from https://doi.org/10.54648/eelr2020032

Berkes, A. (2018, March 28). A New Extraterritorial Jurisdictional Link Recognised by the IACtHR. *EJIL:Talk!*. Retrieved December 31, 2022, from https://www.ejiltalk.org/a-new-extraterritorial-jurisdictional-link-recognised-by-the-iacthr/

Besson, S. (2020). Due Diligence and Extraterritorial Human Rights Obligations—Mind the Gap! *ESIL Reflection, 9*(1). Retrieved December 31, 2022, from https://esil-sedi.eu/esil-reflection-due-diligence-and-extraterritorial-human-rights-obligations-mind-the-gap/

Besson, S. (2021). The Extraterritoriality of the European Convention on Human Rights: Why Human Rights Depend on Jurisdiction and What Jurisdiction Amounts To. *Leiden Journal of International Law, 25*(4), 857–884. Retrieved December 31, 2022, from https://doi.org/10.1017/S0922156512000489

Brown Weiss, E. (1992). In Fairness to Future Generations and Sustainable Development. *American University International Law Review, 8*(1), 19–26.

Burger, M., & Wentz, J. (2015). Climate Change and Human Rights. In M. Faure (Ed.), *Elgar Encyclopedia of Environmental Law* (pp. 198–212). Edward Elgar Publishing.

Clark, P., Liston, G., & Kalpouzos, I. (2020, October 6). Climate change and the European Court of Human Rights: The Portuguese Youth Case. *EJIL:Talk!*. Retrieved December 31, 2022, from https://www.ejiltalk.org/climate-change-and-the-european-court-of-human-rights-the-portuguese-youth-case/

Desierto, D. (2021, October 8). Just Transitions in Climate Change Actions: Are States Respecting, Promoting, and Considering Human Rights Obligations in Setting and Implementing NDCs?. *EJIL:Talk!*. Retrieved December 31, 2022, from https://www.ejiltalk.org/respecting-human-rights-obligations-in-climate-change-actions-are-states-evaluating-ndcs-human-rights-impacts/

Feria-Tinta, M. (2021). Climate Change Litigation in the European Court of Human Rights: Causation, Imminence and Other Key Underlying Notions. *Europe of Rights and Liberties/Europe des Droits and Libertés, 1*(3), 52–71.

Gradoni, L., & Mantovani, M. (2022, March 24). No Kidding!. *Verfassungsblog*. Retrieved December 31, 2022, from https://verfassungsblog.de/no-kidding/

Heri, C. (2020, December 22). The ECtHR's Pending Climate Change Case: What's Ill-Treatment Got To Do With It?. *EJIL:Talk!*. Retrieved December 31, 2022, from https://www.ejiltalk.org/the-ecthrs-pending-climate-change-case-whats-ill-treatment-got-to-do-with-it/

Keller, K., & Heri, C. (2022). The Future is Now: Climate Cases Before the ECtHR. *Nordic Journal of Human Rights, 40*(1), 153–174. Retrieved December 31, 2022, from https://doi.org/10.1080/18918131.2022.2064074

Keller, K., Heri, C., & Piskóty, R. (2022). Something Ventured, Nothing Gained?—Remedies before the ECtHR and Their Potential for Climate Change Cases. *Human Rights Law Review, 22*, 1–26. Retrieved December 31, 2022, from https://doi.org/10.1093/hrlr/ngab030

Kotzé, L. J. (2021). Neubauer et al. versus Germany: Planetary Climate Litigation for the Anthropocene? *German Law Journal, 22*, 1423–1444. Retrieved December 31, 2022, from https://doi.org/10.1017/glj.2021.87

Laukkanen, V., & Laukkanen, V. (2022, November 2). The Point of No Return: The Inadequacy of the Law of State Responsibility in Finding International Responsibility for the Adverse Effects of Climate Change. *International Law Blog*. Retrieved December 31, 2022, from https://internationallaw.blog/2022/11/02/the point-of-no-return-the-inadequacy-of-the-law-of-state-responsibility-in-finding-international-responsiblity-for-the-adverse-effects-of-climate-change/

Leijten, I. (2019). Human Rights v. Insufficient Climate Action: The *Urgenda* Case. *Netherlands Quarterly of Human Rights, 37*(2), 112–118. Retrieved December 31, 2022, from https://doi.org/10.1177/0924051919844375

Mayer, B. (2021a). Climate Change Mitigation as an Obligation Under Human Rights Treaties? *The American Journal of International Law, 115*(3), 409–451. Retrieved December 31, 2022, from https://doi.org/10.1017/ajil.2021.9

Mayer, B. (2021b, June 3). Milieudefensie v Shell: Do Oil Corporations Hold a Duty to Mitigate Climate Change? *EJIL:Talk!*. Retrieved December 31, 2022, from https://www.ejiltalk.org/milieudefensie-v-shell-do-oil-corporations-hold-a-duty-to-mitigate-climate-change/

Milanovic, M. (2011). *Extraterritorial Application of Human Rights Treaties: Law, Principles, and Policy*. Oxford University Press.

Neumann Ciuffo, L. (2022, 7 July). Brazil's High Court First to Declare Paris Agreement a Human Rights Treaty. *Yale Environment, 360*. Retrieved December 31, 2022, from https://e360.yale.edu/digest/paris-agreement-human-rights-treaty-brazil

Nordlander, L., & Monti, A. (2022, June 30). A New Variety of Rights-Based Climate Litigation: A Challenge Against the Energy Charter Treaty before the European Court of Human Rights. *EJIL:Talk!*. Retrieved December 31, 2022, from https://www.ejiltalk.org/a-new-variety-of-rights-based-climate-litigation-a-challenge-against-the-energy-charter-treaty-before-the-european-court-of-human-rights/

Parfit, D. (1984). *Reasons and Persons*. Oxford University Press.

Rajamani, L. (2018). Human Rights in the Climate Change Regime. In J. Knox & R. Pejan (Eds.), *The Human Rights to a Healthy Environment* (pp. 236–251). Cambridge University Press.

Renou-Maissant, P., Abdesselam, R., & Bonnet, J. (2022). Trajectories for Energy Transition in EU-28 Countries over the Period 2000–2019: A Multidimensional Approach. *Environmental Modeling and Assessment, 27*, 525–551. Retrieved December 31, 2022, from https://doi.org/10.1007/s10666-022-09816-7

Sandvig, J., Dawson, P., & Tjelmeland, M. (2021, June 23). Can the ECHR Encompass the Transnational and Intertemporal Dimensions of Climate Harm?

Can the ECHR Encompass the Transnational and Intertemporal Dimensions of Climate Harm?. *EJIL:Talk!*. Retrieved December 31, 2022, from https://www.ejiltalk.org/can-the-echr-encompass-the-transnational-and-intertemporal-dimensions-of-climate-harm/

Savaresi, A., & Setzer, J. (2022). Rights-Based Litigation in the Climate Emergency: Mapping the Landscape and New Knowledge Frontiers. *Journal of Human Rights and the Environment, 13*(1), 7–34. Retrieved December 31, 2022, from https://doi.org/10.4337/jhre.2022.01.01

Setzer, J., & Benjamin, H. (2019). Climate Litigation in the Global South: Constraints and Innovations. *Transnational Environmental Law, 9*(1), 77–101. Retrieved December 31, 2022, from https://doi.org/10.1017/S2047102519000268

Setzer, J., & Higham, C. (2022). *Global Trends in Climate Change Litigation: 2022 Snapshot*. Grantham Research Institute on Climate Change and the Environment and Centre for Climate Change Economics and Policy, London School of Economics and Political Science. Retrieved December 31, 2022, from https://www.lse.ac.uk/granthaminstitute/publication/global-trends-in-climate-change-litigation-2022/

Sharati, A. A., & Tillet, L. (2022, December 19). COP27 Did Not Seize the Opportunity to Open the Debate Around States' Greenhouse Gases Emissions Accountability. *International Law Blog*. Retrieved December 31, 2022, from https://internationallaw.blog/2022/12/19/cop27-did-not-seize-the-opportunity-to-open-the-debate-around-states-greenhouse-gases-emissions-accountability/

Slobodian, L. (2020). Defending the Future: Intergenerational Equity in Climate Litigation. *The Georgetown Environmental Law Review, 32*, 569–589.

Sourgens, F. G. (2022, February 24). Energy Lessons from the Ukraine Crisis. *EJIL:Talk!*. Retrieved December 31, 2022, from https://www.ejiltalk.org/energy-lessons-from-the-ukraine-crisis/

Voigt, C. (2022). The Climate Change Dimension of Human Rights: Due Diligence and States' Positive Obligations. In N. Kobylarz & E. Grant (Eds.), *Human Rights and the Planet* (pp. 152–171). Edward Elgar Publishing.

CHAPTER 7

Just Transition Litigation in Latin America: Striking a Balance Between Economic Development and Environmental and Social Energy Justice

Natalia Urzola, Maria Antonia Tigre, Riccardo Luporini, Lorena Zenteno, Marlies Hesselman, and Pedro Cisterna-Gaete

Introduction

The emergency of the triple planetary crisis (pollution, biodiversity and climate change) is testing the limits of international environmental law. Grounded on the developing scientific knowledge of the limited state of our natural resources and the risks the current developing model places on

N. Urzola (✉)
Global Network for Human Rights and the Environment (GNHRE), White Plains, NY, USA

Elisabeth Haub School of Law, Pace University, NY, USA

D. Iglesias Márquez et al. (eds.), *Legal Challenges at the End of the Fossil Fuel Era*, Global Issues,
https://doi.org/10.1007/978-3-031-61766-9_7

them, the need for an international response to fossil fuel dependency became apparent. The long-term goal to limit global warming to well below 2 °C and preferably to 1.5 °C, compared to pre-industrial levels, guides climate policy-making worldwide (Paris Agreement, 2016). Reducing greenhouse gas (GHG) emissions and mitigating climate change are crucial to reach this goal (IPCC, 2022). Because of their high contribution to GHG emissions, the imperative to reduce emissions relies heavily on extractive industries (Brauch & Toledano, 2022: 2). Yet, decarbonisation strategies are not developed in a vacuum. Socio-economic challenges, including inequality and racial injustice, significantly play a role in developing and implementing policies to move away from fossil fuels. Even though the energy transition could create millions of new (green) jobs worldwide, it could also result in the loss of millions of jobs, overwhelmingly in the energy sector (Renner et al., 2022).[1]

Latin America (LATAM) provides a unique example of these challenges as the region faces a constant dichotomy between ambitious environmental legal frameworks and reliance on extractivism as the foundation for economic development in a complex and multifaceted relationship. On the one hand, there is a real need to increase energy supply and promote economic growth. On the other hand, stakeholders face the obligation to

[1] For a more detailed account of job loss predictions due to a transition away from fossil fuels, see also International Labour Organization, Greening with Jobs, https://www.ilo.org/weso-greening/#Intro-1, accessed on 10 May 2023.

M. A. Tigre
Columbia Law School, Sabin Center for Climate Change Law, New York, NY, USA

R. Luporini
Sant'Anna School of Advanced Studies, Pisa, Italy

L. Zenteno
University of Miami, Miami, FL, USA

M. Hesselman
University of Groningen, Groningen, The Netherlands

P. Cisterna-Gaete
University of Miami, Miami, FL, USA

University of Edinburgh, Edinburgh, UK

pursue environmental protection, achieve climate goals and social justice, and protect human rights. The impulse for increased development, economic growth and the fight against poverty has pushed for large-scale infrastructural and extractive projects in LATAM. This has also evolved based on the private interests of transnational corporations and financial institutions (Little, 2014). Yet despite their obvious economic benefits, these projects severely impact each country's environmental and social dynamics and the region as a whole. Large-scale projects are invasive by nature, leading to environmentally devastating effects, including deforestation and loss of biodiversity (Little, 2014: 60). They also lead to encroachment on indigenous lands, displacement and forced relocation of traditional local communities, as well as inhumane working conditions (Little, 2014: 61). In some cases, nearby towns experience an increase in violence and illicit activities (Little, 2014: 62). A just transition requires minimisation of these impacts, or at the very least, ensuring that they are accounted for when enacting new policies towards decarbonisation.

With some countries in LATAM showing a desire to support a transition towards cleaner energy (*RELAC | IADB*, n.d.), these dynamics are likely to exacerbate. In particular, a new set of distributional issues can potentially emerge. While environmental protection and climate action are the main drivers of the energy transition, the economic impacts on an entire economic sector can jeopardise this transition (OECD, 2022: 23). For example, fossil fuel-dependent communities and workers might be left stranded by cutbacks with limited opportunities to re-enter the workforce (OECD, 2022: 143). These communities and workers, although in favour of cleaner energy, might be forced to carry disproportionate burdens that heighten socio-economic challenges following a just transition. As a result, issues of distributive justice are expected to arise, as the share of benefits and resources, as well as the burdens and negative impacts, will face new challenges. In this case, distributive justice refers to how environmental 'goods' and 'bads' are distributed and shared by and among different groups and communities. This type of justice is strongly related to notions of fairness and equity in the distribution of environmental 'goods' and 'bads' (Stephens & Church, 2017).

Decarbonisation policies will undoubtedly have both positive and negative effects on society. The positive aspect is more apparent since discussions on the benefits of decarbonisation policies abound in academic and public policy scenarios (Savaresi & Setzer, 2022; Carley & Konisky, 2020; Wang & Lo, 2021) and are part of promotion efforts towards a green energy transition. For example, cleaner energy production helps curb GHG emissions, therefore inducing climate action and environmental protection. As for the latter, negative impacts tend to be overlooked or under-discussed. For instance, workers in emissions-intensive jobs and fossil fuel-dependent communities will face livelihoods and income losses and require financial support and re-employment programmes (Saget et al., 2020). Likewise, energy transition projects might continue to affect marginalised communities if their impacts are not adequately considered.

Following recent scholarship on this matter, this chapter treats just transition as a three-fold concept that encompasses the dimensions of (1) a clean, healthy and sustainable environment, (2) decent jobs and livelihoods, and (3) affordable access to essential goods and services for the enjoyment of human rights (Tigre et al., 2023a, 2023b: 6). Ensuring access to decent work and essential goods and services, including energy, transportation and water, as well as moving towards a cleaner environment, is all part of what a just transition should entail.

As governments worldwide—and mainly, within the focus of this chapter, LATAM governments—balance the needs of carbon-dependent communities and industries with international and regional commitments, a new set of litigation cases has emerged. This rise, as identified in the literature, seems to promote *more ambitious* climate change policies across the Global South (Peel & Osofsky, 2015; Peel & Lin, 2020; Setzer & Benjamin, 2020a, 2020b; Osofsky, 2020; Auz, 2022; Alogna et al., 2021; Sindico & Mbengue, 2020). Drawing on that literature, this chapter pinpoints just transition policies and the rise of so-called just transition litigation in LATAM. According to the definition put forth by Tigre et al. (2023a) following the initial conceptualisation of just transition litigation cases by Savaresi and Setzer (2022),

> [j]ust transition litigation questions the distribution of the benefits and burdens of transition policies and activities towards net zero-emission and climate-resilient societies amongst local communities and affected stakeholders. Litigators typically invoke human rights but may rely on other national and international laws and principles. Just transition cases are not necessarily pro- or anti-climate action. Instead, they demand that climate action is undertaken in a just and inclusive manner with attention to those in vulnerable situations (21).

This chapter is based on research on just transition policies and case law in LATAM. It aims to systematically identify and categorise the policies and cases that define just transition, although we do not claim that the method and findings presented here are exhaustive. Through a comparative analysis of a selection of LATAM countries, the chapter questions whether initiatives in energy transformation in the region have relied on just transition dimensions and whether flaws in that process have led to cases at the national level. Overall, this chapter aims to increase the visibility of actions from the Global South to provide practical answers to environmental and social energy justice issues through a just transition towards cleaner energy. The chapter is structured as follows. Section "Just Transition Approaches to Decarbonisation: From International Efforts to National Policy" provides a broad overview of the just transition policies at the international and national levels within selected LATAM countries. Section "An Introduction to Just Transition Litigation in LATAM" provides a snapshot of just transition litigation in LATAM, drawing on existing scholarship, particularly the 2023 Just Transition Litigation report (Tigre et al., 2023a). Section "The Link Between Policy and Litigation: Key Examples of Litigation Challenging Domestic Transition Policies" examines whether LATAM is becoming a hub for just transition litigation due to policies in place. It also examines the cases' impact on human rights and justice in LATAM countries, given their unique circumstances. Finally, section "Conclusions: Lessons Learned" concludes with lessons that place LATAM in an exceptionally prominent position in advancing a just energy transition.

The examples of policies discussed in section "Just Transition Approaches to Decarbonisation: From International Efforts to National Policy" derive from research on progressive attitudes and developments on just transition policy-making in LATAM countries, especially as evident from Nationally Determined Contributions (NDCs), climate and just transition literature, and the experience from the authors, who are in large part from the region. The litigation on just transition included in section "An Introduction to Just Transition Litigation in LATAM" relies on the extensive set of cases outlined in the global database of climate litigation of the Sabin Center for Climate Change Law, the Climate Case Chart and the 2023 Just Transition Litigation report (Tigre et al., 2023a) written by the authors of this chapter, as well as original research into just transition litigation across LATAM countries. While a systematic research analysis of case law in LATAM is beyond the scope of this chapter, our research fills some gaps in the existing data.

Just Transition Approaches to Decarbonisation: From International Efforts to National Policy

The concept of just transition first emerged in labour leader Tony Mazzochi's 1970s campaigns. Mazzochi demanded a 'superfund for workers' in toxic industries that would guarantee their income and educational benefits in transition processes to cleaner industries (Eisenberg, 2019, 285). Almost simultaneously, environmental groups started using the concept of 'just transition' to address the unfair distribution of environmental hazards (Eisenberg, 2019: 275). From there, labour and environmental movements have sponsored the just transition as a concept towards a more equitable future, a demand that continues to evolve, particularly in the climate scenario (Paris Agreement, 2015). International trade unions, for instance, have developed an increasing role in the United Nations Framework Convention on Climate Change (UNFCCC)'s institutional environment, focusing on incorporating the 'just transition' notion into international climate policies (Thomas, 2021). The adaptability of the concept of just transition to different scenarios and contexts has facilitated its development throughout climate discussions.

This section provides an overview of relevant international and regional approaches to just transition policies in the context of climate change, highlighting a few LATAM countries that have included and implemented just transition policies. In addition, it analyses the scenario and the contextual elements that shape the potential development of just transition litigation in LATAM.

International and Regional Approaches

At the international level, the just transition notion was first included in the UNFCCC process at COP15 in Copenhagen. The negotiating text of the COP15 contained an explicit reference to the necessity for a just transition in addressing "the most impacted economic sectors and contributing to building new capacities for both production and service related jobs" (UNFCCC, 2009: 143). Argentina and Norway played a relevant role in including just transition in climate negotiations, supporting the concept in their national mandates (International Trade Union Confederation et al., 2009). In 2010, following COP15, the shared vision of UNFCCC Parties for a future global climate agreement included a reference to the need to ensure "a just transition of the workforce that creates decent work and quality jobs" (Paris Agreement, 2015). Furthermore, the preamble of the Paris Agreement mentioned the need to consider the imperatives of a just transition of the workforce and the creation of "decent work and quality jobs" (Paris Agreement, 2015). The UNFCCC further detailed the necessary steps for achieving a just transition: it requires "a country-specific mix of macroeconomic, industrial, sectoral, social protection, skills, social dialogue and labour policies" as a means to create decent jobs and promote sustainable investment (UNFCCC, 2020: 62).

In parallel, the International Labour Organisation (ILO) offers guidelines for applying the concept of just transition through its Guidelines for a Just Transition (2015). The ILO states that just transition should be "well managed and contribute to the goals of decent work for all, social inclusion and eradication of poverty" (ILO, 2015). Social dialogue is one of the ILO's most relevant elements in developing a just transition. Furthermore, every decision to implement a just transition should be based on "adequate, informed and ongoing consultation" (ILO, 2015). More recently, the UN General Assembly (UNGA) stated that: "the transitions to affordable, reliable, sustainable and modern energy for all should be just, inclusive, equitable and secure, in line with national circumstances" and noted "that an urgent and steep rise" in investments and finance in "clean technologies and quality infrastructure" is necessary, "with a priority focus on the needs of the world's least developed countries and on universal access to electricity" (UNGA, 2022).

In recent years, international and regional human rights law has also increasingly embraced the concept of just transition. For example, in 2020, the UN Special Rapporteur on extreme poverty and human rights devoted an entire report to the topic. This report noted that to comply with just transition standards, states need to implement policies and take measures that can capture the 'triple dividend'. This vision means to capture three vital elements of a just transition: (a) a clean environment; (b) decent job opportunities, especially for vulnerable populations; as well as (c) universal access to affordable goods and services (Human Rights Council (HRC), 2020). In 2021, the Inter-American Commission of Human Rights' (IACmHR) Resolution 21/1 on obligations for the climate emergency equally devoted an entire section to just transition (IACmHR, 2021). Apart from stressing that Organization of American States' (OAS) Member States must protect their vulnerable populations against the adverse effects of climate policies, it also recommended OAS

Member States to actively (re)distribute revenues from climate policies (e.g. carbon pricing) to public investments and social programmes for low-income households and populations most affected by climate change (IACmHR, 2021).

These global and regional human rights approaches are coherent with the call of the Paris Agreement that all States must respect, protect and promote human rights when taking climate action. This implies that States must adequately mitigate GHGs in line with the objectives of the Paris Agreement (HRC, 2019) and ensure that human rights are not violated when implementing mitigation policies and activities. It also resonates with the concept of 'climate justice', which has human rights protection at its core and is emphasised in the Paris Agreement as a means to combat climate change (Paris Agreement, 2015; Skillington, 2017; Schapper, 2018). Therefore, human rights should not be excluded from energy transition analyses (see also Hesselman, 2023; Wewerinke-Singh, 2021; Savaresi & Setzer, 2022). The following subsection presents an overview of the just transition policies adopted in several LATAM countries to comply with international and regional commitments.

Just Transition Policies in LATAM

According to Article 4(2) of the Paris Agreement, all States Parties must maintain, update and take measures to achieve their NDCs under the Agreement. States must adopt such NDCs to mitigate GHG emissions so that the average global temperature rise can stay well below 2 °C, and preferably below 1.5 °C (Paris Agreement, 2015: 4). In addition, NDCs must be set according to the 'highest possible ambition' and imply efforts to adapt to the effects of climate change (Paris Agreement, 2015: 4). Hence, NDCs are vital for meeting global climate targets.

Although LATAM countries are particularly vulnerable to climate change due to poverty and inequality (IPCC, 2022), they account for only 8.1% of global GHG emissions (United Nations, n.d.). All LATAM countries have ratified the Paris Agreement and submitted their respective NDCs, in compliance with Articles 3 and 4 of the Agreement. According to the United Nations Development Programme (UNDP) Quality Checklist, which measures NDCs' quality on three criteria: robustness, feasibility, and ownership and inclusion, the region excels (UNDP, 2022). However, domestic NDCs implementation is not uniform (United Nations, n.d.). Each country has its own limitations, including financing, capacity-building or access to technology (United Nations, n.d.). It is estimated that the region needs USD 77 billion in investments to implement the Paris Agreement's NDCs (Brichetti et al., 2021).

In LATAM, the financial resources needed to implement NDCs are not always available. This is particularly true for those countries that are most vulnerable to the impacts of climate change, including low-lying coastal areas. Many of these countries rely heavily on external financing to implement their NDCs,[2] and the region needs increased public and private sector investment (Natural Resources Defense Council, 2017: 4). Regarding funding, it is also essential to handle proper climate financing strategies to secure financial support. There is a lack of expertise in LATAM countries regarding the strategies to obtain climate funding. Financial assistance is crucial (Pucheta et al., 2021).

A further complicating factor for NDC implementation relates to the economic consequences of the Covid-19 pandemic. The post-Covid economic crisis has restricted funding and made it more difficult for citizens to participate in climate action initiatives (European Commission, 2021). In LATAM, the Covid-19 pandemic had a profound impact, with governments introducing lockdown measures, reducing economic activity and diverting resources to health and social protection. The pandemic led to a significant reduction in their public finances, which has made it difficult for these countries to finance their NDCs. This has been particularly problematic for countries such as Bolivia, Ecuador, Guatemala and Peru, which were already facing significant economic and financial difficulties before the pandemic (Miranda, 2021).

In LATAM, only Argentina, Chile and Colombia have included just transition objectives and measures in their NDCs (United Nations, n.d.). For example, Chile included just transition as one of its 'social pillars', tying it to the decarbonisation of the "power generation matrix and to protect the rights of the most vulnerable in this process" (United Nations, n.d.). Colombia's NDC recognises the impacts of decarbonisation on the labour force, adopting just transition as a leading principle of their climate commitments (United Nations, n.d.). Argentina took a similar view in its last NDC update, incorporating just transition as a way to face the social

[2] For instance, Chile created *Bonos Verdes Soberanos*, which seek to obtain national and international financing for their NDC-related activities. See República de Chile, Marco para Bonos Verdes (2019) https://www.hacienda.cl/noticias-y-eventos/noticias/ministerio-de-hacienda-publica-su-primer-marco-de-bonos-sostenibles-

and economic impacts of the implementation of climate policies (United Nations, n.d.). This section will delve deeper into some of these examples of just transition policies, as well as examples from Brazil and Mexico. In addition to the efforts included in their NDCs as part of their commitment to comply with the Paris Agreement and other international and regional instruments, the first three countries (Chile, Colombia and Argentina) have started to integrate just transition in specific national policy instruments. Moreover, Mexico, despite failing to meaningfully engage with just transition in its NDC, provides a unique perspective on this matter derived from legislation. Lastly, Brazil provides an interesting example of what are just transition policies and how this term can be co-opted if implemented inadequately.

Chile

Chile's NDC dedicates an entire chapter to just transition and sustainable development as part of its social pillars. From the onset, Chile's NDC places a special focus on poverty eradication and treats just transition as an opportunity towards job creation and social justice while promoting a socially and environmentally sustainable economy (Gobierno de Chile, 2020: 23–24). The document identifies just transition as key in the country's goal to transition towards carbon neutrality, bearing in mind the need to minimise potential negative impacts on the most vulnerable and to ensure equal opportunities in the process, including those negatively affected by policies of climate action (Gobierno de Chile, 2020: 24).

In addition to the NDC, the newly elected Chilean government proposed Just Transition Commissions (JTC) to achieve a socio-ecological transition (Chilean Minister of Environment, 2022). These are participatory instances where communities, workers' unions, local and national authorities, and company representatives work together to guarantee just transition processes, ensuring a broad range of representation and participation of all communities involved (Chilean Minister of Environment, 2022). As of March 2023, the policy is in its initial stages and does not have a definitive structure for its implementation yet. However, the proposal has a local emphasis, prioritising the most vulnerable communities and economic sectors to the ecological transition whilst also seeking to develop social and environmental restorative processes for damaged communities (Chilean Minister of Environment, 2022).

Furthermore, Chile enacted the first Framework Law on Climate Change in June 2022 (Law no. 21.455 of June 2022) (Chilean National

Congress, 2022). Whilst the law does not explicitly mention just transition, Article 2 outlines equity and climate justice as fundamental principles (Law no. 21.455 of June 2022). Furthermore, the law requires that public policies consider a fair allocation of burdens, costs and benefits, focusing on gender, vulnerable communities and ecosystems, which can serve as a valuable baseline to implement a just transition (Law no. 21.455 of June 2022).

Colombia

Colombia's updated NDC document also treats just transition as a cross-cutting and foundational aspect of climate action, along with human rights, intergenerational equity, gender equity, differential approach to ethnic communities and vulnerable populations, ecosystems' integrity, food security and poverty eradication (Gobierno de Colombia, 2020: 1). The country's NDC focuses on a just transition for workers that will be affected by both climate change and the actions to fight it (Gobierno de Colombia, 2020: 2). Colombia is working on a national strategy to ensure social and economic inclusion and active participation from relevant stakeholders in the decision-making process.

Colombia first introduced the concept of just transition in the domestic legal framework in November 2019. In partnership with the ILO, workers signed "the Green Jobs and Just Transition Commitment to prepare for the challenges of the energy transition" (Ministerio del Trabajo de Colombia, 2019). The Pledge incorporated a tripartite collaboration mechanism among employers, workers and the government. More recently, in November 2022, at COP27, the then Minister of Energy of Colombia, Irene Vélez, introduced the Just Energy Transition Strategy, which aims to reduce fossil fuel dependence and involve communities in the decision-making process (Deutsche Welle, 2022). There is no doubt that Colombia's strong economic dependence on extractive industries like coal, whose primary destination is Europe, has been an obstacle to the country's just transition process. Nevertheless, promoting energy transition measures in Europe to phase out fossil fuels might be an essential incentive for Colombia's just transition law and policy (Strambo et al., 2018).

Moreover, in 2021, Colombia enacted Law 2169 of 2021 (Climate Action Law), which establishes goals and measures to achieve carbon neutrality. Just transition appears as one of the pillars that guide climate action measures in Colombia (Congress of Colombia, 2021: art. 3). Again, a focus is placed on the workforce, aiming to ensure sustainable production mechanisms and quality green jobs under the banner of human rights, ethnic, gender and intergenerational equity (Congress of Colombia, 2021: art. 7).

Argentina

Argentina's NDC recognises the need to ensure a just transition in the workforce as the country moves towards decarbonisation. The NDC is based upon ambition, justice and equity principles as dimensions that inform national policies and plans to ensure inclusivity in a way that is coherent with Argentina's social and economic context (Presidencia de Argentina, 2021: 8, 14). Furthermore, Argentina recognises that the NDC implementation will require financing from the international community. For this reason, the support provided by Partnership for Action on Green Economies (PAGE), a multi-agency led by the ILO, has been essential in elaborating the country's just transition plans and policies (Pucheta et al., 2021). As part of this effort, assessments and instruments have been developed under the supervision of the Ministry of Labour in Argentina to promote green jobs and social dialogue in support of just transition. These include the National Assessment of Green Jobs, Green Recovery for Argentina and the Road Map towards a National Green Jobs Promotion (The Ministry of Labour of Argentina Promotes Strategies for Green Jobs and Just Transition, 2022).

The adoption of three laws has recently eased the Argentinian transition into a green economy: the Tax Reform Act (Law No 27.430, December 2017) (Republic of Argentina, 2017); Law No 27.604 that reforms Law No 26.865 on fire management (Republic of Argentina, 2020a), and Law No 27.592 called '*Ley Yolanda*' (Republic of Argentina, 2020b) all play an important role in promoting a just transition towards a green economy in Argentina. In particular, the Tax Reform Act introduces a carbon tax on fossil fuels that aims to incentivise a reduction in the use of these fuels, which are major contributors to greenhouse gas emissions and climate change. By taxing these fuels, the government's goal is to encourage businesses and individuals to shift towards more sustainable energy sources. Additionally, the reform of the taxation system in relation to energy use provides a more equitable distribution of the tax burden and helps to ensure that those who can afford to pay more contribute a larger share.

On the other hand, Law No 27.604 helps to prevent the commercial development of land that has been destroyed by fire. This helps to protect the environment and wildlife, as well as promote sustainable land use practices. By preventing the commercial development of such land, the government can ensure that the natural resources in these areas are preserved for future generations. Finally, '*Ley Yolanda*' imposes an environmental training requirement for those in public office. This is important because it ensures that those in positions of power have the necessary knowledge to make informed decisions about environmental protection and sustainable development. By requiring environmental training, the government is taking steps to ensure that future policy decisions are made with the environment in mind, which will help to promote a just transition towards a green economy in Argentina.

Overall, these laws demonstrate the Argentine government's commitment to promoting a just transition towards a green economy, and they provide important tools for achieving this goal. By reducing the use of fossil fuels, protecting the environment, and promoting sustainable land use practices, these laws help to ensure that the transition to a green economy is both environmentally and socially responsible.

Mexico

Unlike Chile, Colombia and Argentina, Mexico's NDC's mention of just transition is bleak. Although the document mentions just transition in the introduction (Gobierno de México, 2022: 6), it does not engage with just transition as a concept nor develop what it means in the Mexican context. Nevertheless, Mexico has enacted relevant legislation to move towards decarbonisation and promotes renewable energy. The 2015 Energy Transition Act and the amendments to the 2014 Electric Industry Law were issued as part of Mexico's plan to comply with its NDC goals of GHG emissions reduction (González Márquez, 2022a). Transition to renewable energy formed a key part of this commitment and the overall objective of energy transition and climate change mitigation action (Gobierno de México, 2022: 12).

The 2015 Energy Transition Act promotes clean energy and establishes prioritisation of renewable energy production (Energy Transition Act, 2015: art. 4–5) (Congress of Mexico, 2015). This law focuses on a gradual increase of clean energy within the country's electric industry as a way to reduce polluting emissions in accordance with the General Climate Change Law (Energy Transition Act, 2015: art. 1–2) (Congress of Mexico, 2015). Similarly, the 2014 Electric Industry Law aims to promote the sustainable development of the electric industry and to ensure compliance with GHG emissions reduction and the clean energy transition (Electric Industry Law, 2014: art. 1) (Congress of Mexico, 2014). In 2021, the Mexican Congress amended this Law ordering, among other measures, the elimination of certain requirements that prioritised clean and renewable energy projects. As noted by González Márquez (2022a: 161–162), the new amendments are "not in line with Mexico's international commitments nor the national energy legal framework currently in force". As a result, several lawsuits, studied in more detail below, have been filed, arguing that these amendments contradict Mexico's commitment towards a clean energy transition.

Brazil

The term just transition in policies or legislation does not necessarily imply its effective implementation. This is the case of the recent Brazilian law on 'just energy transition' (Brazilian Government Commits to Coal Until 2040, 2022). The law creates the 'just energy transition programme' focused on preparing the Santa Catarina region—a coal-reliant subnational state in Brazil—to end fossil fuels (Brazilian Government Commits to Coal Until 2040, 2022). The goal of the programme is to observe environmental, social and economic impacts and align the values of energy and mineral resources to carbon neutrality. Since Brazil had committed to carbon neutrality by 2050 (Brazil, 2022 NDC), the law was adopted to prepare the region to close coal activities by 2040. However, the same just transition law allowed the use of coal energy until 2040, specifically by extending the validity of government authorisations for 15 years (therefore until 2040) and contradicting its own purpose. Furthermore, studies from the Association of Large Energy Consumers in the Market asserted that this permission to continue using coal would make energy costs more expensive, especially for the poorest families (Angeli, 2022). Then this Brazilian law is a clear example of the danger that the cooptation of just transition can have in implementing polluting and coal-friendly policies.

From the previous analysis of selected LATAM's NDCs and policy development on just transition, this chapter concludes that the region's primary focus is on the energy transition and impacts on the workforce, perhaps at the expense of other relevant economic sectors (i.e. agriculture) and stakeholders (i.e. ethnic communities). This strict limit of the just transition approach in LATAM reflects the immense challenges for its implementation and further development. Just transition is a novel concept for LATAM and, as such, is in an initial stage. Just transition policies are still in progress and are subject to modification as the region advances their implementation. As a result of this changing character, the policy frame on which legal conflicts could arise and litigation can be developed, requires ongoing revision of the formulation and implementation of just transition policies. Moreover, just transition efforts in LATAM are being developed against the backdrop of significant levels of inequality and poverty. Perhaps as a result of this structural issue, just transition is vital to make these decarbonising and adaptation routes consistent and fair. Climate litigation, and in particular, just transition litigation, can serve as a crucial instrument for orienting policies and legislation in LATAM.

An Introduction to Just Transition Litigation in LATAM

While LATAM's decarbonisation policies have advanced to pursue the long-term goals of the Paris Agreement, ineffective or inexistent just transition policies—or simply, policies that have no teeth and remain only relevant on paper—represent a risk to governments attempting to decarbonise their economies due to the labour and human rights effects on communities. The effects of decarbonisation on vulnerable groups and communities are multiple. For example, workers will be laid off, and their specialised knowledge acquired over the years could become irrelevant (International Energy Agency, 2021). This requires compensation packages, decent alternatives for high-paying jobs and re-education, which must be an integral part of transition policies (OECD, 2022: 143). These effects are not minimal, as there are entire communities dependent on fossil fuels. Extractive industries may pay taxes and/or fees to local governments that amount to programmes directed to improve the standard of living in the community or support other types of social contributions. For instance, in Colombia, income from the carbon tax is used to finance environmental conservation and sustainability, activities, rural development and peacebuilding (Saget et al., 2020: 87). Furthermore, some projects may implement, as part of their social obligations, programmes and actions directed to improve the conditions of communities through education, basic needs infrastructure improvement, employment of community members and others.

Along those lines, environmental and social justice issues will arise during the energy transition. Years of environmental and human rights harms and exploitation require strong compensation and redress mechanisms that may be left unmet if they are not considered within the just transition policies. Recognising the significant societal implications of both unabated global warming and (unjust) decarbonisation policies and projects, citizens and civil society organisations worldwide have begun to approach courts to challenge the ways in which decarbonisation is implemented. The slow and steady development of the notion of just transition at the international and national levels made way for the beginning of just transition litigation at the national level. This incipient litigation begins to answer this question precisely: does litigation provide pathways to assisting countries in implementing a just energy transition, bridging disagreements over how climate policies affect different communities? Or, on the contrary, does just transition litigation hinder the path to decarbonisation, given claims brought by affected communities that may (even if indirectly) affect climate policies?

As this section shows, national courts in LATAM have slowly started answering these questions, prompted by plaintiffs questioning how just transition policies are developed and implemented. Twenty just transition litigation cases were found in LATAM until the end of 2022 (Tigre et al., 2023a). The sample of cases reinforced the authors' initial assumption that just transition litigation is a phenomenon that has developed in recent years, as all cases identified were filed since 2016 (Tigre et al., 2023a). However, while five cases were filed in the 2017–2019 period, 14 have been filed since 2020. This uptick in cases indicates how just transition litigation is growing as countries respond to the long-term goal of emissions reductions agreed upon in the Paris Agreement. In terms of their geographical distribution, cases were identified in four Latin American jurisdictions: Mexico (10), Chile (5), Brazil (1) and Colombia (1) (Tigre et al., 2023a). In addition, two cases were identified in France, but in relation to decarbonisation policies happening in Mexico that were similarly brought in that jurisdiction. These two cases highlighted the transnational aspects of just transition and were analysed along with the LATAM examples because of their effect on the region.

Just transition brings implicit notions of well-being, equity and justice with the realisation that transitions are 'inherently disruptive' and deliberate effort may be required to avoid disproportionately affecting fossil fuel-dependent communities and industries (IPCC, 2022). Litigation arises in the absence of such policies that focus on injustices within the transition. With climate concerns typically brought as a peripheral or incidental aspect of the legal arguments, just transition cases are grounded on human rights to question the climate goals of policies (Savaresi & Setzer, 2022). At the same time, they challenge the justness or fairness of decarbonisation policies due to their negative effects on carbon-dependent communities or (already) marginalised citizens.

Importantly, the cases identified as examples of just transition litigation all include a "strong justice component", referring to concepts of social justice, equity or general (re)distribution of 'goods' and 'bads' from specific decarbonisation policies or activities, often related to workers, a cleaner environment, indigenous issues or human rights (Tigre et al., 2023a). This research boundary aligns with the conceptualisation of just transition litigation by Savaresi and Setzer, who note that just transition litigation "rel[ies] in whole or in part on human rights to question the distribution of the benefits and burdens of the transition away from fossil fuels and towards net zero emissions" (Savaresi & Setzer, 2022: 3). Within that broader context of 'fairness' in just transition, we observed that some plaintiffs have referred to distributive justice (i.e. intra- and intergenerational equity), as well as procedural justice (i.e. the right to information and participation, and the right to free, prior and informed consent [FPIC]) (Tigre et al., 2023a). Therefore, these complaints question the way climate policies are carried out or their impact on the enjoyment of human rights.

Furthermore, just transition litigation relates, at least marginally, to the energy transition process and challenges policies or activities relating to: (1) the pursuit of the long-term temperature goal of the Paris Agreement, (2) a commitment made by the state at the international level through NDCs and/or (3) the design or implementation of any broader decarbonisation policy or project at the national level. The research on these regional cases suggests that most plaintiffs rely on the Paris Agreement and decarbonisation policies to substantiate their claims.

While several scholars have categorised climate litigation, this categorisation was done while global climate litigation was in its early stages. Legal theories and strategies have developed significantly in the last five years, including with the advent of just transition litigation. In 2012, Markell and Ruhl took note of 'anti' cases in the early stages of climate litigation, categorising those as cases that go in the opposite direction and specifically confront substantive mitigation regulation (Markell & Ruhl, 2012). Similarly, Peel and Osofsky, in 2015, conflated climate cases into two large groups: (1) those supporting climate change regulation (pro-regulatory) and (2) those fighting it (anti-regulatory) (Peel & Osofsky, 2015). According to them, anti-regulatory cases attempt to limit the progress achieved by pro-regulatory lawsuits and fight climate legislation policies—or, in the just transition cases, decarbonisation policies and projects (Peel & Osofsky, 2015). Their analysis shows that most of these cases have been brought by corporations in response to more stringent laws that affect business as usual (Ghaleigh, 2010).

However, as noted by others, just transition litigation cases offer a higher degree of complexity (Setzer & Higham, 2022). They are often classified as 'non-climate aligned' since the plaintiffs question the effects of specific climate policies on vulnerable communities (i.e. decarbonisation plans, renewable energy or mining projects). Nonetheless, they do not necessarily object to climate action in the long term (Setzer & Higham, 2022). Indeed, our research found that while an anti-climate goal informs one-third of cases, two-thirds of cases oppose fossil fuel subsidies, policies or projects, without resisting the energy transition (Tigre et al., 2023a). Even though some of these cases attack climate change mitigation measures, they cannot (or should not) jointly be deemed anti-energy transition. Instead, they present and shed light on issues that are often overlooked. While some just transition cases are 'anti-' energy transition (or 'anti-climate', as the existing scholarship has characterised them), we also found cases that oppose fossil fuel subsidies, policies or projects on the grounds of the climate commitments and the need to promote a just energy transition. As such, they can also be defined as pro-environment, pro-labour or pro-traditional communities (Peel & Osofsky, 2015).

This small subset of climate cases deals with a myriad of seemingly unrelated issues (i.e. labour, indigenous participation, property, poverty relief) that, despite not being against climate action per se, can negatively affect climate legislation. These cases often address complex issues that make it challenging to limit climate cases to a certain yes or no categorisation. With litigation theories and strategies further advancing and litigation arising in the context of a 2 °C or a 3 °C world, these categories will need to be further revised. Just transition cases fall within this small subset of cases. Because these issues have generally been considered disconnected from climate action, mainstream scholarship tends to exclude them in the analysis of climate litigation (Setzer & Vanhala, 2019). Yet, upon closer look, these issues are not only deeply related to climate change but significantly relevant to ensuring a sustainable and decarbonised future. Thus, we suggest just transition litigation as an intermediate category between the 'pro' and 'anti' cases by including cases that promote a just energy transition (and use the just transition framing to promote climate policies) (Tigre et al., 2023a).

Nonetheless, labour, environmental or human rights claims against decarbonisation policies and renewable energy projects can jeopardise a government's push for an energy transition. For example, in the cases analysed, plaintiffs often request: (1) the annulment of a policy or activity; (2) the suspension of a policy or activity or, less frequently (3) the participation of affected groups in the process (Tigre et al., 2023a). These requests can significantly delay the implementation of decarbonisation policies or projects, which are tied up in (costly and time-consuming) litigation. But they can also help ensure the energy transition is, indeed, just. By relying on a wide range of human rights potentially infringed through decarbonisation policies, just transition litigation can provide benchmarks for just processes. For example, in *Company Workers Union of Marítima and Commercial Somarco Limited and Others v Ministry of Energy* (2021), the Chilean Supreme Court ordered the government to adopt measures to ensure that the energy transition incorporates a fair and equitable social and environmental component, with a specific plan to reincorporate workers into the job market. This decision can provide standards for an equitable transition and facilitate the implementation of decarbonisation measures to avoid delays.

The Paris Agreement called for national climate plans to include just transition measures with a centrality of decent work and quality jobs. The transition to a zero-carbon world, according to the ILO, shall therefore be managed by ensuring environmental sustainability, decent work, social inclusion and poverty eradication in accordance with the Sustainable Development Goals (SDGs) (ILO, 2015). At the core of a just transition is ensuring the participation of civil society, workers, unions and communities. These shall be conducted with respect to fundamental rights. As countries develop these plans, it is essential to understand the risks and opportunities of just transition litigation. The few lessons that exist so far—in terms of cases filed and decisions by courts—can provide guidelines in this regard.

These early lessons in just transition litigation complement, in practice, the guidance provided by the Intergovernmental Panel on Climate Change (IPCC). The 2022 IPCC report on mitigation emphasised how inequalities in the distribution of emissions and in the impacts of mitigation policies within countries can affect social cohesion and the acceptability of mitigation and other environmental policies (IPCC, 2022). The IPCC notes with medium confidence that an effective way of integrating equity principles into policies occurs by applying and implementing just transition principles through collective and participatory decision-making processes (IPCC, 2022). With proactive and targeted interventions from governments to reduce any negative social, environmental or economic impact of economy-wide transitions, just transition measures can also maximise the benefits for those disproportionately affected, eradicating poverty, regulating prosperity and creating jobs in 'green' sectors (IPCC, 2022).

Regardless of plaintiffs' motivations, climate change litigation influences regulatory pathways at all levels, affecting government and corporate behaviour (Peel & Osofsky, 2015). More importantly, cases that may initially fall outside the definition of climate litigation can be vital drivers of climate action due to their related consequences. Therefore, as suggested by Bouwer, it is crucial to look beyond the apparent climate cases to assess the role and value of smaller-scale litigation (Bouwer, 2018). The just transition litigation research herein analysed provides an ampler look into a subset of these cases. This research is essential for multiple reasons. First, it contributes to assessing the climate justice angle of climate litigation and whether decarbonisation policies can potentially be at risk by failing to address a wide range of concerns from those affected. In addition, strategic climate litigation—or *Urgenda*-type cases (Maxwell et al., 2022)[3]—could benefit from paying attention to the impacts of litigation occurring outside the traditional boundaries of climate change litigation, especially as it influences climate action, even if indirectly (Bouwer, 2018). As such, questioning the inadvertent effects of decarbonisation policies on local communities, just transition cases question the way that climate policies are carried out or their impact on the enjoyment of human rights. The next section analyses the general threads of the examples found in this research and how these relate to the policies mentioned in section "Just Transition Approaches to Decarbonisation: From International Efforts to National Policy".

[3] That is, strategic cases seeking to compel a State to increase overall mitigation ambition.

The Link Between Policy and Litigation: Key Examples of Litigation Challenging Domestic Transition Policies

To understand the rising trend of just transition climate litigation in LATAM, it is necessary to analyse this momentum within the broader context of climate action. This will inevitably entail discussions of historical injustices faced by the region as well as distributive justice issues. As mentioned above, LATAM countries are among the least contributing to climate change yet are suffering its consequences severely. Pre-existing inequalities are key in determining who gets to reap the benefits of climate change-inducing activities that are environmentally destructive and who carries the burdens of these activities (Preston, 2018; Sinden, 2007). This section dives deeper into the link between the just transition policies many LATAM countries have enacted (as discussed in section "Just Transition Approaches to Decarbonisation: From International Efforts to National Policy") and the main arguments set forth by plaintiffs in the just transition litigation cases (as identified in section "An Introduction to Just Transition Litigation in LATAM"). To this end, the section provides an overview of key examples of just transition litigation cases and how they interplay with the domestic, regional and international legal frameworks.

Scholars have pointed out that human rights and justice underpin the just transition and just transition litigation (Savaresi & Setzer, 2022; Tigre et al., 2023a). We have distinguished in this respect several types of rights that litigants can use, notably: (1) labour rights, (2) indigenous peoples' rights, (3) environmental rights (especially the right to a healthy environment) as well as other human rights (including socio-economic rights related to access to essential services or livelihoods) (Tigre et al., 2023a: 29–34). We also highlighted the special role of procedural rights in achieving justice, including those relevant to indigenous peoples. These rights include access to information, rights to public participation, consultation and FPIC. The cases analysed show that human rights are actively invoked in litigation that seeks to improve, as opposed to harm, transition policies. In just transition litigation, litigants may not always oppose a specific law or project. Still, they may challenge how these are designed or carried out without specific attention to justice considerations. For example, these challenges may rely on a fair(er) distribution of 'burdens and benefits' within and across generations (Savaresi & Setzer, 2022). They could also refer to ownership of the new energy grid and how resources are distributed among the community and workforce. In the case of just transition claims, the right to a healthy environment is frequently invoked to challenge energy, climate and decarbonisation policies and their negative effects on affected groups. This section discusses a number of key examples of just transition litigation and the policy contexts in which they occurred. It is important in this respect to note that all LATAM countries have unique contexts in which transition policies are implemented, which may explain what types of just transition arguments are made.

First of all, the trend of just transition litigation is particularly well visible in Mexico, where several cases have challenged aspects of Mexico's amended Electric Industry Law.[4] The Mexican context of energy transition is characterised by a national energy industry which underwent several controversial changes in recent years, including the opening up of the Mexican energy sector to foreign investment since 2013 (see generally Castrejon-Campos et al., 2022; González Márquez, 2022a). Such investments targeted the renewable energy sector too. In 2017, public auctions for renewable energy production were so successful that they doubled the country's renewable energy capacity and set world records for low-priced renewable energy (Llano et al., 2022). Yet, a change in administration in 2018 redirected the industry's course by restoring state entities' dominance in the energy sector through both "nationalisation and resource nationalism" (Llano et al., 2022). Essentially, the government aimed to "eliminate competition in the energy sector and to reestablish the state monopoly", with significant effects on the involvement of private (renewable) energy companies (González Márquez, 2022b: 36; González Márquez, 2022a). The government's 2021 reforms to the Electric Industry Law and Energy Sector Programs (see *Congreso General de los Estados Mexicanos*, 2021) have been challenged in several cases over the past years, with an emphasis on the right to a healthy environment and on arguing that Mexico is back-tracking on its decarbonisation pathways and compliance with international commitments, including those in the Paris Agreements and the country's own NDC. In fact, it is helpful to note that Mexico's revised NDC is currently directly targeted by climate litigation based on arguments of lowering ambitions on emissions reduction targets (e.g. *Greenpeace v. Instituto Nacional de Ecología y Cambio Climático and Others*, 2021). Despite the country's claims of supporting the Paris

[4] Article 4 of the Constitution of Mexico recognises the right to a healthy environment.

Agreement, several recently issued laws and resolutions contradict its own NDCs and other climate change-related goals.

A range of lawsuits has been launched against the changes in the Mexican Electricity Law specifically, primarily arguing that the amendments effectively prioritise the production and use of non-renewable energy rather than a transition to renewables. Litigants have argued that the new law places higher fees on the generation of renewable energy, thereby disadvantaging this source of energy (e.g. *CEMDA v. Comisión Federal de Electricidad*, 2019). Similarly, lawsuits have challenged the new law as it promotes fossil fuel projects at the expense of renewable energy projects, including by eliminating an obligation to dispatch the cheapest energy first and by giving priority to the Federal Electricity Commission's (CFE) coal and fuel oil-fired power plants in this process. Especially the lawsuits of *Julia Habana* et al. *v. Mexico* (*Julia Habana et al. v. Mexico*, 2021) and *Nuestros Derechos al Futuro y Medio Ambiente Sano* et al. *v. Mexico* (*Nuestros Derechos al Futuro y Medio Ambiente Sano et al. v. Mexico*, 2021) argue that by prioritising electricity generated by coal and fossil fuel plants over cleaner, renewable energy, the statute allows the continuity of projects that have devastating effects on the environment and climate change, which violates the right to a healthy environment. Plaintiffs claim that a just energy transition guarantees the exercise of the right to a healthy environment through the protection of existing ecosystems and a shift in paradigm regarding energy production and consumption. Finally, litigants have challenged the new law because it allegedly altered the rules of the national electric system in terms of energy dispatch and market certification (e.g. *Greenpeace v. Ministry of Energy and Others on the Energy Sector Program 2020 and Electric Industry Law*, 2020). Amongst the different Mexican cases, arguments of 'intergenerational distributive justice' are visible in *Idheas Litigio Estrategico v. Centro Nacional de Control de Energía (CENACE) and Secretaría de Energía (SENER) (Idheas Litigio Estrategico v. Centro Nacional de Control de Energía and Secretaría de Energía*, 2021*)*. In this case, the plaintiffs challenge several government regulations setting tariffs for renewable energy projects under legacy contracts, which would help bolster the position of the State-owned company CFE. Plaintiffs consider that these regulations would increase the use of fossil fuels and hinder the energy transition in the process. The core argument is that these regulations would increase the use of fossil fuels, negatively affecting the energy transition and limiting opportunities for a sustainable future for future generations.

In Chile, the context for just transition litigation is set by its recent development of an ambitious energy transition and decarbonisation policy (Ministerio de Energía, 2021). Due to the country's geographic characteristics, wind and solar projects are attractive. Moreover, Chile is rich in mineral resources that are vital for the development of various renewable energy technologies, such as lithium for batteries (see, e.g. Carrasco & Madariaga, 2022). Factors driving litigation in the country are the (un)fair distribution of benefits of the energy transition and a (lack of) participation of affected communities. Specifically, Chilean litigation involves claims by affected communities based on their human rights to question how they are being actively and fairly included in relevant political processes and negotiations. They may seek to prevent the negative impact of decarbonisation policies and activities on their communities and livelihoods or complain that the benefits of energy transition policies and activities are not accruing to them. Instead, benefits may be diverted to private companies and other communities. In response to lithium mining, indigenous peoples and civil society communities have, for example, (successfully) argued that indigenous consultation and citizen participation have been lacking and that lithium mining projects destroy biodiversity, cause pollution, obliterate indigenous cultural heritage and increase problems with access to water. In 2022, the Chilean Supreme Court granted constitutional relief to plaintiffs and invalidated contracts for the exploration and exploitation of lithium awarded to two foreign companies (see Tigre et al., 2023a: 25; *Atacama Community of Coya v. Ministry of Energy and Others*, 2021).

Chilean litigation has also targeted the implementation of decarbonisation objectives through plans to close coal-fired electricity plants. In *Company Workers Union of Maritima and Commercial Somarco Limited and Others v Ministry of Energy*, three union workers brought a lawsuit arguing that the Energy Ministry of Chile failed to include worker consultation in the reintegration and reconversion plans of coal plants. The case is based on constitutional labour rights. In 2021, the Supreme Court of Chile acknowledged that the closure of coal mines and coal-fired plants was part of Chile's Energy Sector Decarbonisation Plan to achieve carbon neutrality by 2050. Still, it emphasised that carbon neutrality decisions by the government must include a just transition strategy. This strategy must consider the needs and rights of workers affected by the loss of employment, involve their consultation and participation, and account for the loss of services for affected communities due to declining thermoelectric activity. The Supreme

Court ruled in favour of the plaintiffs by ordering government authorities to implement a plan for reintegrating workers into the labour market and to set up mechanisms for control. This case is a paradigmatic example of just transition litigation. As previously argued, plaintiffs did not challenge the decommissioning of the coal activity outright; they questioned the lack of procedures for participation and the indirect effects on their jobs (Tigre et al., 2023a). Unfortunately, the implementation of this judgement is pending as of March 2023. Appellants approached the court to seek an order forcing compliance with the decision in 2022 (Tigre et al., 2023a; Poder Judicial República de Chile, 2022).

This judgement is significant because it highlights the labour dimension of just transition well and some of the LATAM realities associated with this dimension. According to the Inter-American Development Bank and ILO, job loss in some areas of the fossil fuel sector in LATAM economies may be 'limited' compared to the size of the overall labour force or the annual rates of new jobs creation or sizes of GDPs (Saget et al., 2020: 62; Vogt-Schilb and Feng, 2019). For example, an estimated 4000 Chilean jobs—supporting the livelihoods of workers and their dependents—may be lost by the coal-fired power phase-out scenarios by 2030 or 2050. This may not necessarily be a very high number compared to the overall labour force, but for individual workers, the impacts could be devastating. The sectors most affected by job losses as a result of transition policies by 2030 would likely be the fossil fuel-powered electricity and fossil fuel extraction sectors (Saget et al., 2020: 60–62). Even if, according to recent calculations, losses in these sectors would still 'only' represent less than 1%, or less than 0.1% of total employment in LATAM countries (Saget et al., 2020: 62), Vogt-Schilb and Feng (2019) also rightly note that these figures 'mask winners and losers'. Even if tens of thousands of new jobs may be created elsewhere, it is vital those laid-off can access them. This means, inter alia, that newly created jobs must "match the skills available in the geographical location of current coal power plants" (Vogt-Schilb and Feng, 2019).

In Brazil and Colombia, litigation is borne out of different contexts yet again. First of all, in Colombia, plaintiffs combine environmental rights with the special rights of indigenous peoples to demand public policies that embrace energy transition and protect the environment with respect for indigenous peoples' rights to land and natural resources. In the *Wayúu Indigenous community v. Colombia's Ministry of Environment and others* (2019), seeking the annulment of the environmental licence to coal mining activities performed by private companies in the Guajira region (Northern Colombia), where the community lives, applicants directly highlight the incoherence of pretending to address climate change while fostering an extractivist economic model. Plaintiffs do not only make arguments about the negative effects of the project or climate change but draw attention to the broken and 'unjust' model that underlies the problem of climate change generally.

A final example of how litigation targets new 'decarbonisation' laws can be taken from the Brazilian context. As mentioned in section "Just Transition Approaches to Decarbonisation: From International Efforts to National Policy", the Brazilian law on just energy transition created a programme that appears to aim at transitioning heavily coal-dependent regions away from fossil fuels. However, the law is misleading and potentially counter-productive in terms of achieving a transition in the first place. Despite preparing the region for a phase-out of coal-based activities, the programme also includes several rather 'coal-friendly provisions'. In *ADI 7095 (Complexo Termelétrico Jorge Lacerda,* 2022*)*, political parties *Rede Sustentabilidade* (Rede), *Partido Socialismo e Liberdade* (PSOL) e *Partido Socialista Brasileiro* (PSB) filed a Direct Action of Unconstitutionality (ADI) against the Brazilian Congress, including a request for precautionary measures, challenging the law. The complaint addresses both the negative effects of prolonging and subsidising (through passing the costs to consumers) the operation of a severely polluting thermal power complex, as well as the fact that the envisioned Just Transition Council under the Just Energy Transition Program violates important principles of participatory democracy and equality. While the Just Transition Council was created to prepare the region for the 'probable' closure of coal activities, applicants challenge the division of seats proposed, which they claim is unequal.

Moreover, the Council should include the participation of actors related to environmental and labour causes and the sustainable closure of mines. Specifically, the Council is heavily composed of Santa Catarina government representatives—as a state that heavily relies on coal—and civil society organisations that support and incentivise coal use, creating a pervasive incentive for decarbonisation and failing to account for principles of environmental participation and democracy. Instead, workers' unions and other social and environmental non-profit organisations are underrepresented. The case thus calls for an improved energy transition. It supports a faster phase-out of coal while also acknowledging the need to involve various social and environmental stakeholders, including workers' representatives, in the decision-making.

Conclusions: Lessons Learned

The climate crisis is a global crisis with localised impacts: every region in the world experiences it in a differentiated way. The Global South in general and LATAM specifically are unsurprisingly suffering the worst effects of climate change, while their contributions to GHG emissions remain largely minimal. LATAM's context, in particular, is significant in many ways. Many large-scale infrastructural and mining projects, including in the area of renewable hydro-energy, wind energy and biofuel, have been implemented in the region over the past decades, leading to negative human rights impacts and social conflicts. In addition, LATAM societies are already suffering violations of fundamental rights due to climate change. As already recognised by the UNGA, the HRC, the Inter-American Human Rights system and many other international, regional and domestic bodies, the environment is vital for the enjoyment of other human rights, such as the right to life, health, rights of children, among others (UNGA, 2022; HRC, 2021; African Charter of Human and Peoples Rights, 1982; Convention on Access to Information, 1998; San Salvador Protocol, 1988; Escazú Agreement, 2018).

When national decarbonisation policies do not consider the possible harmful effects that such policies create for vulnerable groups, labour markets, the environment or people's ability to access basic services, plaintiffs may face legal challenges in courts. LATAM is no stranger to these legal challenges. Despite trends of ambitious environmental constitutionalism, LATAM still faces a significant number of environmental and human rights violations which have led the region to take a prominent place in the emerging body of climate litigation (Tigre et al., 2023b). Hence, claims grounded on environmental and human rights protection that challenge just transition policies are no surprise.

Most cases examined in relation to LATAM are grounded on environmental rights and, especially, on the right to a healthy environment. LATAM's existing environmental legal framework provides additional support to just transition claims: most LATAM countries, such as Brazil, Mexico, Chile and Colombia, already recognise the right to a healthy environment in their constitutions and domestic legal frameworks, and courts have dealt with issues related to the violation of this right for many years now (Tigre et al., 2023b). Furthermore, the Inter-American System of Human Rights also recognised the right to a healthy environment as an autonomous human right in what has been called 'green' jurisprudence (Tigre & Urzola, 2021).

Without essentialising LATAM's experience, some lessons and general trends can be drawn from how the region has faced the climate crisis from a legal perspective regarding just transition. First, all LATAM countries ratified the Paris Agreement and submitted their NDCs showing the region's openness to climate action. Furthermore, at least three countries (Chile, Colombia and Argentina) explicitly mentioned just transition in their NDCs and have followed up with subsequent domestic legislation that aims towards cleaner energy production and consumption. Additionally, countries like Mexico and Brazil have enacted just transition policies in the context of energy law development, even if some of these policies are insufficient at best.

Moreover, citizens across LATAM are already challenging these laws and putting them to the test through litigation. By bringing these claims to courts, plaintiffs contribute to the overall understanding of just transition policies and what these should entail from an overburdened, overexploited region. Plaintiffs are grounding some of their claims in the region's solid environmental legal framework and the judges' progressive approaches to environmental justice issues (Tigre et al., 2023b). Most of these cases are still pending, though, which means the full impact of LATAM's just transition litigation is yet to be seen. However, one can argue that litigation is yet another tool to prompt countries to implement a just energy transition. Litigation does not yet appear to hinder the path to decarbonisation but seeks to ensure that this path is pursued while respecting and protecting the rights of vulnerable communities and individuals.

This chapter contributes to a burgeoning field within climate litigation in multiple ways. It assesses potential risks to decarbonisation policies and contributes to shaping broader climate litigation strategies, such as strategic climate litigation cases, and to the overall understanding of rights-based climate litigation. Certainly, further research and academic discussion are needed to problematise the way just transition policies and litigation are being developed, including beyond LATAM. For example, even though environmental and climate justice issues seem to be at the core of most just transition efforts in LATAM, the region's over-reliance on extractivism is still strong. Having suffered years of overexploitation, the region still faces challenges in providing a decent standard of living for its citizens while protecting the environment for present and future generations. This dichotomy warrants further research on the real cost of just transition in the Global South and critical approaches to law. However, one thing is certain. LATAM courts now have the opportunity to help shape just transition litigation not only at a domestic or regional level but also at a global level. Since the LATAM region already has a solid environmental legal framework, a wide recognition of the importance of climate action and a history of progressive, bold courts, it could help set the pace of just transition litigation worldwide moving forward.

References

Additional Protocol to the American Convention on Human Rights in the Area of Economic, Social, and Cultural Rights: "Protocol of San Salvador" art. 11, Nov. 17, 1988, OAS. Official records OEA/Ser.A/44.

African Charter of Human and Peoples Rights. (1982). Opened for Ratification June 27, 1981, art. 24, O.A.U. Doc. CAB/LEG/67/3/Rev.5, 21 I.L.M. 58.

Alogna, I., Bakker, C., & Gauci, J. (Eds.). (2021, April 26). *Climate Change Litigation: Global Perspectives.* Brill | Nijhoff. https://doi.org/10.1163/9789004447615

Angeli, M. E. (2022, January 6). *Lei que prorroga uso de térmicas a carvão mineral é criticada por especialistas* Política. Retrieved May 2, 2023, from https://www.correiobraziliense.com.br/politica/2022/01/4975723-lei-que-prorroga-uso-de-termicas-a-carvao-mineral-e-criticada-por-especialistas.html

Atacama Community of Coya v. Ministry of Energy and Others. (2021). Retrieved May 2, 2023, from http://climatecasechart.com/non-us-case/regional-government-of-atacama-v-ministry-of-mining-and-other/

Auz, J. (2022). Human Rights-Based Climate Litigation: A Latin American Cartography. *Journal of Human Rights and the Environment, 13*(1), 114–136. https://doi.org/10.4337/jhre.2022.01.05

Bouwer, K. (2018). The Unsexy Future of Climate Change Litigation. *Journal of Environmental Law, 30*(3), 483–506. https://doi.org/10.1093/jel/eqy017

Brauch, M., & Toledano, P. (2022, June 15). *Integrating Climate Change, Decarbonization, and Just Transition Considerations into Extractive Industry Contracts.* Columbia Center on Sustainable Investment. Retrieved May 2, 2023, from https://ccsi.columbia.edu/news/climate-change-decarbonization-just-transition-mining-oil-gas-contracts

Brazilian Government Commits to Coal Until 2040. (2022, January 17). 350.org. Retrieved May 2, 2023, from https://350.org/brazilian-government-coal-2040/

Brichetti, J. P., Mastronardi, L., Rivas, M. E., Serebrisky, T., & Solís, B. (2021). *The Infrastructure Gap in Latin America and the Caribbean: Investment Needed Through 2030 to Meet the Sustainable Development Goals*. Inter-American Development Bank. https://doi.org/10.18235/0003759

Carley, S., & Konisky, D. (2020). The Justice and Equity Implications of the Clean Energy Transition. *Nature Energy, 5*, 569–577.

Carrasco, S., & Madariaga, A. (2022). The Resource Curse Returns?. *NACLA Report on the Americas, 54*, 445–452. https://doi.org/10.1080/10714839.2022.2145132

Castrejon-Campos, O., Aye, L., & Peng Hui, F. K. (2022). Competition, Coordination, or Institutional Change? A Multi-perspective Analysis of Historical Electricity Transitions in Mexico. *Energy Research and Social Science, 84*, 102362. https://doi.org/10.1016/j.erss.2021.102362

CEMDA v. Comisión Federal de Electricidad. (2019). Retrieved May 2, 2023, from http://climatecasechart.com/non-us-case/cemda-v-rise-of-legacy-transmission-rates/

Chilean Minister of Environment. (2022). *Sesiona por primera vez el Comité de Transición SocioEcológica Justa enfocado en las denominadas zonas de sacrificio*. Retrieved May 2, 2023, from https://mma.gob.cl/sesiona-por-primera-vez-el-comite-de-transicion-socioecologica-justa-enfocado-en-las-denominadas-zonas-de-sacrificio/

Chilean National Congress, Law no. 21.455 of June 2022, Climate Change Law. Retrieved May 2, 2023, from https://www.bcn.cl/leychile/navegar?idNorma=1177286

Company Workers Union of Marítima and Commercial Somarco Limited and Others v Ministry of Energy. (2021). Retrieved May 2, 2023, from http://climatecasechart.com/non-us-case/company-workers-union-of-maritima-commercial-somarco-limited-and-others-with-ministry-of-energy/

Complexo Termelétrico Jorge Lacerda. (2022). Retrieved May 2, 2023, from http://climatecasechart.com/non-us-case/adi-7095-complexo-termeletrico-jorge-lacerda/

Congress of Colombia, Law 2169 of 2021. Retrieved May 2, 2023, from http://www.secretariasenado.gov.co/senado/basedoc/ley_2169_2021.html#:~:text=La%20presente%20ley%20tiene%20por,de%20Colombia%20sobre%20la%20materia

Congress of Mexico, Electric Industry Law of 2014. Retrieved May 2, 2023, from https://www.diputados.gob.mx/LeyesBiblio/pdf/LIElec.pdf

Congress of Mexico, Energy Transition Law of 2015. Retrieved May 2, 2023, from https://www.diputados.gob.mx/LeyesBiblio/pdf/LTE.pdf

Convention on Access to Information. (1998). Public Participation in Decision-Making and Access to Justice in Environmental Matters, art. 1, Oct. 30, 2001, 2161 U.N.T.S. 447.

Decision 1/CP.21 'Adoption of the Paris Agreement'. (2016, January 29). FCCC/CP/2015/10/Add.1,2 (Paris Agreement). Retrieved May 2, 2023, from https://unfccc.int/sites/default/files/english_paris_agreement.pdf

Deutsche Welle. (2022, November 9). *COP27: Colombia anuncia plan de "Transición Energética Justa"*. Retrieved May 2, 2023, from https://www.dw.com/es/colombia-anuncia-en-la-cop27-su-plan-de-transici%C3%B3n-energ%C3%A9tica-justa/a-63697427

Eisenberg, A. M. (2019). Just Transitions. *Southern California Law Review, 92*(2), 273–330. Retrieved May 2, 2023, from https://southerncalifornialawreview.com/2019/01/04/just-transitions-article-by-ann-m-eisenberg/

Electric Industry Law and Energy Sector Programs, *Congreso General de los Estados Unidos Mexicanos*. (2021, March 9). Retrieved December 29, 2022, from https://www.diputados.gob.mx/LeyesBiblio/ref/lielec/LIElec_ref02_09mar21.pdf

European Commission. (2021). *Transición Justa. Un enfoque para vincular las agendas climática, económica y social.* Serie de Estudios Temáticos EUROCLIMA+ n.18. Programa EUROCLIMA, Dirección General de Asociaciones Internacionales. Comisión Europea. Bruselas, Bélgica. 102. Retrieved May 2, 2023, from https://www.euroclima.org/publicacion-euroclima-2/transicion-justa-un-enfoque-para-vincular-las-agendas-climatica-economica-y-social

Ghaleigh, N.-S. (2010). "Six Honest Serving-Men": Climate Change Litigation as Legal Mobilization and the Utility of Typologies. *Climate Law, 1*(1), 31–61. https://doi.org/10.3233/CL-2010-003

Gobierno de Chile. (2020). *Chile's Nationally Determined Contribution: Update 2020.* Retrieved May 2, 2023, from https://unfccc.int/sites/default/files/NDC/2022-06/Chile%27s_NDC_2020_english.pdf

Gobierno de Colombia. (2020). *Actualización de la Contribución Determinada a Nivel Nacional de Colombia (NDC)*. Retrieved May 2, 2023, from https://unfccc.int/sites/default/files/NDC/2022-06/NDC%20actualizada%20de%20Colombia.pdf

Gobierno de México. (2022). *Contribución Determinada a Nivel Nacional Actualización 2022.* Retrieved May 2, 2023, from https://unfccc.int/sites/default/files/NDC/2022-11/Mexico_NDC_UNFCCC_update2022_FINAL.pdf

González Márquez, J. H. (2022a). The New Nationalism of the Mexican Energy Policy in a Turbulent International Context. In C. Banet et al. (Eds.), *Resilience in Energy, Infrastructure, and Natural Resources Law: Examining Legal Pathways for Sustainability in Times of Disruption* (pp. 159–174). OUP. https://doi.org/10.1093/oso/9780192864574.003.0011

González Márquez, J. H. (2022b). The Past, Present and Future of Energy Law in Mexico: Energy Law from 1992 to 2022. *Journal of Energy and Natural Resources Law, 40*(1), 33–41. https://doi.org/10.1080/02646811.2021.2017667

Greenpeace v. Instituto Nacional de Ecología y Cambio Climático and Others. (2021). Retrieved May 2, 2023, from http://climatecasechart.com/non-us-case/greenpeace-v-instituto-nacional-de-ecologia-y-cambio-climatico-and-others/

Greenpeace v. Ministry of Energy and Others (on the Energy Sector Program 2020 and Electric Industry Law). (2020). Retrieved May 2, 2023, from http://climatecasechart.com/non-us-case/greenpeace-mexico-v-ministry-of-energy-and-others/

Hesselman M. (2023). *Human Rights and Access to Modern Energy Services* (PhD Dissertation, University of Groningen).

HRC, Report of the Special Rapporteur on Human Rights and the Environment: Safe Climate (15 July 2019) UN Doc. A/74/161.

HRC, Report of the Special Rapporteur on Extreme Poverty and Human Rights: Just Transition (7 October 2020) UN Doc. A/75/181/Rev.1.

Idheas Litigio Estratégico v. Centro Nacional de control de Energía and Secretaría de Energía, Judgment 19 April 2021. Retrieved May 2, 2023, from http://climatecasechart.com/non-us-case/idheas-litigio-estrategico-v-centro-nacional-de-control-de-energia-cenace-and-secretaria-de-energia-sener/

Inter-American Commission of Human Rights (IACmHR) Resolution 3/2021, 'Climate Emergency: Scope of Inter-American Human Rights Obligations' (31 December 2021).

Intergovernmental Panel on Climate Change (IPCC). (2022). *Sixth Assessment Report.* Retrieved May 2, 2023, from https://www.ipcc.ch/assessment-report/ar6/

International Energy Agency (IEA). (2021). *Roadmap to Net Zero by 2050.* Retrieved May 2, 2023, from https://www.iea.org/reports/net-zero-by-2050

International Labour Organization. (2015). *Guidelines for a Just Transition Towards Environmentally Sustainable Economies and Societies for All.* ILO. Retrieved May 2, 2023, from https://www.ilo.org/wcmsp5/groups/public/@ed_emp/@emp_ent/documents/publication/wcms_432859.pdf

International Labour Organization. (2022, September 30) *The Ministry of Labour of Argentina Promotes Strategies for Green Jobs and Just Transition.* Retrieved May 2, 2023, from https://www.ilo.org/global/topics/green-jobs/news/WCMS_857338/lang%2D%2Den/index.htm

International Trade Union Confederation. (2009). *Trade Unions at the UN Framework Convention on Climate Change.* Retrieved May 2, 2023, from https://www.ituc-csi.org/IMG/pdf/COP15_ITUC_report_final-2.pdf

Julia Habana et al. v. México. (2021). Retrieved May 2, 2023, from http://climatecasechart.com/non-us-case/julia-habana-et-al-v-mexico-unconstitutionality-of-the-reform-to-the-electricity-industry-law/

Little, P. (2014). *Mega-development Projects in the Amazon Region: A Geopolitical and Socio-environmental Analysis with Proposals of Better Government for the Amazon*. Derecho, Ambiente y Recursos Naturales—DAR. Retrieved May 2, 2023, from https://www.dar.org.pe/archivos/publicacion/145_megaproyectos_ingles_final.pdf

Llano, R., Rosenzweig, F., Hamilton, J., & Coulet-Diaz, M. (2022, July 19). *Energy Investors Face Mexico Risks in the Electricity and Lithium Sectors*. White and Case. Retrieved May 2, 2023, from https://www.whitecase.com/insight-alert/energy-investors-face-mexico-risks-electricity-and-lithium-sectors

Markell, D., & Ruhl, J. B. (2012). An Empirical Assessment of Climate Change in The Courts: A New Jurisprudence Or Business As Usual? *Florida Law Review, 64*(15).

Maxwell, L., Mead, S., & van Berkel, D. (2022). Standards for Adjudicating the Next Generation of Urgenda-Style Climate Cases. *Journal of Human Rights and the Environment, 13*(1), 35–63.

Ministerio del Trabajo de Colombia. (2019). *Pacto Por Los Empleos Verdes y Transición Justa En Colombia*. Retrieved May 2, 2023, from https://www.mintrabajo.gov.co/prensa/comunicados/2019/noviembre/pacto-por-los-empleos-verdes-y-transicion-justa-en-colombia-firmaron-oit-y-mintrabajo

Miranda, T. (2021). Nationally Determined Contributions Across the Americas. *Institute of the Americas*. Retrieved May 2, 2023, from https://user-9sjqssx.cld.bz/Nationally-Determined-Contributions-Across-the-Americas

Natural Resources Defense Council. (2017). *Bancos Nacionales de Desarrollo y Bancos de Inversión Verde: movilizando el financiamiento en América Latina y el Caribe para la implementación de contribuciones nacionales*. NRDC. Retrieved May 2, 2023, from https://www.nrdc.org/sites/default/files/bancos-nacionales-de-desarrollo.pdf

Nuestros Derechos al Futuro y Medio Ambiente Sano et al. v. Mexico. (2021). Retrieved May 2, 2023, from http://climatecasechart.com/non-us-case/nuestros-derechos-al-futuro-y-medio-ambiente-sano-et-al-v-mexico-unconstitutionality-of-the-reform-to-the-electric-industry-law/

OECD/ECLAC/Development Bank of Latin America and European Union. (2022). *Latin American Economic, Outlook 2022: Towards a Green and Just Transition*. OECD Publishing. Retrieved May 2, 2023, from https://www.cepal.org/en/publications/48415-latin-american-economic-outlook-2022-towards-green-and-just-transition

Osofsky, H. M. (2020). The Geography of Emerging Global South Climate Change Litigation. *The American Journal of International Law, 114*, 61–66. https://doi.org/10.1017/aju.2020.3

Peel, J., & Lin, J. (2020). Symposium: 'Transnational Climate Litigation: The Contribution of the Global South'. *AJIL Unbound, 114.*

Peel, J., & Osofsky, H. M. (2015). *Climate Change Litigation: Regulatory Pathways to Cleaner Energy.* Cambridge University Press. https://doi.org/10.1017/CBO9781139565851

Poder Judicial República de Chile. (2022). Oficina Judicial Virtual. Retrieved December 29, 2022, from https://oficinajudicialvirtual.pjud.cl/indexN.php#modalDetalleSuprema

Presidencia de Argentina. (2021). *Actualización de la meta de emisiones netas de Argentina al 2030: Octubre 2021.* Retrieved May 2, 2023, from https://unfccc.int/sites/default/files/NDC/2022-05/Actualizacio%CC%81n%20meta%20de%20emisiones%202030.pdf

Preston, B. J. (2018). The Evolving Role of Environmental Rights in Climate Change Litigation. *Chinese Journal of Environmental Law, 131*(2), 132.

Pucheta, M., Alvarez Alonso, C., & Silva Sanchez, P. (2021). Just Transition and Workers' Rights in the Global South: The Recent Argentine and Chilean Nationally Determined Contributions. *Sustainability, 13*(17), 9616. https://doi.org/10.3390/su13179616

RELAC IADB. (n.d.). Retrieved February 27, 2023, from https://hubenergia.org/index.php/en/relac. Accessed on 2 May 2023.

Regional Agreement on Access to Information, Public Participation and Justice in Environmental Matters in Latin America and the Caribbean (Escazú Agreement) (adopted 2018, March 4, entered into force 2021, April 22).

Renner, M., International Renewable Energy Agency, & International Labour Organization. (2022). *Renewable Energy and Jobs: Annual Review 2022.* ILO: International Labour Organization. Retrieved May 2, 2023, from https://policycommons.net/artifacts/2685048/renewable-energy-and-jobs/

Republic of Argentina, Law no 27.430 of December 2017, Tax Reform Act, 2017. Retrieved May 2, 2023, from https://www.boletinoficial.gob.ar/detalleAviso/primera/176831/20171229

Republic of Argentina Law no 27.604 of December 2020a, Fire Management Act. Retrieved May 2, 2023, from https://www.boletinoficial.gob.ar/detalleAviso/primera/239139/20201224

Republic of Argentina Law no 27.592 of December 2020b, Yolanda Law. Retrieved May 2, 2023, from https://www.boletinoficial.gob.ar/detalleAviso/primera/238505/20201215

Saget, C., Vogt-Schilb, A., & Luu, T. (2020). *El empleo en un futuro de cero emisiones netas en América Latina y el Caribe.* Inter-American Development Bank. https://doi.org/10.18235/0002509

Savaresi, A., & Setzer, J. (2022). Rights-based Litigation in the Climate Emergency: Mapping the Landscape and New Knowledge Frontiers. *Journal of Human Rights and the Environment, 13*(1), 7–34.

Schapper, A. (2018). Climate Justice and Human Rights. *International Relations, 32*(3), 275. https://doi.org/10.1177/0047117818782595

Setzer, J., & Benjamin, L. (2020a). Climate Change Litigation in the Global South: Filling in Gaps. *The American Journal of International Law, 114*, 56–60. https://doi.org/10.1017/aju.2020.6

Setzer, J., & Benjamin, L. (2020b). Climate Litigation in the Global South: Constraints and Innovations. *Transnational Environmental Law, 9*(1), 77–101. https://doi.org/10.1017/S2047102519000268

Setzer, J., & Higham, C. (2022). *Global Trends in Climate Change Litigation: 2022 Snapshot 3.* Grantham Research Institute on Climate Change and the Environment and Centre for Climate Change Economics and Policy, London School of Economics and Political Science. Retrieved May 2, 2023, from https://www.lse.ac.uk/granthaminstitute/wp-content/uploads/2022/08/Global-trends-in-climate-change-litigation-2022-snapshot.pdf

Setzer, J., & Vanhala, L. (2019). Climate change litigation: A review of research on courts and litigants in climate governance. *WIREs Climate Change*. Published Online: 4 March 2019. https://doi.org/10.1002/wcc.580

Sinden, A. (2007). Climate Change and Human Rights. *Journal of Land, Resources & Environmental Law, 27*, 255–256.

Sindico, F., & Mbengue, M. M. (Eds.). (2020). *Comparative Climate Change Litigation: Beyond the Usual Suspects.* Springer International Publishing. https://doi.org/10.1007/978-3-030-46882-8

Skillington, T. (2017). *Climate Justice and Human Rights.* Palgrave Macmillan. https://doi.org/10.1057/978-1-137-02281-3

Stephens, C., & Church, C. (2017). In W. Cockerham (Ed.), *International Encyclopedia of Public Health* (2nd ed.). Academic Press. ISBN: 9780128037089.

Strambo, C., González Espinoza, A. C., Puertas Velasco, A. J., & Atteridge, A. (2018). *Privileged Coal: The Politics of Subsidies for Coal Production in Colombia.* Working paper no. 2018-01. Stockholm Environment Institute, Stockholm.

Thomas, A. (2021). Framing the Just Transition: How International Trade Unions Engage with UN Climate Negotiations. *Global Environmental Change, 70*, 102347. https://doi.org/10.1016/j.gloenvcha.2021.102347

Tigre, M., & Urzola, N. (2021). The 2017 Inter-American Court's Advisory Opinion: Changing the Paradigm for International Environmental Law in the Anthropocene. *Journal of Human Rights and the Environment, 12*(24). https://doi.org/10.4337/jhre.2021.01.02

Tigre, M., Zenteno, L., Hesselman, M., Urzola, N., Cisterna-Gaete, P., & Luporini, R. (2023a). *Just Transition Litigation in Latin America: An Initial Categorisation of Climate Litigation Cases Amid the Energy Transition.* Sabin Center for Climate Change Law. Retrieved May 2, 2023, from https://scholarship.law.columbia.edu/sabin_climate_change/197/

Tigre, M., Urzola, N., & Goodman, A. (2023b). Climate Litigation in Latin America: Is the Region Quietly Leading a Revolution? *Journal of Human Rights and the Environment, 14*, 67–93.

United Framework Convention on Climate Change (UNFCCC). (2020, April). *Just Transition of the Workforce, and the Creation of Decent Work and Quality Jobs* [technical paper]. Retrieved May 2, 2023, from https://unfccc.int/sites/default/files/resource/Just%20transition_for%20posting.pdf

United Framework Convention on Climate Change (UNFCCC) FCCC/AWGLCA/2009/INF.1 (2009, June 22). Retrieved May 2, 2023, from https://unfccc.int/resource/docs/2009/awglca6/eng/inf01.pdf

United Nations. (n.d.). *All About the NDCs*. United Nations. Retrieved May 2, 2023, from https://www.un.org/en/climatechange/all-about-ndcs

United Nations Development Programme. (2022). Climate Promise Quality Assurance Checklist for Revising Nationally Determined Contributions, NDC Partnerships. Retrieved August 5, 2024, from https://ndcpartnership.org/knowledge-portal/climate-toolbox/climate-promise-quality-assurance-checklist-revising-nationally-determined-contributions

United Nations General Assembly (UNGA) Res. 77/170, Ensuring Access to Affordable, Reliable, Sustainable and Modern Energy for All. (2022, December 24). Retrieved May 2, 2023, from https://digitallibrary.un.org/record/3954000?ln=en

United Nations General Assembly (UNGA) Res. A/76/L.75. (2022, July 28). Retrieved May 2, 2023, from https://digitallibrary.un.org/record/3982508?ln=en

United Nations Human Rights Council (HRC) A/HRC/48/L.23/Rev.1 The Human Right to a Safe, Clean, Healthy, and Sustainable Environment. (2021, October 5). Retrieved May 2, 2023, from https://undocs.org/a/hrc/48/l.23/rev.1

Vogt-Schilb and Feng. (2019). *Phase Down Scenarios in Chile, IDB Discussion Paper Series, No. IDB-DP-00716*. InterAmerican Development Bank. https://doi.org/10.18235/0001961

Wang, X., & Lo, K. (2021). Just Transition: A Conceptual Review. *Energy Research and Social Science, 82*, 102291. Retrieved May 2, 2023, from https://www.sciencedirect.com/science/article/pii/S2214629621003832

Wayúu Indigenous Community v. Colombia's Ministry of Environment and Others. (2019). Retrieved May 2, 2023, from http://climatecasechart.com/non-us-case/wayuu-indigenous-community-and-others-v-ministry-of-environment-and-others/

Wewerinke-Singh, M. (2021). A Human Rights Approach to Energy: Realising the Rights of Billions Within Ecological Limits. *RECIEL, 31*, 16–20.

PART II

Legal Challenges of a Clean and Just Energy Transition

CHAPTER 8

The Challenge of 'Phasing Out' Fossil Fuels: A Climate-Changing Transition

Guilherme Pratti

Introduction

The slow pace of the current energy transition is not surprising; it is indeed expected (Smil, 2013). Fossil fuels[1] are so entrenched in everyday life that it has become hard to imagine a future exempt from their use. In fact, an exercise of this sort would be deemed naïve if it were done without considering the versatile roles fossil fuels-derived products have

[1] Whenever I mention 'fossil fuels' in the plural form I will be generically referring to coal and hydrocarbons (crude oils and natural gas) and not specifically to one or the other. In the same way, whenever I refer to one type of fossil fuel (gasoline, diesel, kerosene, jet fuel, etc.) I will not be using it as a metonym for the whole of 'fossil fuels'. This distinction is essential to analyse the nuances within the worldwide initiatives that aim to—so to say—'phase out' their incessant use. On the importance of acknowledging the plurality and 'considerable heterogeneity' of the substances encompassed in the concept of 'fossil fuels', please see (Smil, 2008). On the use of the expression 'phase out' in relation to fossil fuels, please see this chapter's section "The Fossil Fuels Propelled International Legal Dialectics".

G. Pratti (✉)
Sant'Anna School of Advanced Studies, Pisa, Italy

D. Iglesias Márquez et al. (eds.), *Legal Challenges at the End of the Fossil Fuel Era*, Global Issues,
https://doi.org/10.1007/978-3-031-61766-9_8

in the contemporary world, which vary much beyond their large share in the global energy system. That is, to name a few such examples: fuelling mechanized crop harvests; transporting raw materials, goods and people throughout all continents incessantly by land, sea and air; and offering a variety of petrochemicals that go from pesticides and fertilizers to medical drugs, synthetic fibres and plastics.

Against this backdrop, this chapter is set out to identify and discuss the main markers of the entrenchment of fossil fuels within international law. It does so by employing an 'integrative approach' to the legal depictions of energy (Viñuales, 2022), especially through the branches of international environmental and energy law.

The socio-legal analysis here developed briefly illustrates, through sections "Introduction" and "Fossil Fuels' Technical Hegemony", the worldwide hegemony of the fossil fuels-based energy technical system and the impacts it has impinged on the Earth System, with special attention to the second half of the twentieth century. Section "'The Great Acceleration' of Impacts on the Earth System" identifies a long-lasting *ab ovo* structural contradiction at the core of the international legal depictions of energy as a legal object, which prioritize fossil fuels-based 'energy production and use' over 'environmental protection' at the same time while pledging to stop offsetting the balance of the Earth's biosphere and of its ecosystems. The fourth and final section discusses how the ongoing national overarching policies towards low-carbon, clean, energy matrices (labelled generically as 'green deals') are enabling the energy transition—whilst facing demanding challenges stemming from the carbon-intensive international normative frameworks.

Fossil Fuels' Technical Hegemony

The first industrial revolution in the mid-eighteenth century is commonly accepted as the main marker of an exponential boost in coal consumption (Smil, 2010b; Steffen et al., 2007). In the same way, the mid-nineteenth century is referred to as the period in which the intense usage of crude oils began to take form—a happening often labelled as the "second industrial revolution" (E. J. Hobsbawm, 1999; Steffen et al., 2007). This division holds good truth if considered geographically bounded to the western northern hemisphere, while bearing in mind that the fossil fuels perennial dependence they created did not happen immediately.

Global rise in the use of fossil fuels, up to becoming the world's main energy source, was a rather lengthy one. Albeit heavily used already in the late eighteenth century, coal only "began to supply more than 5% of all fuel energy around 1840" and by the brink of the twentieth century it "still supplied only about half of demand" (Smil, 2013).

Coal-based energy supply's rise to 50% took roughly 60 years, though it varied from place to place until its use achieved worldwide hegemony.[2] And so the biggest energy source in the twentieth century "was not oil but indeed coal", which only reached its "highest share of global fuel consumption, at about 55%, during the 1910s" (Smil, 2013). The same lengthy process happened to crude oils, which although already in use by then, only surpassed coal in 1964 (Smil, 2013).

One may therefore understand the so-called first and second industrial revolutions to be fundamentally two energy revolutions, which brought upon new energy (re)sources, new modes of production and new overall uses of energy worldwide. These revolutions—or energy transitions—were what powered the 'industrial' in 'industrial revolutions' and basically rendered possible fundamental aspects of contemporary life: from household electricity and heating systems to plastic gadgets and appliances; an unfathomable amount of medical and chirurgical apparatuses; as well as of the global-wide terrestrial, maritime and air transportation systems, which sustain the contemporary trade of commodities and aggregated value goods.

In this sense, at the present moment, all industries and global value chains are virtually fossil fuels-dependent. Either due to the intensive usage of hydrocarbon-derived materials throughout the production systems or to the fossil fuels-based logistics needed to connect world-spanning supply chains. These examples illustrate how fossil fuels-intensive is the hegemonic technical system[3] presently structuring the world: from what we eat to what we wear, from how we commute and build our homes to how the world's industries operate and national energy systems are powered, or yet, from any gadget's smallest plastic component to the kerosene-fuelled military aircrafts. However, as it is now extensively known, this fossil fuel-based technical system spread throughout the globe at great environmental and climate costs.

[2] For instance, Vaclav Smil (2010a, 2013) shows the tipping point in France occurred around 1875; Japan, 1901; the URSS, 1930; China, 1965; and India, the late 1970s.

[3] For my use of 'technical system' please see (Ellul, 2009; Pratti, 2021).

'The Great Acceleration' of Impacts on the Earth System

In the late 1990s, historian Eric Hobsbawm defined the three decades that followed the Second World War (WWII) as a 'golden age', on the grounds of finding this 30-year span a time of "extraordinary economic growth and social transformation" that "probably changed human society more profoundly than any other period of comparable brevity" (Hobsbawm, 1995).

This statement acquired new depths of meaning when atmospheric chemist and Nobel Prize laureate Paul J. Crutzen and limnologist Eugene F. Stoermer popularized the 'Anthropocene' concept. The term describes the "human-dominated, geological epoch", in which anthropogenic emissions of carbon dioxide (CO_2) are causing the global climate to "depart significantly from natural behavior" (Crutzen, 2002; Crutzen & Stoermer, 2000). They argued that the beginning of the current geological epoch can be pinned down to "the latter part of the eighteenth century", which "happens to coincide with James Watt's design of the [coal-intensive] steam engine in 1784" (Crutzen, 2002: 23; Crutzen & Stoermer, 2000).[4] The coincidental factor here being the "beginning of growing global concentrations of carbon dioxide and methane" in the atmosphere (Crutzen, 2002; Crutzen & Stoermer, 2000).

Those 'golden years' not only changed human society, but they also radically intensified human impacts on the Earth's biosphere, more profoundly than any other period of comparable brevity so far. These impacts can be mostly traced back to the incessant emissions of CO_2 and other greenhouse gases (GHGs)—the majority of which are due to the overall massive use of fossil fuels and environmental degradation related to it around the world. This golden period also coincides with crude oils' surpassing of coal as the hegemonic fuel throughout the world (Smil, 2013)—the moment in which fossil fuels became the main propellers of global economic development.

Crutzen and Stoermer's proposal inspired a group of scientists to concentrate efforts and build a "systematic picture of the human-driven changes to the Earth System" as a way to "record the trajectory of the 'human enterprise' through a number of indicators" (Steffen et al.,

[4] Watt's steam engine was defined by Alain Pottage, in a hindsight climate-driven analysis, as "an apocalyptic patent" (Pottage, 2020).

2015).[5] The authors came up with 12 graphs that illustrated the human imprint on the Earth System since 1750, encompassing areas such as population growth, fertilizer consumption, transportation, water and energy use, and urbanization. To the surprise of the scientists involved in the said effort, all graphs showed an exponential acceleration from 1950 onwards (Steffen et al., 2015), therefore confirming the worldwide hegemony of the fossil fuels-based technical system.

In the aftermath of their effort, the period Hobsbawm once called 'golden years' was extended and more appropriately relabelled as 'the great acceleration'[6]—which has certainly not yet come to a stop, in spite of a somewhat noticeable reduction in pace regarding its undesired and unintended consequences (Steffen et al., 2015). This greatly accelerated 'carboniferous'[7] period in the post-mid-twentieth century ultimately reveals that the dynamics of our present time are structured upon the consumption of fossil fuels, as well as upon the overall climate and environmental degradation that comes with it.

The 'great acceleration' concept, as thought out by Steffen et al. (2004), alludes to the ever-growing negative outcomes on the Earth System caused by the steep and exponential increase in the use of fossil fuels from 1950 onwards (with climate change being one among many). Such a concept also offers a valuable context to understand the many social consequences stemming from this 'carboniferous' technical system.[8]

Moreover, whether the beginning of the so-called Anthropocene epoch should be considered the late eighteenth century or the end of WWII, or if it should be even considered a proper geological epoch (or nothing of the sort), is a matter for climate scientists and palaeontologists to settle.[9]

[5] This 2015 article is an updated version of their previous publication on the matter dating back to 2004. For the original graphs, please see (Steffen et al., 2004).

[6] For a comprehensive account on this phenomenon and its consequences, please see (McNeill & Engelke, 2014).

[7] The expression was used by Lewis Mumford when referring to the high level of coal consumption in between the nineteenth and twentieth centuries. Please see (Mumford, 1955).

[8] Such as the "shortening of distances", the acceleration of the pace of contemporary life and of the overall worldwide happenings. On these matters, please see (Ferrarese, 2000; Rosa, 2013; Santos, 2001), respectively.

[9] Formal steps were taken towards the formal recognition of the Anthropocene as a geological time unit as the 'Anthropocene Working Group of the Subcommission on Quaternary Stratigraphy' voted the mid-twentieth century as its official start. See (*Crawford Lake, Canada, Chosen as the Primary Marker to Identify the Start of the Anthropocene Epoch*, n.d.; 'Results of Binding Vote by AWG', 2019; Subramanian, 2019). However, it remains to be seen if the concept relates to a geological 'period', an 'epoch' or simply an 'event' (Gibbard et al., 2022; Head et al., 2023).

This chapter takes into consideration both conceptual possibilities in order to investigate some of the branches of international law that blossomed during this period. From the standpoint hereby undertaken, what matters is to investigate the international normative frameworks composing the fossil fuels-based architecture of the contemporary world.

This normative structure seems to be *ab ovo* marked by a tussle between, on the one hand, the need to secure entitlements over energy resources worldwide and the international trade derived from their exploration, and, on the other hand, the necessity of maintaining environmental and climate stability against the backdrop of an ever more disturbed and unbalanced Earth System. The section that follows describes the main legal traits of this carboniferous normative architecture as seen from the entangled perspective of its structuring tussle.

The Fossil Fuels-Propelled International Legal Dialectics

A Methodological Caveat

The beginning of the so-called 'great acceleration' also happens to coincide with the ever-increasing rate of worldwide environmental degradation, as well as to the unfolding of different realms of international law. However, before advancing the socio-legal analysis scope of this chapter, a methodological caveat is needed as so to narrow down and better define the phenomenon here contemplated.

According to Jorge Viñuales (2022), 'energy' is a rather difficult object to take hold of from the perspective of law for two main reasons. First, it is a legal object that necessarily crosses borders, due to the "substantial mismatch between the States where energy is produced and the States where energy is consumed". Second, because energy is a multifarious object that can be defined as having at least four facets: that of being a resource, a product, an activity and a technology.[10]

[10] This can be summed up, for instance, by imagining the industrial *activities* comprehended from the exploration of an oil well up to the many uses it provides. Crude oil (*resource*) is obtained through a myriad of technological artefacts used in the fracking and refining processes (*technology*). The outcomes (*products*) of the latter phase—fuels, waxes and petrochemicals—are then traded worldwide on a continuous basis to then sustain an infinite number of other endeavours.

It is important to note that Viñuales was, for the most part, referring to a fossil fuels-related conceptualization of energy, especially in his appraisal of its 'resource' facet. In this sense, the 'substantial mismatch' regarding where energy is produced and consumed alludes to the geographical location where reserves of fossil fuels lie around the globe. Therefore, this approach to energy as a legal object works slightly differently when applied to low-carbon renewable energy (re)sources, which, generally, can be produced and consumed within the same State. These two reasons account for the internationalization of energy transactions and the necessarily multidisciplinary nature of the normative frameworks regarding its fourfold facets.

Based on this innovative understanding, Viñuales (2022) sustains that the field of 'international energy law' should actually be understood as encompassing "the entirety of international law", hence why the Cambridge Professor entitled his book as *The International Law of Energy*.[11] Other than multidisciplinary, such an effort is above all transversal, for it does not allow different areas of international law to be considered as separate but as constitutive of a highly complex and multifaceted study object.

Although agreeing with this "integrative approach" (Viñuales, 2022), this chapter mainly focuses on how the fields of international environmental and energy law depict 'energy' as an object. This narrowing down of the available legal sources is justified because these are the—so to say—'branches' of international law that directly address the most fundamental aspects of the above-mentioned 'resource facet' of energy as a legal object.

This purposeful cutback helps to highlight the previously mentioned *tussle* located at the core of the normative architecture of the worldwide fossil fuels-based energy matrix. That is, the normative tussle—or the somewhat contradiction—between securing intense exploitation of fossil fuels, while trying to maintain the Earth's system climate stability. This inconsistency is observed through the dialectics propelled by the international legal sphere's handling of environmental and development concerns from the mid-twentieth century onwards. And so considerations regarding human rights, trade, foreign investment protection under international law as well as the law of the seas are thus willingly put aside due to

[11] Viñuales justifies doing so when he affirms that "*Energy*, not *international law*, is the pivotal term around which an account of the international law of energy must be organized" (2022: xix).

this chapter's length restrictions and to better focus solely on stressing this most fundamental aspect at the core of the legal depictions of energy as an object.

The Ab Ovo Contradiction in Between International Environmental and Energy Law

As mentioned above, the 'great acceleration' of environmental degradation and climate change-inducing activities also happens to coincide with the outset of the field of international environmental law. The 1972 'Stockholm Conference on the Human Environment' can be considered as its 'foundational moment' (Dupuy & Viñuales, 2018), both because it was the first of many conferences of its kind and due to its outcomes: the Stockholm Declaration on the Human Environment (hereinafter Stockholm Declaration), the establishment of the United Nations Environment Program (UNEP) and the Action Plan for the Human Environment are all milestones of the beginning of the global conversations on the entanglement between environmental protection and development through industrialization (Dupuy & Viñuales, 2018).

The Stockholm Declaration's preamble affirms the "major and undesirable disturbances to the ecological balance of the biosphere" result from harmful activities that should be overcome by building, "in collaboration with nature, a better environment", through "an enthusiastic but calm state of mind and intense but orderly work". Five decades (and many gigatons of CO_2 and other GHG thrown in the atmosphere) later, it is somewhat safe to say that 'calm' was not a very soothing word to choose. And that the 'intensity' of the work was not enough to prevent the worsening of the "undesirable disturbances to the ecological balance of the biosphere".

Throughout the Stockholm Declaration's 26 principles, one finds a variety of ways in which natural resources, ecosystems and the Earth's capacity "to produce vital renewable resources" (principle 3) are referred to and hinted at. While on the Action Plan, one notes the focus to be more on the "environmental aspects of natural resources management" (recommendations 19–69), on the "identification and control of pollutants of international significance" (recommendations 70–94) and on the entanglement between "development and environment" (recommendations 102–109).

On the one hand, the Declaration identified the international sphere's concerns with the overall declining health status of the Earth's biosphere, due to environmentally harmful economic-oriented activities, and highlighted the need to find a solution to this unbalanced equation. Whilst, on the other hand, the Action Plan directly linked these concerns "to the environmental effects of energy use and production" (recommendation 57), and urged for a more "effective development of the world's energy resources, with due regard to the environmental effects of energy production and us" (recommendation 59).

As mentioned in the previous section, the contextual backdrop of the 1972 Stockholm Conference is that of Hobsbawm's 'golden age', which can also be referred to as "the great acceleration" of the "human imprint" on the Earth System (Steffen et al., 2015). Or, to put it bluntly, that of the global consumption of crude oil overcoming that of coal (Smil, 2013), therefore firmly establishing fossil fuels as the primary source of energy worldwide. Thus, the Stockholm Action Plan's statement regarding energy resources is to be understood as referring mostly to fossil fuels.

As a matter of fact, this interpretation is strengthened by the Stockholm Declaration's differentiation between, on the one hand, "vital renewable resources" produced by the Earth, which "must be maintained, and, wherever practicable, restored or improved" (principle 3); and, on the other hand, the "non-renewable resources of the earth", which "must be employed in such a way as to guard against the danger of their future exhaustion" (principle 5).

The Stockholm Declaration's references to ecosystems and the "earth's capacity to produce vital renewable resources" implicitly contain the idea of cycles: a series of events that, when completed, return to the initial phase to then repeat themselves on and on, indefinitely. These references contain the assumption of reversibility of the damages caused to the biosphere and ecosystems—possibilities that are fading away with every new Intergovernmental Panel on Climate Change's (IPCC) assessment report.

The 'foundational moment' of the field acknowledges the link between 'energy production and use' and the 'disturbances to the ecological balance of the biosphere', whilst pleading to find harmony between both. Therefore, pleading to accommodate what is irreconcilable and to find a form of equilibrium between an environmental-harmful energy system and the worldwide environment damaged by it. Or, in other words, a system that privileges the exploitation of 'non-renewable resources', even though these are deleterious to the preservation of the 'vital renewable resources' produced by the Earth.

Precisely because this contradiction is constitutive of this so-called foundational moment of international environmental law, it is also present in the developments that followed the Stockholm Conference. And given that 'energy' as a legal object is multifaceted and necessarily transversal, this tussle to accommodate the irreconcilable is structural to the way in which the international legal sphere addresses it in its 'resource facet'.

Moreover, the 1982 World Charter for Nature (WCN), the 1992 Rio Declaration and the 1992 United Nations Framework Convention on Climate Change (UNFCCC) further developed the ways in which natural resources, ecosystems and the Earth's system are referred to and hinted at, reinforcing the idea of reliance on recurring natural cycles. However, while all three instruments link the exploitation of natural resources to the deterioration of the environment and natural systems, only the WCN and the UNFCCC relate the latter to the overall use of non-renewable resources, therefore depicting the dark side of the 'resource facet' of fossil fuels-based energy.

The WCN's preamble states that "life depends on the uninterrupted functioning of natural systems which ensure the supply of energy and nutrients". This means the uninterrupted recurrence of natural cycles. In this sense, the mentioned 'supply of energy' seems to be a reference to the within-nature energies (Pratti et al., 2022), responsible for the cyclic continuance of ecosystemic services, which are at the basis of renewable energy systems such as bioenergy and hydroelectric power.

As a matter of fact, recital ten of the WCN firmly differentiates natural resources as soils, living resources, renewables (including water) and non-renewable resources ("which are consumed as they are used"). Furthermore, it warned that the latter's conversion for consumption should be balanced with their abundance and compatibility with the "functioning of natural systems".

Similarly to (although less assertive than) the WCN, the UNFCCC also relates fossil fuels-based energy to environmental—and, more precisely, climate—degradation. For instance, article 4(10) recognizes that "measures to respond to climate change" are more difficult to implement for the countries whose economies are highly dependent on "fossil fuels and associated energy-intensive products".

The main legal instruments of international energy law 'branch' frame this tussle between the fossil fuels-based energy system and the Earth's environmental and climate stability in a whole different manner.

The preamble to the 1991 European Energy Charter highlights that "efficient energy systems" are of essential importance for environmental protection. It is rather interesting to note that environmental protection is solely mentioned in the context of energy efficiency (as one sees throughout titles I and II of the Charter).

The 1994 Energy Charter Treaty (ECT), designed to fulfil the guidelines of the 'European Energy Charter', is a great example of Viñuales' (2022) "integrative approach" to energy as a legal object. It encompasses a wide range of provisions that go from defining the economic activities of the energy sector and determining investment mechanisms (financing and access to capital, as well as investor protection procedures), to establishing questions of intellectual property and transfer of technology, and offering guidance on international trade and environmental protection. Given this chapter's purposes, however, only the ECT's insufficient handling of environmental protection is under scrutiny at the moment.

In its article 19, entitled "Environmental Aspects", the ECT states that all Parties shall strive to minimize the harmful environmental impacts within the energy cycle in an economically cost-effective manner.[12] Similar wording was already present in the European Energy Charter (Title II, n 7), but the ECT further expanded this strictly economic-oriented conception of environmental protection—and its binding force outweighs the more symbolic aspect of the 1991 instrument.

The 2015 International Energy Charter, a declaration of political intention that updates and expands the European Energy Charter, is in line with this apposing of energy production and environmental protection. Indeed, it recognizes in its preamble the existence of a "trilemma between energy security, economic development and environmental protection".[13]

[12] The Treaty defines 'energy cycle' as "the entire energy chain", including "exploration, production, conversion, storage, transport, distribution and consumption of the various forms of energy".

[13] The World Energy Council has established a 'World Energy Trilemma' in 2010—an index report prepared annually to rank countries energy performances on these three indicators. Please see: *World Energy Trilemma Index* (n.d.).

This 2015 Paris Agreement, designed to enhance the implementation of the UNFCCC and "strengthen the global response to the threat of climate change" (Article 2), did not address this trilemma, nor the energy sector's share of responsibility for the ongoing climate crisis and its dependence on the exploration and use of non-renewable resources. As a matter of fact, the Agreement did not even mention fossil fuels.

Nonetheless, the wording of its article 2(a) states the goal of holding the increase in the global average temperature to well below 2 °C (whilst pursuing efforts to limit it to 1.5 °C) above pre-industrial levels. The mention of 'pre-industrial levels' is a subtle, though evident, reference to the mid-eighteenth-century industrial revolution, which "happens to coincide" (Crutzen, 2002: 23) with the ever-increasing anthropogenic emissions of CO_2 and other GHGs due to the consumption of coal.

In the same way, by establishing the necessity to reach the peak in GHG emissions (article 4), the agreement seemingly hints at the end of the hegemonic era of the fossil fuels-based technical system. However, in spite of the many reasons to celebrate the Paris Agreement as a successful achievement of the international legal sphere, the failure in properly addressing the twenty-first century's structural trilemma is to be acknowledged at least as a missed opportunity—at the very least, from a symbolic perspective.[14]

This was somehow redressed on the Conference of the Parties to the UNFCCC in 2023, during the so-called COP 28, which happened in Dubai, in the United Arab Emirates, and also held the first Global stocktake under the Paris Agreement. Regarding the latter, its final text is characterized by a strong worded phrasing in reference to fossil fuels. More specifically, its recital 28, which mentions the need to accelerate the "phase down of unabated coal power", recognizes the need of "transitioning away from fossil fuels" in energy systems, as so to achieve net zero by

[14] I by no means intend to diminish the astonishing importance of the Paris Agreement. I fully agree with Professor Lavanya Rajamani (2016) when she firmly notes that "the Paris Agreement represents the most ambitious outcome possible in a deeply discordant political context".

2050, and calls for the "phasing out of inefficient fossil fuels subsidies that do not address energy poverty or just transitions as soon as possible".[15]

The (Un)balance Between Environmental Protection and the Energy Cycle

An unsettling problem has been present throughout the major developments of the international environmental and energy law fields in the past 80 years. At their intersection, a contradictory structural trait may as well be defined as the main distinctive legal marker of the so-called great acceleration: that of the relation between 'energy production and use' and overall environmental consequences.

On the one hand, from its foundational moment onwards, international environmental law acknowledges the link between 'energy production and use' and the 'ecological disturbances to the balance of the biosphere'. The solutions hinted at, however, urge for the harmonization of both, given that worldwide ecological balance is widely important, among other reasons, for the maintenance of ecosystemic services and the sustenance of 'energy supply'.

On their part, the WCN and the UNFCCC both narrowed down the negative outcomes of this link to the usage of 'non-renewable' energy resources, with the latter directly addressing fossil fuels. Therefore, indicating the way out of this negative-sum game of 'more energy production' equalling 'more biosphere degradation' to be a shift towards renewable energy sources. While, as previously highlighted, the 2015 Paris Agreement

[15] Although the direct references to 'fossil fuels' is a paradigmatic happening and should be celebrated as a victory, one must acknowledge it was—to some extent—a 'Pyrrhic victory'. The expression "phase down of unabated coal power" is an urge to diminish what should be urgently abandoned; "transitioning away from fossil fuels" in energy systems, is a reference to energy matrices and not to the GHG-intensive infrastructures of global logistics and of urban mobility; lastly, the "phasing out of inefficient fossil fuels subsidies that do not address energy poverty or just transitions" is simply a too narrow, sector-specific call to action. This being so, although the direct references to fossil fuels are a great symbolic victory, one can say that from a practical perspective the word 'victory' does not apply—given that the range of impacts has been narrowed down rather significantly. In this sense, the Paris Agreement—that does not even mention fossil fuels but put in place important mechanisms directed at facing climate change—can be even further appreciated. On the Paris Agreement's mechanisms, see section "The Fossil Fuels Propelled International Legal Dialectics". For the full text of the global stocktake, see (UNFCCC, 2023).

did not straightforwardly mention, this entanglement between fossil fuels and the degradation of the biosphere.

On the other hand, although the field of international law of energy[16] also recognizes the unsurmountable relation between the energy field and the environment—as well as with many other areas of international, regional and domestic law—it does so in an insufficient and one-sided manner. This is so because environmental protection is seen as a matter of the efficiency of the energy cycle, understood in cost-effective terms throughout the whole energy chain. That is, environmental protection is something to be done *through* and *for* the benefit of the energy production cycle, and not to prevent or redress the many different forms of ecological impact the activities from 'well to wheels' generate. This ignores the fact that the 'energy cycle' of renewable energy (re)sources, such as hydropower and biofuels, is largely dependent on the ecosystemic services disturbed by the use of fossil fuels.

Oddly enough, the international legal sphere's overall approach to the 'resource facet' of energy as a legal object seems to have a tendency to prioritize 'energy production and use' over its manifold environmental consequences—hence, rendering the protection of the ecological balance of the biosphere almost as an afterthought, when it comes to the securing of the fourfold facets of the fossil fuels-based energy matrix.

All things considered, given the current health status of the planet and the dire and ever-worsening consequences of global warming and climate change, this multidisciplinary and transversal socio-legal interpretation of the entanglement between energy production and environmental protection does not seem to be too far off.

In this regard, it is possible to say, paraphrasing Viñuales (2013), that *environmental protection* is seen as an 'immigrant' in the land of *energy* law and as such, it would only have the scope of action that is consistent with *energy production*. If *environmental protection* was to collide with *energy cycle* considerations, the latter would likely prevail,[17] especially so in regard to the fossil fuels-based energy cycle.

[16] On the meaning of the expression 'international law of energy', please see note 11 above.

[17] In the original, Viñuales was analysing the relationship between 'environmental protection' and 'development/growth', respectively. I replaced the latter by 'energy'.

The Present Energy Transition: A Race to the Top on Climate Actions?

The Worldwide Quest towards Renewable Energy Sources

As previously stated, there are many reasons to celebrate the 2015 Paris Agreement. Among these, two particularly stand out—considering the multidisciplinary and transversal perspective taken in this chapter.

First, the agreement can be considered as a hard-fought landmark in the UN climate negotiations, for it managed to establish ambitious and aspirational goals, extensive binding obligations of conduct, alongside rigorous oversight mechanisms (Rajamani, 2016) on matters relating to the global efforts against climate change.

As a matter of fact, the mechanisms contained in articles 3, 4 and 14 institute a due diligence-based standard of care (Voigt & Ferreira, 2016), requiring States' efforts to achieve the goal set on article 2 to progress over time, with the newest efforts necessarily going beyond the previous ones. According to Cristina Voigt and Felipe Ferreira (2016), these mechanisms have "the potential to function as a catalyst for a race to the top on climate action".

Second, by defining that the international community's response to climate change must aim at the goal of holding "the increase in the global average temperature to well below 2°C above pre-industrial levels", whilst "pursuing efforts to limit the temperature increase to 1.5°C", the Paris Agreement did more than initially meets the eye. It bluntly introduced a temporal frame of reference and a temperature rise threshold that shall not be exceeded, in contrast to which the efforts to stop offsetting the equilibrium of the Earth's biosphere can be measured against. In other words, it embedded the climate emergence, heir of the great acceleration, in the realm of international law.

The 1.5 °C mark was not merely symbolic nor arbitrarily chosen. It indicates a tipping point for the Earth's climate system that will, in case it is exceeded, set in motion irreversible systemic damages to the Earth's ecosystems, as the IPCC climate models show (Hoegh-Guldberg et al., 2018). Needless to say, if the 2 °C threshold is reached or overshot, the consequences will worsen catastrophically.

Among the many after-Paris efforts to adapt and mitigate climate change, the so-called green deals seem to be the most paradigmatic initiatives.[18] These are overarching national policy programmes (except in the case of the European Union's, which is not strictly 'national') with their own schedules regarding the peak and subsequent reduction of CO_2 and other GHG emissions, therefore moving their energy matrices away from fossil fuels.

Examples of the many worldwide comprehensive initiatives currently taking form are, for instance, the European Union's 2019 *Green Deal*, China's 2020 *Action Plan for Carbon Dioxide Peaking Before 2030*, South Korea's 2021 *New Deal*, as well as the recently announced US *Inflation Reduction Act* (IRA)[19] and Japan's *Green Transformation* programme, both in 2022.

The differences within these initiatives provide guidance concerning the social transformation their respective countries aim to achieve. On the one hand, the transversal goals of the ecological and just transitions pervading all policy areas of the European Green Deal exemplify the weight environmental and social matters have acquired within the EU.[20] On the other hand, the market-oriented and pro-competition perspective of the US IRA shows the primacy given to large-scale implementation of clean technologies over underlying social matters.

However, among the myriad of regulatory and financing policies these national initiatives institute, they all share the 'net zero goal' as the main, across-the-board, objective. China's plan to peak and then start reducing its carbon footprint from 2030 onwards, as well as the US and the EU's respective plans to become carbon neutral by 2050 are the most important examples of this common objective. That is, reducing CO_2 and other GHG emissions to as close as possible to zero, in such a way that the remaining emissions shall be absorbed from the atmosphere by, for instance, oceans and forests (United Nations, 2022). In other words, 'net zero' means 'carbon neutrality', which necessarily implies moving away from fossil fuels.

[18] Formally, only the European Union's initiative is called 'Green Deal'. However, given the similar goal the worldwide policies have—that of the net zero and the phasing out of fossil fuels—I have been generally referring to all as 'green deals'. For more on my use of the expression and analysis on what these initiatives entail to the realm of the law, please see (Pratti, 2021).

[19] The IRA can be understood as complementing and strengthening the United States officially rejoining the Paris Agreement.

[20] On how the EU has been handling its 'ecological transition', please see (Chiti, 2022).

Achieving the net zero goal means putting a stop to the rise of the global average temperature, stabilizing and establishing the conditions for the reduction of the overall concentration of atmospheric GHG. However, as indicated by the 2022 IPCC sixth assessment report (Pathak et al., 2022: 58), the global temperature is set to reach the 1.5 °C mark by the end of the next couple of decades.

These overarching national policy programmes currently being shaped worldwide are, at their core, the transition of their respective energy matrices from fossil fuels to low-carbon renewable (re)sources. In this sense, they represent the many different national journeys towards the fulfilment of the Paris Agreement's article 2 goal. Ultimately, they represent the transition of the global energy system. And given the advantages being a forerunner on new and clean technologies brings (Pratti, 2021), these initiatives seem to be competing with each other, which seems to be giving form to a 'race to the top' on the establishing of normative frameworks regarding the energy transition.[21]

This worldwide quest towards renewable energy sources faces multiple challenges and requires distinct areas of expertise to be tackled. One may say that the feasibility of timely large-scale implementation of carbon neutral technologies, the financing mechanisms to do so and the legal adaptations these require from all levels of law are the most pressing ones. However, from the socio-legal perspective of this chapter, a multifaceted and transversal challenge deserves special attention. In particular, because possible solutions necessarily evoke the entirety of international law and concern the legal architecture the present energy transition is arising in and struggling against.

'Phasing Out' Fossil Fuels: What Does It Legally Entail?

As seen throughout sections "Introduction", "Fossil Fuels' Technical Hegemony", and "'The Great Acceleration' of Impacts on the Earth System", from the mid-eighteenth century's rise of massive use of coal to the present non-stop drilling, refining, international trading and transporting of crude oils and their derived products, a fossil fuels-based

[21] There is, of course, a geopolitical dimension inherent to these initiatives, especially so regarding the access to the 'mineral foundation' of the energy transition, as well as to be the establish who will be the forerunners on the deployment of low-carbon, clean technologies and on the establishing of legal performance standards for these. On the matter, please see (Siddi, 2023a, 2023b).

socio-technical system of worldwide reach emerged—and brought about with it an international normative architecture consistent to it.

As a matter of fact, fossil fuels are so entrenched in the contemporary world that one is able to trace them throughout "the entirety of international law" (Viñuales, 2022). Be that in the international normative frameworks aiming at granting access to energy resources, investment in their exploration and securing their trade, or in the ones conforming the international community's actions against environmental and climate negative consequences caused by their consumption.

Evidently, the same happens with the overarching policies currently being shaped around the world to meet the Paris Agreement's temperature goal. Unlike the structural tussle at the core of the international legal depictions of energy as an object, however, these national policies seem to have a tendency to prioritize environmental and climate protection over 'energy production and use', especially so if the latter is based on non-renewable (re)sources.

Here, it becomes possible to highlight two fundamental facts regarding the present energy transition. First, it is giving rise to a multitude of new normative frameworks within national legal spheres. This is so because the so-called green deals must necessarily pervade all areas of contemporary life, given that fossil fuels are somehow virtually embedded in everything. For instance, from a non-exhausting bird's eye view and solely focused on the European experience, the legal areas currently being substantially modified or restructured anew are: regulations on access to resources and secondary raw materials; on waste management and recycling; on the efficiency, adaptation, and transitioning of the energy matrix towards renewable sources; on public and private financing and investment mechanisms in regard to intensive CO_2 and other GHG emitting activities; on carbon pricing, emissions trading system (ETS) and carbon border adjustment mechanisms (CBAM); on emission performance standards and discontinuation programmes for vehicles based on their polluting capacities; and on the vast field of agriculture, forestry and other land use (AFOLU).[22]

[22] For instance, ever since the Green Deal Communication in 2019 (European Commission, 2019), the EU has had an impressive normative output on key areas of the energy transition without falling short with its environmental and social ambitions. This has been the case, for instance, with the 'Fit for 55 legislative package', the Clean Hydrogen Alliance (European Commission, 2022b), the Battery Alliance (European Commission, 2022a), the Climate Law (Reg 2021/1119, 2021), the RePower EU strategy (European Commission, 2022c), the Biodiversity Strategy (European Commission, 2020), and the CBAM (Regulation (EU) 2023/956, 2023), the amendment of its ETS (Directive (EU) 2023/959, 2023).

With overarching objectives like this having multifarious impacts on all areas of the EU polity—other than its energy matrix and its ecology—the EU, its "regulatory space", and its administrative system are expected to undergo a remarkable transformation process (Giorgi, 2023). At the same time, each of these normative frameworks pushes forward the global objective of disentangling, of eviscerating, fossil fuels from the structural core of daily life. Of course, doing so will neither be easy nor fast and this "is expected" (Smil, 2013). However, it will probably still not mean zero fossil fuels consumption, especially given the infinity of uses petrochemicals have. Moreover, this interpretation is in line with the previously mentioned outcome of the UNFCCC's COP 28, which called for "transitioning away" from fossil fuels *in energy systems*—as opposed to a fully fledged abandonment of it.[23]

Regarding these points, the European Green Deal (European Commission, 2019) offers a primary example through the way it articulates the 'net zero' and the 'phasing out fossil fuels' goals: the former is to be achieved by 2050,[24] whereas to the latter no specific time frame was established. This possibly indicates that the energy transition may not mean the absolute end of fossil fuels (especially of crude oils) but most likely the emancipation of the world from its dependence on them.

Second, there is an apparent conflict emerging between national policies that prioritize environmental protection and climate stability over fossil fuels-based 'energy production and use' and the depiction of energy as a legal object by international law—which has been seemingly doing the opposite, as discussed in section "'The Great Acceleration' of Impacts on the Earth System". Duly noted, this aspect is connected to the tussle between the many normative frameworks currently being designed, as well as the 'carboniferous' international normative architecture in place.

In this regard, the energy-oriented international normative framework in place may represent an obstacle to the energy transition. On the one hand, the original intent of securing investments and access to natural resources—at the roots of the internationalization of energy transactions (Viñuales, 2022)—may actually hamper national initiatives to move away

[23] On this matter, please see note 15 above.

[24] The European Union established the carbon-neutrality goal to be achieve by 2050 and set an intermediate target of reducing 55% of its emissions already by 2030 with its climate law (European Climate Law (Reg 2021/1119), 2021). However, as this chapter is being written, there is no 2040 goal set yet (Keating, 2023).

from the exploration of fossil fuels. The 20-year sunset clause of the ECT's article 47 is a primary example of such, as its entrenchment effect and "limits imposed on the agenda of policymakers" are unparalleled (Kouroutakis et al., 2022).[25]

On the other hand, because this framework has been designed while having as a main concern the geographical mismatch of where energy is produced and consumed, as well as the fourfold facets of fossil fuels-energy, it may not apply as intended to the critical minerals, rare earths and overall raw materials that are fundamental to the energy transition.

The main fact that stands out in this regard is that there is not a fully fledged overarching international normative architecture when it comes to mining activities—neither an UN-led international organization[26] nor even an international treaty similar to the ECT—which may become a problem in the future, given the worldwide consumption of raw materials is increasing ever faster. Although already having an impact on a variety of supply chains around the world, the "mineral foundation" (Bazilian, 2018) of the present energy transition is yet to fully develop its more disruptive consequences.

Nonetheless, it remains to be seen what impacts the rather different dynamics of the 'geographical mismatch' of non-renewable energy (re) sources will have on the national, regional and international normative frameworks of the energy sector. Not due to matters concerning availability and access to solar, wind or geothermal energy power though, but because the technologies to explore these energy sources are themselves subject to a rather different and multifaceted 'geographical mismatch'. The same applies to the core technologies at the basis of electric vehicles.

These are all fundamentally dependent on critical raw materials, which are largely concentrated in a few countries. And, on top of that, the capacity to refine and process these elements is largely dominated by even fewer countries—mostly not those in which such resources are found, with the exception of China.

[25] As a possible answer to this problem, the European Commission proposed a planned, joint withdrawal from the ECT. Please see (Bektasheva, 2023; European Commission, 2023).

[26] In this regard, however, there is the International Seabed Authority, established under the 1982 United Nations Convention on the Law of the Sea (UNCLOS), who is expected to decide by 2025 about the possibility of deep sea mining on areas beyond national jurisdictions.

Be that as it may, the indispensable mineral-based materials and technologies of the ongoing energy transition must be transported many times throughout the world before reaching their final clean low-carbon uses. And this will certainly be done by fossil fuels-intensive means of transportation for the foreseeable future, until these are finally disentangled from the structure of contemporary life and the global value chains underlying it.

By shedding light on these two fundamental facts regarding the ongoing energy transition and on the inescapable challenges they impose, it is possible to infer that prior to being a scientific or a technological matter, the ongoing energy transition's goal of 'phasing out' fossil fuels is a legal challenge. As such, it has the potential of reshaping some deeply rooted structures of international law while pushing forward the redesigning of national normative frameworks around the world. In sum, we seem to be witnessing the beginning of a new and hopefully more sustainable dimension of the transformative, form-giving function of the realm of the law (Kjaer, 2022). In the meantime, we ought to bear with the climate-changing consequences of the presently enduring carboniferous socio-technical system.

Conclusion

The previous energy transitions saw the emergence of fossil fuels-based energy resources and it gave rise to a world-spanning carboniferous socio-technical system, which is reflected within the core structures of "the social practice we call international law" (Viñuales, 2022). By delving into this structure whilst having the 'resource facet of energy' as a legal object, this chapter identified the unsettling, and somewhat contradictory, privilege that both international environmental and energy law have been assigning to fossil fuels-based 'energy production and use' over 'environmental protection' matters.

This problematic and structural mishandling of the inseparable link between energy and the environment poses a fundamental challenge to the efforts of phasing out fossil fuels. On the one hand, as seen through the example of the ECT's article 47, this international normative architecture may bluntly hamper national initiatives towards the hegemonization of renewable energy (re)sources. On the other hand, the lack of fully fledged international normative frameworks to secure investment and access to the 'mineral foundations' of the ongoing energy transition, alongside the geographical mismatches of the countries in which these are located and refined before their final use, are still to present problems whose complete extensions and solutions remain to be seen.

The most immediate and promising answers to these challenges are stemming from the so-called green deals, which are redesigning innumerous national normative frameworks to eviscerate their dependences on fossil fuels and lay down the regulatory pathway for the implementation of low-carbon, renewables-based energy matrices. These national overarching normative frameworks are rebalancing the scale, which has been prioritizing 'energy production and use' from non-renewable sources, in favour of environmental and climate protection and renewables. These are to be watched closely for they seem capable of having a spillover effect on the legal spheres beyond the State—transnational, international and global law—rendering these more consistent with the needs of the times. It remains to be seen how the *ab ovo* contradiction this chapter identified will be dealt with as these likely developments unfold.

This is not the first—and most likely will not be the last—energy transition. But it certainly is the most important one up until now. Through the legal dialectics currently brewing, the realm of the law—in all of its spheres (national, regional, international and global)—seems to be going through a watershed happening regarding the enabling of its form-giving, transformative functions. May it thus not lose sight of this energy transition's *raison d'être*: to maintain a human-favourable balance of the Earth's biosphere and of its ecosystems, so as to ultimately avoid the climate catastrophe that awaits by the end of this century. To this end, one can only hope the law will have the significant role it is expected of it (Pratti, 2021).

Acknowledgments I dearly thank Nelson Aloysio Reis de Almeida Passos for the constant exchange of ideas and the constructive criticism offered throughout the writing of this chapter. A heartfelt acknowledgement is also due to Emmalucia Virardi, for the kind words and support concerning the main ideas of this paper. I would also like to acknowledge the anonymous reviewers for their helpful notes and suggestions. Any mistakes, misunderstandings and ambiguities are the author's responsibility.

References

Bazilian, M. D. (2018). The Mineral Foundation of the Energy Transition. *The Extractive Industries and Society, 5*(1), 93–97. https://doi.org/10.1016/j.exis.2017.12.002

Bektasheva, A. (2023, June 29). Withdrawal from the Energy Charter Treaty without its Modernization, or Modernization of the ECT Treaty without Withdrawal by EU Members: What is the Impact of the Two Scenarios? *International Law Blog*. Retrieved from https://internationallaw.blog/2023/06/29/withdrawal-from-the-energy-charter-treaty-without-its-modernization-or-modernization-of-the-ect-treaty-without-withdrawal-by-eu-members-what-is-the-impact-of-the-two-scenarios/

Chiti, E. (2022). Managing the Ecological Transition of the EU: The European Green Deal as a Regulatory Process. *Common Market Law Review, 59*(1), 19–48.

Crawford Lake, Canada, Chosen as the Primary Marker to Identify the Start of the Anthropocene Epoch. (n.d.). ScienceDaily. Retrieved July 31, 2023, from https://www.sciencedaily.com/releases/2023/07/230711133248.htm

Crutzen, P. J. (2002). Geology of Mankind. *Nature, 415*(6867), 6867. https://doi.org/10.1038/415023a

Crutzen, P. J., & Stoermer, E. F. (2000, May). The "Anthropocene." *IGBP Newsletter: A Study of Global Change of the International Council for Science, 41*, 17–18.

Directive (EU) 2023/959 of the European Parliament and of the Council of 10 May 2023 amending Directive 2003/87/EC establishing a system for greenhouse gas emission allowance trading within the Union and Decision (EU) 2015/1814 concerning the establishment and operation of a market stability reserve for the Union greenhouse gas emission trading system (Text with EEA relevance), EP, CONSIL, 130 OJ L (2023). Retrieved from http://data.europa.eu/eli/dir/2023/959/oj/eng

Dupuy, P.-M., & Viñuales, J. E. (2018). *International Environmental Law* (2nd ed.). Cambridge University Press.

Ellul, J. (2009). *Il sistema tecnico: La gabbia delle società contemporanee* (G. Carbonelli, Trans.). Editoriale Jaca Book.

European Commission. (2019). *The European Green Deal—COM(2019)640 Final—Communication from the Commission to the European Parliament, the European Council, the Council, the European Economic and Social Committee and he Committee of the Regions.*

European Commission. (2020). *Biodiversity Strategy. COM(2020)380 Final—Communication from the Commission to the European Parliament, the European Council, the Council, the European Economic and Social Committee and he Committee of the Regions.*

European Commission. (2022a). *European Battery Alliance*. Retrieved from https://single-market-economy.ec.europa.eu/industry/strategy/industrial-alliances/european-battery-alliance_en

European Commission. (2022b). *European Clean Hydrogen Alliance*. Retrieved from https://single-market-economy.ec.europa.eu/industry/strategy/industrial-alliances/european-clean-hydrogen-alliance_en

European Commission. (2022c). *REPowerEU: Affordable, Secure and Sustainable Energy for Europe* [Text]. European Commission—European Commission. Retrieved from https://ec.europa.eu/info/strategy/priorities-2019-2024/european-green-deal/repowereu-affordable-secure-and-sustainable-energy-europe_en

European Commission. (2023). *European Commission Proposes a Coordinated EU Withdrawal from the Energy Charter Treaty*. Retrieved from https://energy.ec.europa.eu/news/european-commission-proposes-coordinated-eu-withdrawal-energy-charter-treaty-2023-07-07_en

Ferrarese, M. R. (2000). *Le istituzioni della globalizzazione: Diritto e diritti nella società transnazionale*. Il Mulino.

Gibbard, P., Walker, M., Bauer, A., Edgeworth, M., Edwards, L., Ellis, E., Finney, S., Gill, J. L., Maslin, M., Merritts, D., & Ruddiman, W. (2022). The Anthropocene as an Event, not an Epoch. *Journal of Quaternary Science, 37*(3), 395–399. https://doi.org/10.1002/jqs.3416

Giorgi, A. (2023). The EU Green Deal and the Transformations of the European Administrative System: Does the "Epistemic Leadership" of the Scientific Advisory Board Push the Agency Model Over the Sunset Boulevard? *European Papers, 8*(2).

Head, M. J., Zalasiewicz, J. A., Waters, C. N., Turner, S. D., Williams, M., Barnosky, A. D., Steffen, W., Wagreich, M., Haff, P. K., Syvitski, J., Leinfelder, R., McCarthy, F. M. G., Rose, N. L., Wing, S. L., An, Z., Cearreta, A., Cundy, A. B., Fairchild, I. J., Han, Y., et al. (2023). The Anthropocene is a Prospective Epoch/Series, not a Geological Event. *Episodes Journal of International Geoscience, 46*(2), 229–238. https://doi.org/10.18814/epiiugs/2022/022025

Hobsbawm, E. (1995). *Age of Extremes: The Short Twentieth Century 1914–1991*. Abacus.

Hobsbawm, E. J. (1999). *Industry and Empire: From 1750 to the Present Day*. The New Press.

Hoegh-Guldberg, O., Jacob, D., Taylor, M., Bindi, M., Brown, S., Camilloni, I., Diedhiou, A., Djalante, R., Ebi, K., Engelbrecht, F., Zhou, G., Joel, G., Hijioka, Y., Mehrotra, S., Payne, A., Seneviratne, S., Thomas, A., Warren, R., Halim, S., & Guillén Bolaños, T. (2018). Impacts of 1.5oC Global Warming on Natural and Human Systems. In *Global Warming of 1.5°C. An IPCC Special Report on the Impacts of Global Warming of 1.5°C Above Pre-industrial Levels and Related Global Greenhouse Gas Emission Pathways, in the Context of Strengthening the Global Response to the Threat of Climate Change, Sustainable Development, and Efforts to Eradicate Poverty* (pp. 175–311).

Keating, D. (2023). Does the EU need a 2040 climate target? *Energy Monitor*. Retrieved from https://www.energymonitor.ai/policy/does-the-eu-need-a-2040-climate-target/#catfish

Kjaer, P. F. (2022). What is Transformative Law? *European Law Open, 1*(4), 760–780. https://doi.org/10.1017/elo.2023.1

Kouroutakis, A., Pavy, E., & Van Der Elst, F. (2022). *Sunset Clauses in International Law and their Consequences for Eu Law*. Policy Department for Citizen's Rights and Constitutional Affairs. European Parliament.

McNeill, J. R., & Engelke, P. (2014). *The Great Acceleration: An Environmental History of the Anthropocene since 1945*. The Belknap Press of Harvard University Press.

Mumford, L. (1955). *Technics and Civilization* (7th ed.). Routledge and Kegan Paul Ltd.

Pathak, M., Slade, R., Shukla, P. R., Skea, J., Pichs-Madruga, R., & Ürge-Vorstaz, D. (2022). Climate Change 2022: Mitigation of Climate Change. *Contribution of Working Group III to the Sixth Assessment Report of the Intergovernmental Panel on Climate Change.*

Pottage, A. (2020). An Apocalyptic Patent. *Law and Critique, 31*(3), 239–252. https://doi.org/10.1007/s10978-020-09278-4

Pratti, G. (2021). Bad Moon Rising: The Green Deals in the Globalization Era. *Rivista Quadrimestrale Di Diritto Dell'Ambiente, 1*, 177–196.

Pratti, G., Putzer, A., & de Marinis, L. (2022). Advancing Legal and Practical Recognition of the Non-Human Right to Energy. In K. Araújo (Ed.), *Routledge Handbook of Energy Transitions* (1st ed., pp. 86–101). Routledge.

Rajamani, L. (2016). Ambition and Differentiation in the 2015 Paris Agreement: Interpretative Possibilities and Underlying Politics. *International and Comparative Law Quarterly, 65*(2), 493–514. https://doi.org/10.1017/S0020589316000130

Regulation (EU) 2021/1119 of the European Parliament and of the Council of 30 June 2021 establishing the framework for achieving climate neutrality and amending Regulations (EC) No 401/2009 and (EU) 2018/1999 ('European Climate Law'), 243 OJ L (2021). Retrieved from http://data.europa.eu/eli/reg/2021/1119/oj/eng

Regulation (EU) 2023/956 of the European Parliament and of the Council of 10 May 2023 establishing a carbon border adjustment mechanism (Text with EEA relevance), 130 OJ L (2023). Retrieved from http://data.europa.eu/eli/reg/2023/956/oj/eng

Results of Binding Vote by AWG. (2019, May 21). *Working Group on the 'Anthropocene' Subcommission on Quaternary Stratigraphy*. Retrieved from http://quaternary.stratigraphy.org/working-groups/anthropocene/

Rosa, H. (2013). *Social Acceleration: A New Theory of Modernity* (J. Trejo-Mathys, Trans.). Columbia University Press.

Santos, M. (2001). *Por uma outra globalização: Do pensamento único à consciência universal* (6th ed.). Editora Record.

Siddi, M. (2023a). *Europe's Policies for a Green Transition: The European Commission's Geopolitical Turn and Its Pitfalls*.

Siddi, M. (2023b). The Geopolitics of Energy Transition: New Resources and Technologies. In J. Berghofer, A. Futter, C. Häusler, M. Hoell, & J. Nosál (Eds.), *The Implications of Emerging Technologies in the Euro-Atlantic Space: Views from the Younger Generation Leaders Network* (pp. 73–85). Springer International Publishing. https://doi.org/10.1007/978-3-031-24673-9_5

Smil, V. (2008). *Oil: A Beginner's Guide*. Oneworld Publications.

Smil, V. (2010a). *Energy Transitions: History, Requirements, Prospects*. Praeger.

Smil, V. (2010b). *Prime Movers of Globalization—The History and Impact of Diesel Engines and Gas Turbines*. The MIT Press.

Smil, V. (2013). The Long Slow Rise of Solar and Wind. *Scientific American, 310*(1), 52–57. https://doi.org/10.1038/scientificamerican0114-52

Steffen, W., Sanderson, R. A., Tyson, P. D., Jäger, J., Matson, P. A., Moore, B., III, Oldfield, F., Richardson, K., Schellnhuber, H.-J., Turner, B. L., & Wasson, R. J. (2004). *Global Change and the Earth System: A Planet under Pressure* (1st ed.). Springer.

Steffen, W., Crutzen, P. J., & McNeill, J. R. (2007). The Anthropocene: Are Humans Now Overwhelming the Great Forces of Nature? *Ambio, 36*(8), 614–621.

Steffen, W., Broadgate, W., Deutsch, L., Gaffney, O., & Ludwig, C. (2015). The Trajectory of the Anthropocene: The Great Acceleration. *The Anthropocene Review, 2*(1), 81–98. https://doi.org/10.1177/2053019614564785

Subramanian, M. (2019). Anthropocene Now: Influential Panel Votes to Recognize Earth's New Epoch. *Nature*. https://doi.org/10.1038/d41586-019-01641-5

United Nations. (2022). *Net Zero Coalition*. United Nations; United Nations. Retrieved from https://www.un.org/en/climatechange/net-zero-coalition

United Nations Framework Convention on Climate Change. (2023). *Outcome of the First Global Stocktake under the Paris Agreement (Draft Decision -/CMA.5)*. United Nations. Retrieved from https://unfccc.int/documents/636608

Viñuales, J. E. (2013). The Rise and Fall of Sustainable Development. *Review of European Community and International Environmental Law, 22*(1), 3–13.

Viñuales, J. E. (2022). *The International Law of Energy*. Cambridge University Press. https://doi.org/10.1017/9781108235273

Voigt, C., & Ferreira, F. (2016). 'Dynamic Differentiation': The Principles of CBDR-RC, Progression and Highest Possible Ambition in the Paris Agreement. *Transnational Environmental Law, 5*(2), 285–303. https://doi.org/10.1017/S2047102516000212

World Energy Trilemma Index. (n.d.). World Energy Council. Retrieved July 31, 2023, from https://www.worldenergy.org/transition-toolkit/world-energy-trilemma-index

CHAPTER 9

The Role of Hydrogen for a Sustainable Energy Transition in the European Union: Between a Fossil-Based and a Renewable Path

Aloia López Ferro

Introduction

In recent years, a new wave of interest in hydrogen has emerged. Russia's invasion of Ukraine has accelerated the momentum for hydrogen development in the European Union (EU), and several large hydrogen projects are being developed or proposed for the production and transport of this so-called game-changer molecule. However, since hydrogen-related market regulation is largely absent in the EU, a proper regulatory framework for hydrogen is needed (Lavrijssen & Vitez, 2020).

Currently, 25% of the total energy consumption in the EU is provided by fossil gas. However, to meet the climate goals, fossil gas consumption must be reduced and decline sharply before 2050 (Commission, 2020a).

A. López Ferro (✉)
Rovira i Virgili Unviersity, Tarragona, Spain
e-mail: aloia.lopez@urv.cat

D. Iglesias Márquez et al. (eds.), *Legal Challenges at the End of the Fossil Fuel Era*, Global Issues,
https://doi.org/10.1007/978-3-031-61766-9_9

Therefore, creating a future-proof regulatory framework for hydrogen represents a major challenge that policymakers must carefully address "to regulate hydrogen in a manner that serves decarbonisation goals and minimises costs to consumers" (Anderson & Jahn, 2022: 8).

Not all hydrogen production methods, applications and infrastructure are completely decoupled from fossil gas use. Firstly, under the umbrella term of low-carbon hydrogen, projects for generating hydrogen from fossil fuels using carbon capture techniques are enabled. Secondly, fossil gas infrastructures could also be maintained over time by blending hydrogen with gas. Thirdly, failure to apply a proper prioritisation of hydrogen applications and the omission of the additionality principle in the production of renewable hydrogen increase the risk of delaying the end of fossil fuels in the EU and the fulfilment of the climate objectives for a sustainable energy transition.

The main objective of this chapter is to assess how the EU is carrying out hydrogen planning and regulation with the goal of achieving a reduction of consumption and dependence on fossil gas. To this end, it analyses the main hydrogen-related policies, regulatory frameworks and proposals, and the role of methane gas that they envisage.[1] In addition, it identifies legal sustainability challenges arising from the EU's approach to hydrogen deployment. The research method employed is inductive, starting with the analysis of existing and proposed policies and regulations, and then identifying the legal challenges that must be addressed.

Fossil-Based Hydrogen and Renewable Hydrogen in the EU Policies

In recent years, the declining costs of renewable energies, technological breakthroughs and the urgency to drastically reduce greenhouse gas (GHG) emissions have led to a growing interest in hydrogen. In Europe, policymakers have given hydrogen an increasingly prominent role, especially since 2020. This section assesses what is the role for methane gas that the hydrogen-related policies forecast.

From "A Clean Planet for All" to the European Green Deal

The role of hydrogen in the decarbonisation of Europe's energy system and industries has been linked to the use of methane as a bridging solution

[1] This chapter was completed in October 2023. Therefore, the analysis of regulations approved after this date has not been included.

since 2018, when the Commission presented the Communication "A clean planet for all. Europe's long-term strategic vision for a prosperous, modern, competitive, and climate-neutral economy" (Commission, 2018). This strategy foresaw fossil-based hydrogen with carbon capture in a transitional phase.

In 2019, the Communication on the European Green Deal (Commission, 2019) set out an initial roadmap of key policies and measures to, among other objectives, achieve net emissions of GHG in 2050. Regarding the gas sector, actions involved supporting the development of decarbonised gases, designing a decarbonised and competitive gas market, and solving the problem of methane emissions. It also pointed out that the regulation of energy infrastructures should be modified to allow for the development of clean hydrogen and carbon capture. Although the Communication does not define the concept of clean hydrogen and does not contain any mention of low-carbon hydrogen, references can be found to decarbonising gas—as opposed to phasing out coal—which is indicative of the recognition of the Commission's vision on the role of natural gas on the path to a carbon neutral future.

The EU's Hydrogen Strategy

The specific piece of hydrogen policy, EU Hydrogen Strategy, was published by the Commission in 2020 (Commission, 2020b). This strategy lays the foundation for hydrogen to become an intrinsic part of an integrated European energy system. It identifies the challenges to be overcome and includes a roadmap of actions for the coming years. These actions relate to the promotion and stimulation of public and private investment, the development of a regulatory framework to enable the creation of a new hydrogen market, the necessity of research and innovation, the changes and improvements on the infrastructure network and cooperation with third countries.

The strategy envisages the role of hydrogen in three phases: until 2024, the EU will support the installation of at least 6 GW of renewable hydrogen electrolysers and the production of up to 1 million tonnes of renewable hydrogen in all Member States; from 2025 to 2030, hydrogen should become an intrinsic part of the energy system, with at least 40 GW of renewable hydrogen electrolysers and the production of up to 10 million tonnes (MT) of renewable hydrogen in the EU; and from 2030, renewable hydrogen will be deployed on a large scale in all sectors that are difficult to decarbonise.

The Communication considers that the large-scale deployment of clean hydrogen at a fast pace is critical for the EU. According to this strategy, clean hydrogen refers to renewable hydrogen, that is hydrogen derived from a production pathway with near-zero GHG emissions. However, low-emission hydrogen also plays a role in the overall strategy: while the EU's priority in the long term is to develop renewable hydrogen, in the short to medium term, the Commission highlights that other forms of low-carbon hydrogen are needed. Low-carbon hydrogen will serve to rapidly reduce emissions from current hydrogen production, contributing to scaling-up production and the hydrogen market, as well as enabling a gradual hydrogen development pathway, considering different speeds in different sectors and regions.

Fossil gas and its associated infrastructures will also play a role in the envisaged phases of large-scale hydrogen development. In this respect, the strategy announces that some of the existing hydrogen production plants should be retrofitted with carbon capture and storage technologies; the existing gas network could be partially reused for the transport of renewable hydrogen, and non-network-based transport options, as liquefied natural gas (LNG) terminals may be adapted for hydrogen. In this regard, according to the strategy, LNG terminals could be used for the transportation of hydrogen or other hydrogen carriers, such as ammonia by trucks or ships in the future; however, it is explicitly stated that this should be "insofar as technically feasible". The strategy also considers that blending hydrogen into the natural gas grid at a limited percentage could contribute to the development of hydrogen in a transitional phase; and, finally, that infrastructure to support carbon capture use and storage may be needed to produce low-carbon hydrogen and synthetic fuels.

The regulatory framework that the strategy envisages aims at establishing a well-functioning European hydrogen market and incentivising the supply and demand for renewable and low-carbon hydrogen. As the policy support framework for hydrogen scale-up will be directed not only towards renewable hydrogen but, in a transitional period, towards low-emission hydrogen, the strategy outlines specific actions. In particular, it states that it will be necessary to define what should be considered renewable and low-carbon hydrogen. In particular, it recognises that there is a need for a common threshold or standard to support low-carbon and renewable hydrogen installations, considering their full lifecycle GHG performance.

In addition, it announces that terminology and criteria for the certification of renewable and low-carbon hydrogen will be developed at the European level and promoted internationally.

Moreover, recognising the environmental impacts associated with both renewable and low-carbon hydrogen, the strategy delineates specific actions. Regarding fossil hydrogen with carbon capture, methane emissions will be addressed as part of the methane strategy. Concerning electrolytic hydrogen, the Commission emphasises the necessity of enabling hydrogen demand during periods of abundant renewable electricity supply in the grid.

The Energy System Integration Strategy

The Energy System Integration Strategy (Commission, 2020c) proposes concrete policy and legislative measures to gradually shape a new integrated energy system. It encompasses three complementary and mutually reinforcing ideas: a more circular energy system, with energy efficiency at its core, in which the least energy-intensive choices are prioritised, unavoidable waste streams are reused for energy purposes and synergies are exploited across sectors; higher direct electrification of end-use sectors; and use of renewable and low-carbon fuels, including hydrogen, for end-use applications where direct heating or electrification are not feasible, not efficient or have higher costs. In addition, it highlights the potential of hydrogen to provide storage solutions for energy produced from variable renewable sources, which could serve to create synergies between the power, gas and end-use sectors.

Again, low-carbon hydrogen is considered necessary during a transition phase to replace existing hydrogen and kick-start an economy of scale. However, the "Energy System Integration Strategy anticipates gas as a solution of last resort, coming in only where other solutions do not fit" (Anderson et al., 2022: 16), that is in the hard-to-decarbonise sectors, such as the production of ammonia, fertilisers, methanol, the direct reduction of iron ore, high-temperature processes (including the production of cement, steel, glass, ceramics) or refining processes where hydrogen is used to lower the sulphur content of diesel fuel (EEB, 2023).

From the "Fit for 55 Package" to the REPowerEU Plan

In July and December of 2021, the Commission adopted the legislative initiatives of the "Fit for 55 package" (Commission, 2021a). The package proposes revising several pieces of EU climate and energy legislation to achieve the objectives targeted in the climate regulation of the EU. Although the Commission Communication accompanying this package does not mention low-carbon hydrogen, the regulatory proposals do, as this chapter discusses.

In May 2022, the Commission published the REPowerEU Plan (Commission, 2022a). Building on the "Fit for 55" package, the REPowerEU Plan proposed additional actions to reduce GHG emissions and strengthen energy supply security. These actions aimed to end Europe's dependence on Russian fossil fuels as soon as possible, as well as to accelerate the energy transition (Conti & Kneebone, 2022). Hydrogen receives special attention in the REPowerEU Plan, as the Commission believes that the EU needs to accelerate the ramp-up of both the demand and supply of renewable hydrogen. In this sense, while the 2030 target for renewable hydrogen in "Fit for 55" was 5.6 MT, the new plan increases the target to 20 MT (of which 4 MT is for ammonia). It also sets the target of producing 10 MT of renewable hydrogen, and the same amount is added for imports. Notably, they are anticipated to originate from regions in the Global South, including Africa and Latin America, as highlighted by Müller et al. (2022). As regards the demand, it increases the expected quotas for the use of renewable fuels for transport and industry. Alongside the acute increase in supply and demand for renewable hydrogen, the Commission also outlined a need to accelerate the deployment of renewable capacity to produce it. Consequently, REPowerEU increases the EU's renewable ambitions specifically for hydrogen production to 80 GW.

As far as fossil gas is concerned, it explicitly recognises that its reduction will be accelerated, limiting its consideration as a transition fuel. In fact, the REPowerEU does not mention low-carbon hydrogen. However, this does not mean that fossil gas consumption will disappear in the EU: the plan announces an increase in imports of LNG to diversify the gas supply. Furthermore, the Commission considers it a priority to analyse whether measures and investments are needed in hydrogen-ready gas infrastructure and interconnections to overcome bottlenecks to fully use the EU's LNG capacity. To identify the remaining infrastructure gaps, the Commission requested the European Network of Transmission System Operators for Gas (ENTSOG) to support the assessment.

Ultimately, the plan does not alter the regulatory basis for low-carbon hydrogen deployment that had begun to be developed. In fact, the REPowerEU calls upon the European Parliament and the Council to "rapidly conclude the revision of the Hydrogen and Gas Market package" (Commission, 2022a: 7). In addition, the Commission expects that 1.3 MT of hydrogen will be blended into the fossil gas network by 2030 (Commission, 2022b).

Fossil Gas in the Approved and Proposed Hydrogen-Related Regulatory Framework

Hydrogen regulation is currently in the spotlight. Most of the proposals concerning hydrogen regulation were included in the 2021 "Fit for 55 package" and in the hydrogen and decarbonised gas market package. In this section, an analysis is conducted on the EU legislative acts and proposals adopted since the European Green Deal to explore the role of fossil gas in hydrogen regulation development.

Taxonomy Delegated Acts

In 2021, the Commission approved the Delegated Regulation (EU) 2021/2139 of 4 June 2021 supplementing Regulation (EU) 2020/852 of the European Parliament and of the Council by establishing the technical screening criteria for determining the conditions under which an economic activity qualifies as contributing substantially to climate change mitigation or climate change adaptation and for determining whether that economic activity causes no significant harm to any of the other environmental objectives.

In Section 3.10, it outlines the technical criteria for hydrogen production to be considered as substantially contributing to climate change mitigation in section. While the delegated act does not explicitly define low-carbon hydrogen, there is broad acceptance that the threshold that it establishes allows for the classification of low-carbon hydrogen as sustainable (ECOS, 2022). This assertion is further supported by the explicit acknowledgement of the potential inclusion of carbon sequestration in hydrogen production.

The delegated act also includes, in Section 4.14, the construction, conversion, repurposing or retrofitting of gas networks for low-carbon gases as sustainable activities. In particular, it refers to the following activities: construction or operation of new transport and distribution networks for hydrogen or other low-carbon gases; conversion/reconversion of existing natural gas networks for 100% hydrogen transport and distribution; and renewal of gas transmission and distribution networks to enable the integration of hydrogen and other low-carbon gases into the gas transmission and distribution networks, including leak detection and repair of existing pipelines and other network elements to reduce methane leakage. When describing the above-mentioned activities related to the transport and distribution of hydrogen, the delegated act does not distinguish between low-carbon or renewable hydrogen and does not provide any definition about these concepts. In fact, it mentions hydrogen and other low-carbon gases, without including any distinction, considering that all these activities related to hydrogen are sustainable. Moreover, the inclusion of retrofitting gas networks as a sustainable activity, facilitating the integration of hydrogen through blending in Section 4.14, is particularly controversial.

In 2022, the Commission adopted the Delegated Regulation (EU) 2022/1214 of 9 March 2022 amending Delegated Regulation (EU) 2021/2139 as regards economic activities in certain energy sectors. This complementary act was approved despite the opposition from the Platform on Sustainable Finance and legal action against it by some environmental organisations as well as Austria. It contains the criteria to qualify nuclear energy and gas as significantly contributing to climate mitigation.

This complementary delegated act considers fossil gas and nuclear energy as transitional activities, changing and broadening the concept previously defined in the taxonomy regulation. Accordingly, it includes as sustainable activities the construction and operation of nuclear power plants for hydrogen production; electricity generation from fossil gaseous fuels; high-efficiency cogeneration of heat, cooling and power from fossil gaseous fuels; and production of heat and cooling from fossil gaseous fuels in efficient district heating and cooling systems. In these cases, fossil fuel use is allowed as a transitional measure, as long as the facility is designed and constructed to use renewable and/or low-carbon gaseous fuels by 2035. Ensuring fuel switching relies on a commitment and a verifiable plan approved by the management body of the undertaking. Furthermore, it also depends on verification from a third party that should not have any conflict of interests and which should transmit to the Commission an annual report of whether the activity is on a credible trajectory to fulfil the commitment.

These aspects of the complementary delegated act have been highly criticised and described as a "recipe for a greenwashing disaster" as "even if the commitments for shifts to low-carbon or renewable gases were realised, the proposal risks gas power plants becoming a black hole for renewable energy, which would be more efficiently used elsewhere" (Bellona, 2022b: 1). In any case, a similar formula, based on the credibility of the climate commitments and future gas shift, can also found in Communication from the Commission "Guidelines on State aid for climate, environmental protection and energy 2022". These guidelines enable investments in fossil gas refuelling infrastructures if the Member State demonstrates a credible pathway towards the phasing out of hydrogen that is not renewable or low-carbon to supply the refuelling infrastructure by 2035.

Revision of the Regulation on Guidelines for Trans-European Energy Infrastructure

In 2022, it was passed a new Regulation (EU) 2022/869, of the European Parliament and the Council of 30 May 2022, on guidelines for trans-European energy infrastructure. One of the main purposes of the revision consisted of removing fossil gas and oil projects from the lists of projects of common interest (PCI) to adapt it to the climate objectives. According to the European Parliament, it should represent the ending of policy and financial support to fossil infrastructure (European Parliament, 2022).

This revised regulation not only addresses renewable hydrogen infrastructure but also low-carbon hydrogen. In this sense, it enables the development of low-carbon hydrogen production and projects by making carbon capture as well as CO_2 transport projects eligible for PCI status and to receive EU financial assistance. For this reason, it enables a "hybrid" decarbonisation, in which both renewable and low-carbon hydrogen play a role (Yafimava, 2022: 4).

Regarding fossil gas infrastructures and blending, point 59 of the explanatory memorandum of the Regulation points out that "blending hydrogen with natural gas or biomethane could be used in the scaling up of the hydrogen production capacity and facilitating the transport of hydrogen". Accordingly, article 2.10 of the Regulation includes in the concept of smart grid any equipment and installation enabling blending hydrogen with methane in natural gas networks. Moreover, article 31 of the Regulation contemplates an exemption clause, according to which, for a transitional period until 2029, gas networks or infrastructure could be used to transport or store a blend of hydrogen with natural gas or biomethane. However, eligibility for EU financial assistance for such projects will cease on 31 December 2027. In these cases, ensuring fuel switching relies on the demonstration of the project promoter of how, by the end of the transitional period, the natural gas assets will become dedicated hydrogen assets and how the use of hydrogen will be enhanced during the transitional period. Such pieces of evidence shall include an assessment of the supply and demand of renewable or low-carbon hydrogen, as well as a calculation of the GHG emission reduction attributable to the project. Furthermore, the Agency for the Cooperation of Energy Regulation should carry out the verification of timely transition. In addition, recital 59 also includes that funding agreements for these projects should establish a condition for the repayment of the funds in case of delay in the timely transition.

This Regulation was hardly criticised by environmental organisations, such as Ecologistas en Acción, Greenpeace, Climate Action Network, Global Witness, Friends of the Earth and E3G. The main points of discrepancy consist of the role ENTSOG still has in the new Regulation, which could lead to overinvestment in the gas network, and the grandfathering clause, that is the transitional period related to methane infrastructure (Couffon, 2021). In particular, they criticised that the new gas-related categories create a serious risk that the Regulation on guidelines for trans-European energy infrastructure will continue to subsidise investments in fossil fuel infrastructure, as the hydrogen category includes projects for low-carbon hydrogen. In addition, they considered that hydrogen infrastructure development should have a local approach, locating hydrogen production facilities as close to demand centres as possible.

The Revised Renewable Energy Directive (RED III) and the Delegated Acts on Renewable Hydrogen

The revised Renewable Energy Directive (RED III) includes targets to incentivise the demand of renewable fuels of non-biological origin (RFNBOs), including renewable hydrogen and ammonia, methanol or e-fuels when produced from renewable hydrogen. Accordingly, it includes a binding target in industry to procure at least 42% of its hydrogen from RFNBOs by 2030 (article 22 bis). However, an exception is incorporated to benefit those countries that aim to promote hydrogen produced with nuclear energy (commonly known as pink hydrogen), such as France. In this sense, for those countries that can achieve a fossil-free hydrogen mix of at least 77% by 2030, the target is reduced by 20% (article 22 ter). In transport, fuel suppliers must achieve either a 14.5% reduction GHG emissions associated with their fuels or achieve at least a 29% renewable share in the fuel supply. In addition, at least 5.5% of the fuel mix must be composed of advanced biofuels and RFNBOs. Although fuel suppliers will be free to choose their preferred fuel, they must guarantee at least 1% is sourced from RFNBOs. As regards the maritime sector, Member States with maritime ports should endeavour to ensure that from 2030 the share of renewable fuels of non-biological origin in the total amount of energy supplied to the maritime transport sector is at least 1.2% (article 25).

In addition, the REPowerEU Plan announced two proposals for delegated acts regarding hydrogen. These proposals are inextricably linked to the Renewable Energy Directive, as they clarify the rules applicable to renewable hydrogen under this Directive. The proposals of delegated acts have been adopted by the Commission in February 2022. On the one hand, the Commission adopted the Delegated Regulation supplementing Directive (EU) 2018/2001 of the European Parliament and of the Council by establishing a Union methodology setting out detailed rules for the production RFNBOs. Its objective is to ensure that these fuels can only be produced from "additional" renewable electricity generated at the same time and in the same area as their own production (temporal and geographical correlation).

On the other hand, it was adopted the Delegated Regulation supplementing Directive (EU) 2018/2001 of the European Parliament and of the Council by establishing a minimum threshold for greenhouse gas emissions savings of recycled carbon fuels and by specifying a methodology for assessing greenhouse gas emissions savings from renewable liquid and gaseous transport fuels of non-biological origin and from recycled carbon fuels. It establishes the methodology for GHG savings and puts forward a detailed scheme to calculate the lifecycle emissions of renewable hydrogen as well as recycled carbon fuels to meet the GHG emission reduction threshold set in the Renewable Energy Directive.

These regulations are of the utmost importance to hydrogen development and reducing fossil gas consumption. From a climate change point of view, they should seek to establish a direct link between the production of renewable hydrogen and the production of renewable electricity, ensuring the principle of additionality. The objective of this principle consists of avoiding the production of renewable hydrogen to cannibalise renewable energy, which means leading to an overall increase in emissions by redirecting the renewable electricity from other end uses that also need renewable electricity to decarbonise (ERCST, 2022). From a consumer's point of view, if the additionality requirement is not met, the additional electricity demand in the short term will need to be covered by additional fossil fuel power production and, consequently, will increase consumers' electricity prices. Furthermore, inadequate regulation of the additionality principle also raises state aid concerns as justification relies on the premise that it will effectively reduce GHG emissions (Agora Energiewende, 2022).

The initial proposals of the Commission have triggered an intense debate around if they ensure strict regulation for the additionality principle (ClientEarth, 2022a; Bellona, 2022a). Finally, the acts adopted by the Commission include some measures that, for some environmental organisations, risk an emission increase in the short term (Bellona, 2023a). The most critical aspects refer to the entering in to force of the principle of additionality, which is delayed until 2028, allowing for all the electrolysers connected to the grid in the coming years to not fulfil the additionality principle until 2038. In addition, hydrogen producers will be able to match their hydrogen production with their contracted renewables on a monthly basis until 1 January 2030, instead of an hourly matching.

The Alternative Fuel Infrastructure Regulation (AFIR), the ReFuelEU Aviation Regulation and the FuelEU Maritime Regulation

In the last semester of 2023, three new regulations were approved: the Regulation (EU) 2023/1804 of the European Parliament and of the Council of 13 September 2023 on the deployment of alternative fuels infrastructure, and repealing Directive 2014/94/EU (AFIR), the Regulation (EU) 2023/1805 of the European Parliament and of the Council of 13 September 2023 on the use of renewable and low-carbon fuels in maritime transport, and amending Directive 2009/16/EC (FuelEU maritime regulation), and the Regulation on ensuring a level playing field for sustainable air transport (ReFuelEU Aviation). These three new regulations will impact the development of hydrogen in the transport sector and determine to what extent low-carbon or renewable hydrogen will be promoted in these sectors.

On the one hand, the REFuel Aviation regulation requires aviation fuel suppliers to ensure that all fuel provided to aircraft operators at EU airports contains a minimum share of sustainable aviation fuels. In the case of synthetic fuels, the minimum share is 1.2% in 2030, increasing to 35% in 2050. These targets can also be achieved using renewable hydrogen for aviation and low-carbon aviation fuels. However, in this case, low-carbon aviation fuel encompasses only hydrogen derived from non-fossil non-renewable sources that meets a lifecycle emissions savings threshold of 70%.

On the other hand, the FuelEU Maritime regulation incorporates a range of measures to support the adoption of sustainable RFNBOs in the maritime transport. Specifically, a 2% RFNBO subtarget is set to be implemented from 2034 if, following market monitoring, the Commission determines that the share of RFNBO in the maritime bunker fuels used by ships covered by this Regulation is below 1% by 2031. However, this subtarget does not apply to a ship demonstrating that an equivalent share of the yearly energy used on board is achieved through other fuels offering comparable GHG emissions savings and certified under Article 10 of this Regulation. This provision allows for the use of low-carbon hydrogen certified in accordance with Union legal acts governing the internal markets in renewable and natural gases, establishing a GHG emissions savings threshold and an associated methodology for calculating GHG emissions from the production of such fuels. Consequently, the key distinction between the shipping and aviation regulations is that the former permits the use of fossil-based low-carbon fuels, whereas the latter does not.

Finally, the Regulation on the deployment of alternative fuels infrastructure (AFIR) establishes compulsory deployment objectives for electric recharging and hydrogen refuelling infrastructure in the road sector, as well as for shore-side electricity supply in maritime and inland waterway ports, and stationary aircraft electricity supply. The regulation deems crucial the development of refuelling infrastructure to drive the adoption of hydrogen-powered vehicles. In this regard, it establishes mandatory targets for the deployment of public hydrogen refuelling points (article 6). Furthermore, it mandates that Member States must develop a national action framework for market development in the alternative fuels sector, specifically addressing, among other aspects, measures envisaged or adopted to promote the establishment of infrastructure for alternative fuels for captive fleets (article 14).

Proposals for a Recast Regulation and Directive on EU Gas and Hydrogen Markets

In December 2021, the Commission adopted a legislative proposal to recast the 2009 EU Gas Regulation and a proposal for a Directive on common rules for the internal markets in renewable and natural gases and in hydrogen, as part of the hydrogen and decarbonised gas markets package. In the package, the Commission considers that low-carbon hydrogen can play a role in the short and medium term to reduce emissions from existing hydrogen production and support the future uptake of renewable hydrogen. In fact, "it does not make any meaningful distinction between the treatment of low-carbon and renewable fuels or gases" (Kneebone, 2022).

As regards the differentiation about hydrogen production methods, the explanatory memorandum of the Regulation proposal admits that there is no common EU terminology and certification system for low-carbon fuels and gases. For that reason, the package includes a proposal for the establishment of terminology and certification of low-carbon hydrogen and low-carbon fuels. On the one hand, it defines low-carbon hydrogen as hydrogen derived from non-renewable sources that produces at least 70% less greenhouse gas emissions than fossil natural gas across its full lifecycle. But, despite its great importance, the situation of low-carbon hydrogen is not expected to be clarified until 2024, when a delegated act shall be approved, setting the methodology for assessing GHG emissions savings. Once this methodology is established, it will reveal whether hydrogen produced through methane reforming with subsequent carbon capture can meet the 70% threshold. On the other hand, concerning renewable hydrogen, article 2 of the Gas Directive refers to Directive (EU) 2018/2001 of the European Parliament and the Council of 11 December 2018 on the promotion of the use of energy from renewable sources.

With the intention of facilitating the integration of renewable and low-carbon gases into the existing gas network, the package establishes some enabling rules equally applicable to both types of gases. In this sense, the Regulation proposal seeks to ensure that such gases have access to the gas wholesale market, to abolish costs for cross-border tariffs, to facilitate trade and reduce injection costs, as well as to allow exemptions from general tariff rules, third-party access obligations and unbundling rules.

The gas package proposal also includes connections between hydrogen and fossil gas in the following aspects: firstly, article 4 of the Regulation proposal allows cross-subsidies between users of gas and hydrogen, although under conditions. For some authors, this would seem to be appropriate in the early development stages of the hydrogen markets with the purpose of facilitating the hydrogen infrastructure rollout (Tanase & Herrera Anchustegui, 2023: 17). However, there is an intense debate about that, as some argue that it is "is likely to result in gas or electricity users funding the development of hydrogen infrastructure they will never use". Additionally, there are concerns about the risk of financial resources being allocated to inefficient uses of hydrogen (EEB et al., 2022: 7).

Secondly, hydrogen blending is also allowed. In this sense, article 20 of the Regulation proposal states that transmission system operators (TSOs) must accept gas flows with a hydrogen content (irrespective of whether renewables or fossil-based) of up 6 to 5% by volume at interconnection points from October 2025. In fact, allowing blending is one of most controversial aspects of the proposal, as it is seen as a waste of renewable hydrogen that should be used in other more suitable applications (Anderson et al., 2022; Bellona, 2023b).

Thirdly, the proposal of Directive, in its articles 63 and 64, adopts a legal and account horizontal unbundling (but without functional unbundling). For some authors, this would risk creating an incentive for gas network operators advocate for hydrogen blending in existing gas pipelines, upgrade existing pipelines to transport hydrogen instead of gas and overbuild new infrastructure. For this reason, there have been proposals for stringent horizontal unbundling provisions to prevent conflicts of interest that could result in an oversized hydrogen infrastructure and hinder the decommissioning of fossil gas infrastructure (EEB et al., 2022; Anderson & Jahn, 2022).

Regarding hydrogen governance, the proposals contemplate the creation of a new entity, the European Network of Network Operators for Hydrogen (ENNOH), dedicated to overseeing the EU hydrogen network and facilitating the trading and supply of hydrogen across EU borders. However, there is concern that the proposed regulation lacks sufficient mechanisms to ensure ENNOH's independence from ENTSOG, as there is not a robust horizontal unbundling provisions (EEB et al., 2022).

Some authors, such as Anderson et al. (2022), have expressed strong criticism of the proposal. They perceive the exemption included in the proposed regulation as loopholes that foster the development of unnecessary gas and hydrogen infrastructure. Furthermore, they pointed out that it lacks strict obligations for decommissioning gas distribution networks, fails to address the establishment of a no-regret hydrogen infrastructure and merely replicates the existing gas infrastructure "by framing the problem as one of gas decarbonisation, instead of system decarbonisation" (Anderson et al., 2022: 18). Additionally, it does not provide decision-makers with sufficient information to consider measures that would limit the growth of the system, such as infrastructure upgrades. In fact, while the proposal restricts long-term contracts for unabated fossil gases beyond 2049, it lacks specific targets or detailed rules regarding the phase-out of fossil gas (Kneebone, 2022). Moreover, the inclusion of low-carbon hydrogen in the proposal is highly controversial, and some of the most representative environmental organisations in the EU do not support any reference to "low-carbon" hydrogen in the gas package (EEB et al., 2022).

Proposal for a Recast of the Energy Taxation Directive

The proposal for a Directive restructuring the Union framework for the taxation of energy products and electricity, proposed by de Commission in July 2021, seeks to harmonise the taxation of energy products with energy and climate policies. This involves eliminating exemptions and lowering rates that currently incentivise the use of fossil fuels. The Commission's initial proposal contemplated applying a reduced tax rate to electricity, advanced sustainable biofuels, biogas and renewable fuels of non-biological origin, including renewable hydrogen. Nevertheless, it also extended the same reduced rate to low-carbon hydrogen for a transitional period of ten years.

Regulatory Challenges for Hydrogen Development on the Path to Climate Neutrality

Hydrogen can play a vital role as a complementary energy vector in decarbonisation efforts and the phasing out of fossil fuels. However, for this to happen, the current legal framework governing hydrogen development in the EU must consider various limitations that affect its potential. Below, we identify crucial challenges for hydrogen regulation and assess how they have been addressed within the current EU hydrogen regulatory framework.

Absence of Clear Legal Definitions

One of the most critical questions in the past and current policy debate on regulating hydrogen in the EU is whether and to what extent renewable hydrogen should be promoted in combination with low-carbon hydrogen as a bridging solution. This approach would include fossil-based hydrogen with carbon capture, or even nuclear energy, as some countries, such as France, have been advocating. However, to decide which type of hydrogen should be prompted, it is necessary to properly classify and define hydrogen types.

A colour-coded system is commonly used to refer to different hydrogen production methods. For example, renewable hydrogen is known as green hydrogen, and hydrogen derived from natural gas through the process of steam methane reforming is known as blue hydrogen. However, this is not a precise classification as it does not specify how hydrogen is produced and the impacts that are caused (EEB, 2022a). Hence, currently, there is a lack of regulation on the production methods and classification of hydrogen types (European Parliament, 2021).

The current regulatory framework of the EU lacks proper definitions of hydrogen types. In the case of low-carbon hydrogen, the EU Hydrogen Strategy defines low-carbon as "fossil-based hydrogen with carbon capture and electricity-based hydrogen, with significantly reduced full life-cycle greenhouse gas emissions compared to existing hydrogen production", but does not define "significantly reduced". The gas package states that low-carbon hydrogen must have a GHG reduction threshold of 70%, but the details for this calculation will be determined in a future delegated act.

This clarification of the renewable and low-carbon hydrogen concepts is one of the most pressing issues of hydrogen regulation, needed "for business plans to move forward and for investment decisions to be made" (Marcu et al., 2022). In addition, since hydrogen is only an energy carrier and not a primary energy source, "any strategy that relies on hydrogen will first have to determine how it can be produced in a way that results in an overall reduction of GHGs" (Anderson et al., 2022: 20). Furthermore, clarifying the legal concepts "will allow for more informed choices when deciding on policy options at the EU or national level" (Commission, 2020c), considering the environmental impacts associated with hydrogen production methods.

The Sustainability Risks of Different Hydrogen Production Methods

Despite the absence of clear definitions, it is widely recognised that hydrogen production methods are associated with certain risks, which hydrogen regulation must adequately address. In any case, although renewable hydrogen has a lower carbon footprint than the fossil fuels it would replace, no hydrogen production method is free of environmental impact (Monterrubio et al., 2022).

As regards low-carbon hydrogen, it is a subpart of fossil-based hydrogen. Nonetheless, in this production method, GHG emitted as part of the hydrogen production process are captured (Commission, 2020b). Some of the concerns around this production method involves the following: it creates risks of methane leakage (Howarth & Jacobson, 2021), which is considered a far more powerful climate forcer than CO_2 (Commission, 2021a); it could produce even higher emissions than grey hydrogen, because of the increased use of fossil gas to power the carbon capture (Bauer et al., 2022); there is a lack of certainty about the effectiveness and possible rates of carbon capture methods (IRENA, 2022a; Rosenow & Lowes, 2021); there are costs and impacts associated with transporting and storing CO_2, and it requires monitoring of stored CO_2; it does not support the goals of energy security, as it uses fossil fuels, which expose the production to price fluctuations or volatilities and dependences on producer countries (IRENA, 2022b); and it increases the risks of creating stranded assets (IRENA, 2022b).

The so-called blue hydrogen is inextricably linked to fossil gas. According to IRENA (2022b: 93), "blue hydrogen is sometimes portrayed as a safe bet because it allows producer countries to monetise natural gas resources and pipelines that might otherwise become stranded". As some authors have stated, the push for blue hydrogen would be a clear example of carbon lock-in, as "blue hydrogen is primarily supported by the fossil fuel industry, which has a vested interest in maintaining its existing business model" (Rosenow & Lowes, 2021: 1527). It is also argued that the "bridge technology" or the "transition fuel" narrative is used to legitimise investments in natural gas infrastructure and natural gas usage in line with the economic interests of natural gas advocates (Kemfert et al., 2022).

On the other hand, renewable hydrogen is mainly produced by the electrolysis of water in an electrolyser powered by electricity from renewable sources. In addition, Commission also considers biogas reforming and biochemical conversion of biomass as renewable hydrogen if certain sustainability requirements are met (Commission, 2020b). Some of the concerns around this production method involve the following: electrolytic hydrogen requires large amounts of electricity to be produced, implying a massive installation of renewable production infrastructure with its associated impacts. It is expected that a quarter of the renewable electricity will be used for renewable hydrogen production by 2050 (Commission, 2020b). To fulfil hydrogen demand of electricity, Europe is expected to deploy 7 GW of dedicated renewable capacity for hydrogen production during 2022–2027, with Spain in the lead position, accounting for half of Europe's growth (IEA, 2022). In that regard, the development of renewable hydrogen largely depends on the effective increase in renewable energy and the possibility of allocating a large amount of it to hydrogen production (Ferrari, 2022). Therefore, the development of hydrogen is largely unpredictable at present. Moreover, the need to increase the amount of renewable energy available leads to regulatory pressure to speed up authorisation procedures, including streamlining the environmental assessment of plans and projects that could significantly affect Natura 2000 sites. In that regard, it was adopted the Council Regulation (EU) 2022/2577 of 22 December 2022 laying down a framework to accelerate the deployment of renewable energy. For some environmental organisations, this proposal represents a "backdoor" amendment of environmental legislation (ClientEarth, 2022b). As regards the rapid increase in the generation of renewable energy that is expected to occur in the coming years, numerous environmental associations and members of civil society point out the need for renewable development not to result in impacts on biodiversity, and claim the importance to take into account the fair distribution of the benefits and burdens derived from the different territorial conditions, the need to reduce the demand for energy and the limitations represented by the scarcity of raw materials (EEB, 2022b).

Another challenge related to renewable hydrogen is that despite renewable hydrogen being primarily produced using electricity, its production method could still be associated with fossil fuels. Consequently, electrolytic hydrogen is only as environmentally friendly as the electricity used in its production (Lovisolo et al., 2021). That means that even when hydrogen is derived from electricity generated by renewable sources, it does not guarantee sustainability in every case. Without regulations ensuring that the electricity for hydrogen production originates from new or additional renewable energy generation, there is a risk of increased reliance on fossil fuels (Pototschnig, 2021). Therefore, as cautioned by IRENA, "failure to adhere to the principle of additionality could also impede progress towards expanding access to energy for those who lack it today if countries prioritize the deployment of renewables for green hydrogen export" (2022c: 105).

Finally, there are shared challenges between fossil-based hydrogen and renewable hydrogen, such as the scarcity of the raw materials essential for implementing these technologies (IEA, 2021). Furthermore, both green and blue hydrogen require substantial amounts of pure water for production, prompting the recognition that "a growing number of countries may need to consider whether hydrogen production is suitable in the longer term" (IRENA, 2022b: 97). Given this, and in light of increasing water stress, the selection of water sources for hydrogen production should be explicitly addressed in strategies and further detailed in project planning, especially concerning water consumption in connection with the energy and food sector (IRENA, 2020; Monterrubio et al., 2022). Lastly, concerns exist regarding the impact of hydrogen leakage on climate change (Arrigoni & Bravo Díaz, 2022). Hydrogen leaks influence the lifespan of other greenhouse gases, including methane, ozone and water vapour. Consequently, they alter atmospheric chemistry and contribute to global warming.

The current and proposed regulation of the EU regarding hydrogen allows for the use of fossil gas in hydrogen production and considers low-carbon hydrogen as a production pathway that could help to decarbonise the energy system. This is the case of the taxonomy delegated acts, the proposal for a recast Regulation and Directive on EU gas and hydrogen markets, the proposal for a recast of the Energy Taxation Directive and the FuelEU maritime regulation. As regards the impacts associated with renewable hydrogen, the delegated acts on renewable hydrogen are of utmost importance to guarantee that hydrogen do hinder electrification. However, the applicability of the principle of additionality is delayed until 2028, allowing for all the electrolysers connected to the grid in the coming years to not fulfil the additionality principle until 2038.

A Non-regret Hydrogen Strategy

Hydrogen can be potentially used for the decarbonisation of different sectors and applications: it can be used as a feedstock, fuel or energy carrier in industrial and chemical processes; in air, maritime and heavy-duty road transport; in heating applications as well as in energy storage. However, it is widely recognised that it should only be employed on those specific applications for which its added value is clear and proven. For that reason, hydrogen should be avoided in applications such as building-level heating (Rosenow, 2022), low-temperature heat in industry, and cars and light-duty vehicles (Agora Energiewende and Industry, 2021). In particular, blending faces multiple challenges, as it leads to limited CO_2 benefits, increases energy costs and risks wasting the limited amounts of hydrogen to sectors which could have more effective and efficient solutions—that is direct electrification—(IRENA, 2022c; Bard et al., 2022). Regarding hydrogen refuelling stations for personal mobility, there is high controversy about personal mobility as an acceptable use for hydrogen, considering its low efficiency (Agora Energiewende and Industry, 2021).

In addition, there are still uncertainties about the future demand for hydrogen and its technological maturity to be deployed at a large scale. Hydrogen, as a decarbonisation option, faces the risk of lock-in if investments are undertaken despite the significant uncertainties (Kranzl et al., 2022). On the one hand, prioritising high-demand applications or non-regret options, for which hydrogen is clearly the best alternative, is more likely to be cost-effective and less susceptible to the risks associated with nascent markets (IRENA, 2022c). On the other hand, the electricity generated from renewable sources for productive uses should be prudently assessed before diverting to produce green hydrogen (IRENA, 2020). As IRENA (2022d: 109) warns, despite hydrogen's great potential, it must be kept in mind that its production, transport and conversion require energy, as well as a significant investment. Therefore, "indiscriminate use of hydrogen could therefore slow down the energy transition".

If the demand of hydrogen is not correctly estimated and other solutions are not considered to cover the end-uses, such as, for example, direct electrification when feasible, the resulting infrastructure network would be expensive and could lock in hydrogen use in applications where it is not the most beneficial option. Thus, regulators must clearly prioritise uses, correctly assess the need for new or repurposed infrastructure and consider that hydrogen infrastructure will differ from fossil, as their employment should not be in the same sectors. Furthermore, infrastructure planning and governance of hydrogen should establish mechanisms to avoid pressure and lack of information provoking the network to expand unnecessarily, replicating the fossil model. Proper prioritisation of hydrogen end-uses and infrastructure planning is essential as the effects of policies and regulations we adopt today will have an impact for decades. Thus, a precautionary approach based on the principle of energy efficiency first (Commission, 2021a, 2021b) should be applied to avoid lock-in effects, as well as to lead to higher emissions. As Anderson and Jahn (2022: 8) point out:

> Adopting the structure of fossil gas network development, without questioning whether those structures are useful, would only be another mistake with long-lasting consequences.

Regarding the recently approved text of the Renewable Energy Directive (RED III), it has established a binding target aimed at decarbonising current uses of hydrogen in the industrial sector. However, Bellona (2023b) notes that whilst these regulations create a hydrogen market with a focus on priority sectors, the steel industry falls outside the scope of this mandate.

Turning to the transport sector, the Alternative Fuel Infrastructure Regulation acknowledges the nascent stage of market implementation in this domain and the existing uncertainty about the types of vehicles and prevalent technologies. Nevertheless, it included specific targets for the development of a hydrogen refuelling infrastructure. Some entities have contended that the regulation should have restricted the deployment of those stations to strategically locations like ports and industrial clusters. In addition, they suggested maintaining indicative targets, at least until the regulation undergoes revision in 2026 (Transport and Environment, 2021; Bellona, 2022c).

Moreover, fossil gas and its associated infrastructures will play a role in the envisaged phases of large-scale hydrogen development, not only by using low-carbon hydrogen but also by retrofitting fossil infrastructure or even maintaining it for a transitional period. In this sense, the taxonomy delegated acts consider blending as a sustainable activity and the Regulation on guidelines for trans-European energy infrastructure includes an exemption clause allowing the transport or storage of a blend of hydrogen with fossil gas until 2029. Furthermore, taxonomy delegated acts adopted an "hydrogen-ready" approach, which determines that investments in fossil infrastructures and facilities are sustainable if designed and constructed to use renewable or low-carbon hydrogen by 2035. However, short-term solutions in fossil gas assets risk not fitting in a decarbonised future. Decisions made with technological uncertainties and a lack of confidence in their future development present risks. In the case of LNG infrastructures, although the Commission believes that they should start by being hydrogen-ready—as the terminals that Germany has already announced—the question remains if they will end up as stranded assets, given the technical difficulties involved in guaranteeing their future use for hydrogen (Heidecke et al., 2022; Kemfert et al., 2022; Riemer et al., 2022).

Hydrogen's Potential to Decarbonise Hard-to-Abate Industries and the Need for a Strategic Step-Change Approach

For the decarbonisation of fossil fuel or fossil-dependent industries, where grey hydrogen is already used as an energy vector or as a feedstock, renewable hydrogen could improve their carbon footprint. That is the case of hydrogen currently used as a feedstock for the petrochemical industry, which includes crude oil refining, ammonia synthesis (primarily for fertiliser production) and methanol production for a wide variety of products, including plastics (IRENA, 2022c). In addition, hydrogen is suitable for decarbonising the most energy-intensive industries and transport, such as iron, steel and cement manufacturing, as well as shipping and aviation (EEB, 2022a). Since the iron, steel and chemical sectors represent about 8% and 5% of the global energy and process-related greenhouse emissions, achieving their reduction is a positive outcome (IRENA, 2022a).

However, even for those hard-to-abate industrial sectors where hydrogen is suitable, regulation must consider that fuel or feedstock switching is only one part of a bigger picture. A broader decarbonisation portfolio for industry should include, among other options: energy efficiency, circular economy, electrification, heat use and waste reduction (IRENA, 2022a). In particular, as the best energy is the one not produced because there is no need to use it, reducing demand should also be preferred to making energy from climate-neutral sources (Commission, 2021a, 2021b).

Furthermore, when considering which industrial sectors policymakers should prioritise for investments and how the public sector should allocate funds to boost the production and demand of renewable hydrogen, it is crucial to also consider whether a gradual or a step-changing approach is necessary. As stated by IRENA (2022a: 28), "the goal of reaching zero emissions requires a very different mindset compared to an objective of gradually reducing emissions". An approach based on setting gradual policies entails risks of lock-in emissions, as not all the hard-to-abate industries may be proactive in decarbonising their processes and could decide to continue investing in fossil technologies. For that reason, a step approach with bans and mandates to leave fossil fuels is appropriate measures to avoid the construction of new industrial assets highly dependent on fossil fuels. These assets can have decades-long lifetimes, even exceeding the 2050 deadline (IRENA, 2022a).

Conclusions

Fossil gas has no place in a decarbonised future. Because efficiency losses during the conversion of hydrogen involve a massive growth in the energy required, the role of fossil gas in today's energy system cannot be replaced by hydrogen. Consequently, planning and hydrogen regulation must provide for the decline of fossil gas without trying to replicate the role of fossil gas in the current energy system.

The EU is at a crucial point concerning hydrogen regulation. Decisions taken today will determine the future of the energy system and its ability to cope with the climate crisis. In shaping the hydrogen legal framework for the coming decades, discussions are ongoing on how to design it with due regard to the decreasing role of fossil gases in the energy system, to establish the most efficient uses of hydrogen and to avoid reproducing the current gas energy infrastructure and system, which are highly centralised and dependent on fossil fuels imports.

The EU Hydrogen Strategy acknowledges low-carbon hydrogen as a transitional or bridging solution, influencing regulations and proposed measures related to hydrogen development. Despite fossil gas consumption should significantly decrease in the forthcoming decades, the currently adopted and proposed regulations in the EU allow for the utilisation of fossil gas in hydrogen production and treat it equivalently to renewable hydrogen, consider sustainable the blending of hydrogen with fossil gas and permit the fostering of investments in fossil infrastructure during a transitional period. As a result, the EU's regulatory approach to hydrogen development is firmly anchored in the perception of fossil gas as a bridging technology, adopting an approach characterised by incremental policies that heighten the risk of creating stranded assets and fossil lock-in.

However, the pending regulatory decisions and its implementation will determine how fast the system is decarbonised, for how long the dependence on fossil gases is prolonged, the scale of deployment of renewable energy production facilities and whether there is a risk of creating stranded assets. Notably, the challenges in shaping a hydrogen regulation that is resilient for the future include clearly defining the role and establishing stringent sustainability criteria for low-carbon hydrogen, deciding on the maintenance or decommissioning of fossil gas infrastructure, and managing the influence and power of the fossil gas governance system.

To minimise the risks of fossil lock-in and stranded assets, the regulatory approach to hydrogen development should ensure that hydrogen use and the allocation of public funds are focused on no-regret hydrogen applications. To this end, all available options to reduce energy demand under the principle of energy efficiency first need to be considered to determine which are the most beneficial targeted uses. In addition, the production of renewable hydrogen should strictly be linked to additional renewable generation capacity. Otherwise, overinvestment and misallocation of public funds in hydrogen projects could undermine the effectiveness of the energy transition and hinder the achievement of climate goals.

Acknowledgments This chapter was financed by the Secretaria d'Universitats i Recerca del Departament d'Empresa i Coneixement de la Generalitat de Catalunya (Pla de Doctorats Industrials) and has been carried out within the research group of the Rovira i Virgili University "Territory, Citizenship and Sustainability", recognised as a consolidated research group by the Generalitat de Catalunya and which has the support of the Department of Research and Universitat of the Generalitat de Catalunya (2021 SGR 00162).

References

Agora Energiewende. (2022). *Feedback Submission on the RED Delegated Acts for the Production of RFFNBOs and the Methodology for GHG Emissions Accounting for RFNBOs*. Retrieved December 9, 2022, from https://ec.europa.eu/info/law/better-regulation/have-your-say/initiatives/12713-Renewable-energy-method-for-assessing-greenhouse-gas-emission-savings-for-certain-fuels/F3315048_en

Agora Energiewende and Industry. (2021). *12 Insights on Hydrogen*. Retrieved December 9, 2022, from https://static.agora-energiewende.de/fileadmin/Projekte/2021/2021_11_H2_Insights/A-EW_245_H2_Insights_WEB.pdf

Anderson, M., & Jahn, A. (2022). *Getting the Hydrogen Network We Need for Decarbonisation*. Retrieved December 9, 2022, from https://www.raponline.org/knowledge-center/getting-the-hydrogen-network-we-need-for-decarbonisation/

Anderson, M., et al. (2022). *The Clash with Gas: Should it Stay or Should it go? Principles to Address the Changing Role of Gas in a Decarbonised Energy System*. Retrieved December 9, 2022, from https://www.raponline.org/wp-content/uploads/2023/09/rap-anderson-etal-gas-transition-2022-aug.pdf

Arrigoni, A., & Bravo Díaz, L. (2022). *Hydrogen Emissions from a Hydrogen Economy and Their Potential Global Warming Impact*. Publications Office of the European Union. Retrieved December 9, 2022, from https://publications.jrc.ec.europa.eu/repository/handle/JRC130362

Bard, J., et al. (2022). *The Limitations of Hydrogen Blending in the European Gas Grid*. Retrieved December 9, 2022, from https://www.iee.fraunhofer.de/content/dam/iee/energiesystemtechnik/en/documents/Studies-Reports/FINAL_FraunhoferIEE_ShortStudy_H2_Blending_EU_CommissionF_Jan22.pdf

Bauer, C., et al. (2022). On the climate Impacts of Blue Hydrogen Production. *Sustainable Energy & Fuels, 2022*, 6. Retrieved December 9, 2022, from https://pubs.rsc.org/en/content/articlelanding/2022/se/d1se01508g

Bellona. (2022a). *Explainer on the Delegated Acts on RFNBOs and GHG Emissions Reductions*. Retrieved December 9, 2022, from https://network.bellona.org/content/uploads/sites/3/2022/06/Explainer-on-the-Delegated-Acts-on-RFNBOs-and-GHG-emissions-reductions-2022.pdf

Bellona. (2022b). *Leaked Taxonomy Proposal: Fossil Gas "Sustainable" Label Relies on Promises in Bad Faith—Still Risks Wasting All Our Renewable Energy*. Retrieved December 9, 2022, from https://bellona.org/news/renewable-energy/2022-01-leaked-taxonomy-proposal-fossil-gas-sustainable-label-relies-on-promises-in-bad-faith-still-risks-wasting-all-our-renewable-energy

Bellona. (2022c). *Bellona's Main Asks for AFIR*. Retrieved December 9, 2022, from https://bellona.org/news/energy-systems/2022-03-bellonas-main-asks-for-afir

Bellona. (2023a). *Renewable Hydrogen Rules Delay Clear Link Between Hydrogen and Renewable Electricity until after 2030*. Press Release: Renewable hydrogen rules delay clear link between hydrogen and renewable electricity until after 2030—Bellona.org. Retrieved 21 February 2023.

Bellona. (2023b). *Europe's Hydrogen Journey: It's Time to Move from Hype to Prudent Planning*. Retrieved November 15, 2023, from https://bellona.org/news/climate-change/2023-11-europes-hydrogen-journey-its-time-to-move-from-hype-to-prudent-planning

ClientEarth. (2022a). *RePower European Union Proposal for a REDII Amendment*. Retrieved December 9, 2022, from https://www.clientearth.org/media/ite-jqaby/clientearth_repowereu-permitting-amendments-briefing_july-2022.pdf

ClientEarth. (2022b). *Delegated Acts on Hydrogen Methodologies to Assess GHG Emissions Savings and for Production of RFNBOs*. Retrieved December 9, 2022, from https://www.clientearth.org/media/vr4pmmjp/clientearth-comments-on-proposed-delegated-acts-relating-to-rfnbos-220617.pdf

Commission. (2018). *Communication "A Clean Planet for All A European Strategic Long-term Vision for a Prosperous, Modern, Competitive and Climate Neutral Economy", COM/2018/773 final*. Retrieved from https://eur-lex.europa.eu/legal-content/EN/TXT/?uri=CELEX%3A52018DC0773

Commission. (2019). *Communication "The European Green Deal", COM/2019/640 Final*. Retrieved December 9, 2022, from https://eur-lex.europa.eu/legal-content/ES/TXT/?uri=CELEX%3A52019DC0640

Commission. (2020a). *Staff Working Document Impact Assessment Accompanying the Communication "Stepping Up Europe's 2030 Climate Ambition", SWD(2020) 176 Final*. Retrieved December 9, 2022, from https://eur-lex.europa.eu/legal-content/EN/TXT/?uri=CELEX:52020SC0176

Commission. (2020b). *Communication "A Hydrogen Strategy for a Climate-Neutral Europe", COM/2020/301 Final*. Retrieved December 9, 2022, from https://eur-lex.europa.eu/legal-content/EN/TXT/?uri=CELEX%3A52020DC0301

Commission. (2020c). *Communication "An EU Strategy for Energy System Integration", COM/2020/299 Final*. Retrieved December 9, 2022, from https://eur-lex.europa.eu/legal-content/EN/ALL/?uri=COM:2020:299:FIN

Commission. (2021a). *Communication "Fit for 55": Delivering the EU's 2030 Climate Target on the Way to Climate Neutrality, COM(2021) 550 Final*. Retrieved December 9, 2022, from https://eur-lex.europa.eu/legal-content/EN/TXT/?uri=CELEX%3A52021DC0550

Commission. (2021b). *Recommendation on Energy Efficiency First: From Principles to Practice*. Retrieved December 9, 2022, from https://eur-lex.europa.eu/legal-content/EN/TXT/?uri=CELEX:32021H1749

Commission. (2022a). *Communication "REPowerEU Plan", COM(2022) 230 Final*. Retrieved December 9, 2022, from https://eur-lex.europa.eu/legal-content/ES/TXT/HTML/?uri=CELEX:52022DC0230&from=EN

Commission. (2022b). *Commission Staff Working Document. Implementing the Repower EU Action Plan: Investment Needs, Hydrogen Accelerator and Achieving the Bio-methane Targets, SWD(2022) 230 Final*. Retrieved December 9, 2022, from https://eur-lex.europa.eu/legal-content/EN/TXT/PDF/?uri=CELEX:52022SC0230&from=EN

Conti, & Kneebone, J. (2022). *A First Look at REPowerEU: The Commission's Plan for Energy Independence from Russia. Florence School of Regulation*. Retrieved December 9, 2022, from Https://Fsr.Eui.Eu/First-Look-at-Repowereu-Eu-Commission-Plan-for-Energy-Independence-from-Russia

Couffon, E. (2021). *Is the TEN-E Regulation Fit for a Decarbonized Future? A Battle to Shape the European Energy Transition, IFRI*. Retrieved December 9, 2022, from https://www.ifri.org/sites/default/files/atoms/files/couffon_ten-e_regulation_2021.pdf

ECOS. (2022). *EU Climate Taxonomy: The Good, the Bad, and the Ugly*. Retrieved December 9, 2022, from https://ecostandard.org/news_events/good-bad-ugly-taxonomy/

EEB, et al. (2022). *NGO and Stakeholder Briefing on the EU Gas Package*. Retrieved December 9, 2022, from https://www.rescoop.eu/toolbox/ngo-and-stakeholder-briefing-on-the-eu-gas-package

EEB. (2022a). *Face to Face with Hydrogen: The Reality Behind the Hype*. Retrieved December 9, 2022, from https://eeb.org/wp-content/uploads/2021/05/EEB-Hydrogen-Position-Paper-FINAL.pdf

EEB. (2022b). *EEB Policy Brief: Policy Measures Towards Nature-Positive Renewable Energy in the EU using PAC Scenario Results*. Retrieved December 9, 2022, from https://eeb.org/wp-content/uploads/2022/07/Policy-Brief_Nature-positive-renewable-energy.pdf

EEB. (2023). *Policy Brief: A Sustainable Hydrogen Strategy for the EU*. Retrieved February 28, 2023, from https://eeb.org/library/policy-brief-a-sustainable-hydrogen-strategy-for-the-eu/

ERCST. (2022). *Feedback to the Commission RED Delegated Acts on RFNBOs (Renewable Fuels of Non-Biological Origin) and RCFs (Recycled Carbon Fuels)*. Retrieved December 9, 2022, from https://ercst.org/ercst-feedback-to-red-das/

European Parliament. (2021). *Resolution of 19 May 2021 on a European Strategy for Hydrogen*. Retrieved December 9, 2022, from https://www.europarl.europa.eu/doceo/document/TA-9-2021-0241_EN.html

European Parliament. (2022). *Revision of the TEN-E Regulation EU Guidelines for New Energy Infrastructure*. Retrieved December 9, 2022, from https://www.europarl.europa.eu/RegData/etudes/BRIE/2021/689343/EPRS_BRI(2021)689343_EN.pdf

Ferrari, G. F. (2022). La regulación del hidrógeno y de los gases renovables en la Unión Europea, en Alenza García, J.F. Mellado Ruiz, L. (coords.), Estudios sobre cambio climático y transición energética, Marcial Pons (pp. 223–243).

Heidecke, L., et al. (2022). *The Revision of the Third Energy Package for Gas, Publication for the Committee on Industry, Research and Energy (ITRE), European Parliament.* Retrieved December 9, 2022, from https://www.europarl.europa.eu/RegData/etudes/STUD/2022/734009/IPOL_STU(2022)734009_EN.pdf

Howarth, R. W., & Jacobson, M. Z. (2021). How Green is Blue Hydrogen? *Energy Science and Engineering, 9*(10), 1676–1687. Retrieved December 9, 2022, from https://onlinelibrary.wiley.com/doi/full/10.1002/ese3.956

IEA. (2021). *The Role of Critical Minerals in Clean Energy Transitions.* Retrieved December 9, 2022, from https://www.iea.org/reports/the-role-of-critical-minerals-in-clean-energy-transitions

IEA. (2022). *Renewables 2022. Analysis and Forecast to 2027.* Retrieved December 9, 2022, from https://iea.blob.core.windows.net/assets/64c27e00-c6cb-48f1-a8f0-082054e3ece6/Renewables2022.pdf

IRENA. (2020). *Green Hydrogen Cost Reduction: Scaling Up Electrolysers to Meet the 1.5°c Climate Goal.* Retrieved December 9, 2022, from https://www.irena.org/-/media/Files/IRENA/Agency/Publication/2020/Dec/IRENA_Green_hydrogen_cost_2020.pdf

IRENA. (2022a). *Green Hydrogen for Industry: A Guide to Policy Making.* Retrieved December 9, 2022, from https://www.irena.org/publications/2022/Mar/Green-Hydrogen-for-Industry

IRENA. (2022b). *Global Hydrogen Trade to Meet the 1.5°c Climate Goal. Part II: Technology Review of Hydrogen Carriers.* Retrieved December 9, 2022, from https://www.irena.org/publications/2022/Apr/Global-hydrogen-trade-Part-II

IRENA. (2022c). *Geopolitics of the Energy Transformation: The Hydrogen Factor.* Retrieved December 9, 2022, from https://www.irena.org/-/media/Files/IRENA/Agency/Publication/2022/Jan/IRENA_Geopolitics_Hydrogen_2022.pdf?rev=1cfe49eee979409686f101ce24ffd71a

IRENA. (2022d). *Accelerating Hydrogen Deployment in the G7: Recommendations for the Hydrogen Action Pact.* Retrieved December 9, 2022, from https://mc-cd8320d4-36a1-40ac-83cc-3389-cdn-endpoint.azureedge.net/-/media/Files/IRENA/Agency/Publication/2022/Nov/IRENA_Hydrogen_deployment_G7_2022.pdf?rev=ef39150b82754588b2f31bd5afc74c01

Kemfert, C., et al. (2022). The Expansion of Natural Gas Infrastructure Puts Energy Transitions at Risk. *Nature Energy, 7*, 582–587. Retrieved December 9, 2022, from https://www.nature.com/articles/s41560-022-01060-3

Kneebone, J. (2022). *A First Look at the EU Hydrogen and Decarbonised Gas Market Package.* Retrieved December 9, 2022, from https://fsr.eui.eu/a-first-look-at-the-eu-hydrogen-and-decarbonised-gas-markets-package/

Kranzl, L., et al. (2022). *Renewable space heating under the revised Renewable Energy Directive*. Publications Office of the European Union. Retrieved December 9, 2022, from https://data.europa.eu/doi/10.2833/525486

Lavrijssen, S., & Vitez, B. (2020). Make Hydrogen Whilst the Sun Shines: How to Turn the Current Momentum into a Well-Functioning Hydrogen Market? *Carbon and Climate Law Review, 14*(4), 266–280. Retrieved December 9, 2022, from https://cclr.lexxion.eu/article/CCLR/2020/4/6

Lovisolo, M., et al. (2021). *Electrolysis Hydrogen Production in Europe*. Bellona. Retrieved December 9, 2022, from https://bellona.org/publication/electrolysis-hydrogen-production-in-europe

Marcu, A., et al. (2022). *Taking Stock of the EU Hydrogen Regulatory Framework, ERCST*. Retrieved December 9, 2022, from https://ercst.org/taking-stock-of-the-eu-hydrogen-regulatory-framework/

Monterrubio, A., et al. (2022). *Hidrógeno verde como combustible, Oficina de Ciencia y Tecnología del Congreso de los Diputados*. Retrieved December 9, 2022, from https://www.oficinac.es/informes-c/hidrogeno-verde

Müller, et al. (2022). *Hydrogen justice. Environmental Research Letters, Volume 17*. Retrieved March 2, 2023, from https://iopscience.iop.org/article/10.1088/1748-9326/ac991a/meta

Pototschnig, A. (2021). *Renewable Hydrogen and the "Additionality" Requirement: Why Making it More Complex than is Needed?*. Retrieved December 9, 2022, from https://fsr.eui.eu/publications/?handle=1814/72459

Riemer, M., et al. (2022). *Conversion of LNG Terminals for Liquid Hydrogen or Ammonia. Analysis of Technical Feasibility and Economic Considerations*. Fraunhofer Institute for Systems and Innovation Research ISI. Retrieved December 9, 2022, from https://www.isi.fraunhofer.de/content/dam/isi/dokumente/cce/2022/Report_Conversion_of_LNG_Terminals_for_Liquid_Hydrogen_or_Ammonia.pdf

Rosenow, J. (2022). Is Heating Homes with Hydrogen All but a Pipe Dream? An Evidence Review. *Joule, 6*(10), 2225–2228. https://doi.org/10.1016/j.joule.2022.08.015

Rosenow, J., & Lowes, R. (2021). Will blue hydrogen lock us into fossil fuels forever? *One Earth, 4*(11), 1527–1529. Retrieved December 9, 2022, from https://www.sciencedirect.com/science/article/pii/S2590332221006047

Tanase, L., & Herrera Anchustegui, I. (2023). EU Hydrogen and the Decarbonized Gas Market Package: Unbundling, Third-Party Access, Tariffs and Discounts Rules at the Core of Transport of Hydrogen. Íñigo del Guayo Castiella, Lorenzo Mellado Ruiz (Directores), "Retos Regulatorios de los Gases Renovables en la Economía Circular" (2023). https://doi.org/10.2139/ssrn.4431113.

Transport and Environment. (2021). *AFIR: Providing Infrastructure to Make Transport Fit for 55*. Retrieved November 14, 2023, from https://www.transportenvironment.org/wp-content/uploads/2021/11/20211004_AFIR_Briefing.pdf

Yafimava, K. (2022). *The TEN-E Regulation: Allowing a Role for Decarbonised Gas*. Oxford Institute for Energy Studies. Retrieved December 9, 2022, from https://www.oxfordenergy.org/wpcms/wp-content/uploads/2022/08/The-TEN-E-Regulation-allowing-a-role-for-decarbonised-gas-NG174.pdf

CHAPTER 10

The Legal Protection of the Territory of Catalonia Against Specific Renewable Energy Projects

Ivan Hortigüela Bolsa

Introduction

The sources of energy used over the last century are one of the main causes of the current climate crisis. In 1965, the report 'Restoring the quality of our environment' warned that the extraction and consumption of fossil fuels were leading to the release into the atmosphere of carbon dioxide that had been trapped in the natural environment for millions of years (Environmental Pollution Panel of the President's Science Advisory Committee, 1965).

Concern about greenhouse gas emissions, global warming, and climate change has grown exponentially over time (Jackson, 2007). Burning fossil fuels is one of the main causes of climate change. These

I. Hortigüela Bolsa (✉)
INSTA, La Canonja, Spain

Rovira i Virgili University, Tarragona, Spain
e-mail: ihortiguela@instajuridic.com

D. Iglesias Márquez et al. (eds.), *Legal Challenges at the End of the Fossil Fuel Era*, Global Issues,
https://doi.org/10.1007/978-3-031-61766-9_10

fuels account for 80% of global primary energy demand, and the energy system is the source of approximately two-thirds of global CO_2 emissions (Foster & Elzinga, 2015).

To mitigate the worst impacts of climate change, these emissions must be reduced. In October 2014, the European Council agreed on the European Union's 2030 climate and energy framework (General Secretariat of the Council, 2014), looking ahead to the 21st Conference of the Parties to the United Nations Framework Convention on Climate Change. The targets set with the 2030 horizon in mind were to reduce greenhouse gas emissions by 40% compared with 1990 levels and for renewable energy to make up 27% of all energy consumption.

The Directive (EU) 2018/2001 of the European Parliament and of the Council of 11 December 2018 on the promotion of the use of energy from renewable sources would increase later to 32% the target for renewable energy consumption. To achieve these targets, dependence on fossil fuels must be reduced, and one of the tools for this is the use of renewable energies such as solar and wind. According to the Directive's fourth *Whereas*, the use of renewable energy in the electricity sector, the heating and cooling sector, and the transport sector are effective tools, together with energy efficiency measures, for reducing greenhouse gas emissions in the Union and the Union's energy dependence.

Article 3.2 of the Directive (EU) 2018/2001 establishes that Member States shall set national contributions to meet, collectively, the binding overall target on the share of energy from renewable sources in the Union's gross final consumption of energy in 2030 is at least 32%. These contributions shall be included in their integrated national energy and climate plans in accordance with Articles 3–5 and 9–14 of Regulation (EU) 2018/1999 of the European Parliament and of the Council of 11 December 2018 on the Governance of the Energy Union and Climate Action, amending Regulations (EC) No 663/2009 and (EC) No 715/2009 of the European Parliament and of the Council, Directives 94/22/EC, 98/70/EC, 2009/31/EC, 2009/73/EC, 2010/31/EU, 2012/27/EU, and 2013/30/EU of the European Parliament and of the Council, Council Directives 2009/119/EC and (EU) 2015/652, and repealing Regulation (EU) No 525/2013 of the European Parliament and of the Council.

In Spain, the Integrated National Energy and Climate Plan 2021–2030 (*Plan Nacional Integrado de Energía y Clima 2021–2030*) sets out the path for the country to achieve 100% renewable electricity generation by

2050 (MITECOGOB, 2020). Achieving this objective will require major efforts to deploy renewable energies within the timeframe of the Integrated National Energy and Climate Plan: by 2030, Spain must have reduced greenhouse gas emissions by one-third, and electricity from renewable energy should account for 42% of the final energy use and 74% of all electricity generated.

Spain's Council of Ministers declared a climate emergency in January 2020. The Declaration on the Climate and Environmental Emergency in Spain (La Moncloa, 2020) sets out 30 priority lines of action for combating climate change. These lines of action are largely aimed at decarbonisation and linked to generating energy from renewable sources (Gobierno de España, 2020).

The Spanish competencies in energy are shared between the state and the autonomous communities that conform it. The Spanish Constitution (*Constitución Española*, 2022) provides in its Article 149.1.25° that it is the responsibility of the State to develop the foundations of the energy system. Likewise, the exclusive competence for the authorisation of electrical installations falls under the responsibility of the State when their use affects more than one autonomous community or the transportation of energy extends beyond the territorial scope of the community where the installation is located (Article 149.1.22°). Taking this into account, the authorisation of electricity generation projects that do not exceed the territorial limits of an autonomous community may correspond to it.

In the case of Catalonia (an autonomous community located in the northwest of the country), the Statute of Autonomy, reformed by Organic Law 6/2006 of 19 July (*Ley Orgánica 6/2006, de 19 de julio, de reforma del Estatuto de Autonomía de Catalunya*), provides in its Article 133.1 that the Generalitat (institutional system in which the self-government of Catalonia is politically organised) has shared competence in energy matters. This competence includes the regulation of activities related to the production, storage, and transportation of energy, the authorisation for facilities that are entirely located within the territory of Catalonia, and the inspection and control of all existing facilities in its territorial limits, as well as the promotion and management of renewable energies and energy efficiency, among other aspects. By virtue of the shared competence on energy, several Catalan policies and laws have been adopted to mitigate the adverse effects of climate change and promote the development of renewable energies. Prime examples of these are Law 16/2017, of 1 August, on Climate Change (*Ley 16/2017, de 1 de agosto, del cambio climático*), and

Decree-Law 16/2019, of 26 November, on Urgent Measures for the Climate Emergency and to Encourage Renewable Energies (*Decreto Ley 16/2019, de 26 de noviembre, de medidas urgentes para la emergencia climática y el impulso a las energías renovables*).

This chapter will briefly examine the social context for the implementation of renewable energies in Catalonia, their legal framework in this autonomous community, and a case study consisting of the deployment of three wind farms in land unsuitable for urban development. To preserve professional secrecy, when the case study will not include names of the companies and the business group linked to it nor exact locations where the wind farms were supposed to be located.

Social Context for the Deployment of Renewable Energies in Catalonia

Society has greeted the implementation of renewable energies in Catalonia with varying degrees of acceptance. Developer companies approved the facilities to develop renewable energy projects. Nevertheless, some citizens and environmental NGOs have been reluctant to accept specific projects and their massive way of deployment due to environmental and social impacts.

The Catalan Decree-Law 16/2019 establishes a streamlined administrative process for renewable energy projects of up to 50 megawatts (MW) within the autonomous community of Catalonia. Since Decree-Law 16/2019 entered into force, developers have submitted plans for hundreds of projects involving solar photovoltaic plants and wind farms. According to data from the Catalonia Renewable Energy Observatory (*Observatori Català de les Energies Renovables*) (2020), 167 applications for wind farms and 467 applications for solar photovoltaic plants were submitted to the Renewable Energies Committee.

Most of the wind turbines already installed in Catalonia are located on land classified as unsuitable for development in the provinces of Tarragona and Lleida. Based on district (*comarca*)-level calculations of official data published on the website of the Government of Catalonia, approximately 64% of the wind turbines in Catalonia are in the province of Tarragona and 22% in the province of Lleida (Generalitat de Catalunya, Medi Ambient i Sostenibilitat, 2022). If we look at wind energy generated across the whole autonomous community, 60.5% is generated in the province of Tarragona and 25.1% in the province of Lleida (Observatori de les Energies Renovables, 2022).

Most of the solar photovoltaic plants are similarly located on land unsuitable for development in the province of Lleida. This can be inferred from the Environmental Viewer for Renewable Energies in Catalonia (*Visor Ambiental de las Energías Renovables de Cataluña*) web app (Generalitat de Catalunya, 2022). If we look at photovoltaic solar energy generated across the whole autonomous community, 40.7% is generated in the province of Lleida, and 33.8% in the province of Barcelona (Observatori de les Energies Renovables, 2022).

The Decree-law 16/2019 has been the subject of harsh criticism for preventing public participation in the process and for not providing a planification for the deployment of renewable energy projects, among other issues (De la Varga Pastor, 2022). The concentration of renewable energy projects in specific areas, the potential social and environmental impacts, and the fact that the energy produced is not necessarily consumed in the vicinity of the affected area have generated strong opposition from certain sectors of the community. The fact that energy generation facilities are located far away from the consumption areas makes this renewable energy deployment model inefficient, with potential energy losses of over 15% (Gil Casión, 2022). Even local and supra-local authorities have displayed opposition to particular renewable energy projects (Hidalgo & Aymí, 2021).

From a demographic perspective, 77% of municipalities (30 out of 39 municipalities) that have wind farms in Catalonia have less than 1000 inhabitants, and 80% (31 out of 39) have lost population between 2008 and 2018. The *comarca* (district level administrative demarcation) of Terra Alta, which concentrates 25% of the installed wind power in Catalonia, is the region that has lost the highest percentage of population between 2008 and 2018, a 9.7% (1251 inhabitants) (Saladié Gil, 2021).

In recent years, a recurring complaint made in the affected Catalan areas regarding the deployment of renewable energies has been about the lack of participation (Faneca, 2021). Other objections put forward by the public are the risk of depopulation of rural areas (Agència Catalana de Notícies, 2022) and the increasing concentration of energy projects in already overcrowded areas (Rocaselva Soler, 2022).

As regards environmental impacts, the deployment of infrastructure for generating and transmitting renewable energy requires large areas of land, if a massive, concentrated, and oversized model is followed. The requirements of this model of implementation imply increased pressure on the territory and on biodiversity (Nació, 2021).

In the case of wind farms and above-ground feed-in infrastructure, the effects on birdlife provide another example of environmental impacts, as will be shown in the case study. Other impacts could include photovoltaic solar projects taking up agricultural land. The deployment of hectares of solar panels on agricultural land involves the degradation of the land's agricultural, ecological, and productive value for as long as the installation exists.

As one might expect, large-scale renewable energy projects concentrated in specific areas have a greater impact on territory and biodiversity. In contrast, small, scattered generation projects tend to be more compatible with land conservation, as they require less space and less infrastructure.

The village of Bellprat provides an example of head-on opposition to perceptions of powerlessness: residents were concerned about the social and environmental impacts of some projects and blocked access to the village to prevent machinery from constructing renewable energy macro-projects (Escolà, 2022).

The developers of renewable energy projects have a different perspective. In their view, political, regulatory, and administrative measures should be implemented to favour the development of renewable energies, especially projects of more than 5 MW. According to the sector, these large projects must be encouraged in order to reach 700 MW of onshore wind power per year by 2050 to meet the targets set by the Government of Catalonia (El Periódico de la Energía, 2022). It is worth mentioning that according to data from the Energy Outlook for Catalonia 2050 report (*Prospectiva Energètica de Catalunya per a l'any 2050/PROENCAT 2050*), it will also be necessary to deploy photovoltaic solar energy projects, which will require the production of thousands of megawatts to meet the targets set (Institut Català de l'Energia, 2022).

The developers also believe the administrative processing of procedures for establishing renewable energy projects needs to be streamlined (El Periódico de la Energía, 2022). As a result, developers have started to merge projects in order to exceed the 50 MW threshold, above which the administrative processes follow national procedures rather than Catalan ones (Sala, 2022).

The Rules Governing the Deployment of Renewable Energy and Participation in Administrative Procedures

The legal framework for the administrative procedure for deploying renewable energy projects below 50 MW in Catalonia is Decree-Law 16/2019, of 26 November, on Urgent Measures for the Climate Emergency and to Encourage Renewable Energies. This chapter will describe the version of the Decree Law before being modified as well as the most important aspects of subsequent amendments, which had a significant impact on public participation in the administrative process.

The purpose of Decree-Law 16/2019 is as follows (Article 1):

(a) to take urgent measures to address the climate emergency through an ecological and energy transition that enables the objectives set out in Law 16/2017, of 1 August, on Climate Change, to be achieved as rapidly as possible;
(b) to amend the Recast Text of the Urban Planning Law, approved by Legislative Decree 1/2010, of 3 August (*Decreto Legislativo 1/2010, de 3 de agosto, por el que se aprueba el Texto Refundido de la Ley de Urbanismo*), in order to facilitate and simplify the implementation of installations for exploiting solar and wind energy;
(c) to determine the requirements for the authorisation of wind and solar photovoltaic power generation installations; define the energy, environmental, urban planning, and landscape rules that should govern their implementation; and simplify the administrative procedure for authorising them.

Article 6.1 of Decree-Law 16/2019 sets out the scope of application of the administrative procedure for the deployment of renewable energy projects. The latter applies only to projects on land unsuitable for development that meet certain criteria. In accordance with the second paragraph of Article 6 of Decree-Law 16/2019, projects located on land suitable for development are governed only by the provisions of Chapter IV, on the authorisation regime for renewable energy installations with regard to energy authorisation and environmental impact assessment. The urban planning authorisation must be governed by the applicable urban planning legislation.

Wind farms must be installations for producing electricity from wind power with a capacity greater than 100 kW but equal to or less than

50 MW (with or without self-consumption). The following are also considered to form part of a wind farm: the electricity feed-in infrastructure, the wind farm's substation, newly built access points, and modifications of previously existing access points. Solar photovoltaic plants must be installations for producing electricity from solar energy using the photoelectric effect with a capacity greater than 100 kW but equal to or less than 50 MW (with or without self-consumption). The following are also considered to form part of a photovoltaic solar plant: the inverters, the plant's substation, the electricity feed-in infrastructure, newly built access points, and modifications of previously existing access points.

Article 7 sets out some general rules for the implementation of renewable energy projects. In their original version, these projects must have no significant impact on the area of influence, natural heritage, biodiversity, or cultural heritage; must comply with guidelines and objectives for regional land-use planning and landscape; must minimise the territorial impact of new access points to the facilities or the modification of existing access points; and must minimise the territorial impact of power lines connecting to the electricity grid, seek optimum proximity to the electrical grid, and prevent them from running through areas of high natural value.

When Decree-Law 16/2019 was amended by Decree-Law 24/2021, on Accelerating the Deployment of Distributed and Shared Renewable Energies, the following rule was added: 'Improve social acceptance in the territory, enabling local participation in projects' (current version of Article 7.1.e of Decree-Law 16/2019).

In the first version of Decree-Law 16/2019, Article 8 set out some specific rules for the implementation of wind energy projects. These projects must minimise the impact on land of high natural value, endangered species or species particularly vulnerable to wind farms, and migratory stopover sites for birds, and avoid critical areas for endangered birds of prey; must avoid sites of high impact on the landscape and of high significance or relevance to society according to official landscape catalogues; and must take account of the cumulative impact of the concentration of wind farms in particular parts of the territory. In addition, natural areas with special protection, special protection areas for birds, and natural areas with a surface area of less than 1000 ha included in Catalonia's Plan for Areas of Natural Interest (*Plan de Espacios de Interés Natural*) are considered unsuitable for the installation of wind farms. However, this last rule can be modified and clarified through specific studies and analyses, which must be reflected in a territorial land-use plan. At the time of writing this contribution, the sectorial land-use plan for the implementation of renewable energies in Catalonia is still being developed and has not been definitively approved yet.

These rules were subsequently slightly amended by Decree-Law 24/2021 and Law 2/2021, of 29 December, on Fiscal, Financial, and Administrative Measures for the Public Sector (*Ley 2/2021, de 29 de diciembre, de medidas fiscales, financieras y administrativas del sector público*). One notable change was the requirement to maintain a minimum distance of one kilometre between the site of the wind turbines and urban centres (amendment introduced by Decree-Law 24/2021): with the amendment of Law 2/2021, this minimum distance was reduced to 500 metres.

Similarly, Article 9 of Decree-Law 16/2019 sets out a number of specific conditions for the deployment of solar photovoltaic plants. These projects must respect the biophysical nature of the land, taking account of the rule of proportionality to the local environment and the existing layout of plots of land; must take account of the suitability for the terrain on which they are sited; must maintain the routes of existing roads and not making significant changes to them; must respect the configuration of the perimeters and other existing elements such as trees of interest, water courses, irrigation channels, and similar items, even if this means that the footprint of the plant has to be discontinuous; must minimise earthworks so that the panels are placed primarily directly on the ground, without continuous bonding; must maintain a minimum distance between fences and roads and more frequented areas; and must not have a significant impact on soils of high agrological value or high agricultural interest.

This last condition didn't establish a specific definition of this type of soil, which highlighted the need to define more precisely the criteria applicable to land use, with a special need regarding the planning of agricultural land uses that promote compatibility of food sovereignty with the energy transition (Gil Casión, 2022). It wasn't until the modification established by Decree-Law 24/2021, that it was established that soils of classes I, II, III, and IV defined in the system for evaluating soil agrological capabilities in the official cartographic information of Catalonia are considered high agrological value and of high agricultural interest. Certain soil occupancy limitations were also added based on their agrological capabilities. Furthermore, natural areas included in the Natura 2000 Network are considered not to be suitable for the installation of solar photovoltaic plants, except if the plants are intended for self-consumption, or occupy less than 3 ha, or are located in a municipality that does not have any other type of land. However, this rule can be modified and clarified through specific studies and analyses, which must be reflected in a territorial land-use plan. As exposed before, the sectorial land-use plan for the implementation of renewable energies in Catalonia has not been definitively approved yet.

As with wind farms, these rules have since been amended. One of the most relevant changes has been setting a more demanding rule for the maximum area that solar photovoltaic plants can occupy on Natura 2000 Network land, reducing it from 3 ha to 1 ha. Rules were also included for the installation of solar photovoltaic plants on land of high agrological value and forest land.

The version of Decree-Law 16/2019 before being modified stated that the Renewable Energies Committee should be consulted in advance on the viability of the location of wind farms and solar photovoltaic plants (Article 11). The consultation had to be carried out through the Government of Catalonia's Business Management Office and the project developer had to provide a preliminary draft defining the characteristics and specific location of the facilities, associated infrastructures, access and service roads, and the project's deadlines; an environmental and land-use diagnostic study of the area affected by the project (the study presented to the Committee had to include compliance with the rules on the deployment of renewable energies described earlier); and a study describing the main alternatives considered and including an analysis of the potential impacts of each of them.

After receiving this documentation, the Committee had to provide notice of the consultation to the local councils of the municipalities where the activity was to be carried out. It was also supposed to notify other public administrations considered to be affected.

The Committee had to give its decision on the viability of the proposal and indicate the conditions under which the activity could take place within three months. An unfavourable pronouncement, resolving that the site was not viable, prevented the construction of the facility. If the developer did not receive a response within three months from the consultation submission date, project approval procedures could be initiated.

It is worth mentioning that for the first few months after Decree-Law 16/2019 came into force, the documentation submitted for prior consultation by the developers was published on a web portal belonging to the Government of Catalonia. The publication of the documentation led to a high level of participation by citizens, third-sector bodies, and administrations. This participation took the form of submitting comments on the documentation. The documentation subsequently ceased to be published on the web portal, and instead, only those administrations and individuals considered to have an interest in the procedure were notified. With the entry into force of Decree-Law 24/2021, Article 11 of Decree-Law 16/2019 was repealed, and the prior consultation of the Renewable Energies Committee on the viability of the project site ceased to be carried out. The elimination of the prior consultation also eliminated a moment in the administrative process in which interested parties could participate by submitting comments.

According to Article 12 of Decree-Law 16/2019, the authorisation of a wind farm requires the intervention of the departments responsible for energy, urban planning, landscape, and environment, and includes a prior official authorisation for the construction of the wind farm; the approval of a specific development plan of public interest on land unsuitable for development (urban planning procedure); and an environmental impact assessment of the project, if required by national legislation.

The departments responsible for energy, urban planning, landscape, and the environment must check the adequacy and suitability of the documentation provided by the developer. The environmental department must then notify the department responsible for energy of the list of bodies and administrations that must be consulted for the environmental impact assessment (Article 15 of Decree-Law 16/2019). Within one month of having verified that the documentation submitted is adequate, the department responsible for energy must put the documentation through the public consultation procedure, for a minimum of 30 working days.

At the same time, the department responsible for energy must hold a hearing with the administrations and bodies that may be affected and ask organisations and companies involved in public services or in services of general interest to issue a report on parts that may affect them.

After the public consultation period, the representations and reports received must be sent to the developer, who has 30 working days to respond. The department responsible for energy must provide notification of the representations and reports and the developer's response to those responsible for urban planning, landscape and environmental assessment, and the affected town councils. Those departments may submit representations within a month.

After the project's environmental impact assessment has been carried out, the Regional Urban Planning Commission must give final approval to a specific development plan on land unsuitable for development within one month (Article 17 of Decree-Law 16/2019).

Within one month of obtaining the outcome of the environmental and urban planning process, the department responsible for energy must process the application for prior official authorisation, the Declaration of Public Interest if the developer has requested it, and the construction licence for the project. After this, the operating licence must be processed and entered into Catalonia's Register of Electricity Production Facilities (Article 20 of Decree-Law 16/2019).

The rules governing the deployment of renewable energy represent a fast way to increase the production of energy from non-fossil sources but also implied an authentic avalanche of projects in certain areas of Catalonia. From the author's point of view, and based on their experience in post-public consultation report preparation, this regulatory framework, probably perceived as simple and agile, has led to the submission of many project documents to the Renewable Energies Committee with low technical quality. Some examples are project documents that contain manifest urban incompatibilities or environmental and land-use diagnostic studies that neglect potential environmental impacts. Another clear example of deficiencies in the documentation can be seen in some alternative studies that present as an option a mere adjustment in the number of wind turbines, locations of the park that are practically identical, or alternatives located in the Natura 2000 network, and therefore destined to not be a real and eligible alternative. All these aspects could cause the location selected by the developer to be declared as non-viable by the Renewable Energy Committee when the location consultation step still existed.

Case Study

This case study is based on the experience of the author as an environmental lawyer. In January 2021, our legal team, specialised in environmental law, was contracted by a supra-municipal administration to draw up a post-public consultation report regarding three projects submitted to the Renewable Energies Committee's prior consultation process. This procedure, covered by Article 11 of Decree-Law 16/2019, was abolished in the subsequent amendment in October 2021, resulting in one less stage in which citizens, interested entities, and affected administrations could participate.

The projects submitted by the developer were three wind farms. Each of the projects was submitted for prior consultation by a different limited company, but all three companies belonged to the same business group, and there were even cross-references in the documentation submitted to the Committee. To preserve professional secrecy, the names of the companies and the business group will not be disclosed. Project A was composed of nine wind turbines, with a total power of 49.5 MW; project B was composed of five wind turbines, with a total power of 27.5 MW; and project C was composed of nine wind turbines, with a total power of 49.5 MW. The intended locations were adjacent municipalities in the province of Lleida, not far from the province of Tarragona. To maintain professional secrecy, these municipalities will not be named; instead, they will be referred to by numbers (1, 2, 3, and 4). In addition to the nine wind turbines, Project A also included a high-voltage line, which was intended to transmit the electricity generated by wind farms A, B, and C from a transformer substation to the province of Barcelona.

The report prepared for the supra-municipal administration examined factors that could compromise the viability of the chosen site. These are set out below.

Compatibility with Regional Land-Use and Urban Planning Requirements

The post-public consultation report examined the degree of compatibility of the wind turbines and the planned roads with regional land-use and urban planning requirements. The wind turbines and roads were planned to be located on land unsuitable for development, that is land on which urban development cannot take place. Urban development may not be permitted for a variety of reasons, such as features that require special protection or the terrain being unsuitable for building on.

Wind Turbines

Project A planned nine wind turbines, of which six were to be located in Municipality 1 and three in Municipality 2. The six wind turbines of Project A in Municipality 1 were to be sited on land categorised as agricultural land unsuitable for urban development. Article 122 of Municipality 1's urban planning regulations defined this land as 'land that has agricultural value, is barren, or has historically been cultivated as rain-fed land, that qualifies for protection because of its value as an agricultural, environmental, or landscape reserve, the transformation of which into land unsuitable for development is considered harmful to the necessary balance between strictly urban areas and areas of special agricultural or natural value.'

On the permitted uses of land categorised in this way, the same Article 122 states that

> in general, its use for agriculture, livestock, and forestry is permitted. With regard to its use for housing, only single-family dwellings are permitted, if they comply with the conditions established in Article 118–4, which regulates housing on land unsuitable for development. In addition, existing farmhouses may be used for hotel, health, socio-cultural, educational, catering, recreational, and sporting purposes, according to the conditions and procedures established in Article 48 of the Law on Urban Planning.

In the post-public consultation report, it was argued that wind turbines and their associated infrastructure were unsuitable for land categorised in this way, because this type of facility is not described as an appropriate use and also because they did not align well with the reasons why this categorisation had been applied. The deployment of wind turbines and evacuation infrastructures could be harmful to its value as an agricultural, environmental, or landscape reserve.

The other three wind turbines of project A were planned to be located in Municipality 2. One of them was planned to be located on specially protected land: land of natural value and protection, according to the Partial Regional Plan for the Ponent area. Partial territorial plans are land-use planning outlines that establish a land-use model for the different areas of Catalonia: *Alt Pirineu i Aran*, *Àmbit metropolità de Barcelona*, *Camp de Tarragona*, *Comarques Gironines*, *Comarques Centrals*, *Penedès*, *Ponent*, and *Terres de l'Ebre*. They set out decisions on the three basic land-use systems that are structured in plans: the one for open spaces, which is a priority reference for urban planning; the one for settlements, which sets out guidelines that urban plans must put into practice; and the one for mobility infrastructure, to create a network that structures the territory.

Article 2.5 of the regulations of the Partial Regional Plan for the Ponent area states 'in regard to specially protected land that this category includes spaces with features warranting a level of protection that would severely restrict the potential for transformations that might affect them. It includes land that forms part of the areas of protection established in the sectoral regulations (such as the Plan for Areas of Natural Interest and the Natura 2000 Network) and land that the Plan considers necessary to preserve because of its value as an object or connectivity corridor of natural interest and also because of its specific role in environmental balance, as is the case with aquifer recharge areas.'

While it is true that the deployment of renewable energy plants is not expressly prohibited, this category of land is nevertheless subject to special protection. Such projects are considered inappropriate when they may clearly affect elements requiring special protection (Article 2.6 of the regulations of the Partial Regional Plan for the Ponent area).

According to the urban planning regulations of Municipality 2, the land on which the developer planned to locate this wind turbine was forest land unsuitable for development. Article 114 of these regulations defines this category of land as 'rural land covered with trees or shrubs, bushes, and grasses, barren land on the edges of forests that is required for the protection of those forests, and barren land which, because of its characteristics, is suitable for afforestation or reforestation.' On conditions of use, the same article stipulates that the general use of this category of land is forestry.

Despite the fact that Municipality 2's regulations on regional land use and urban planning do not expressly prohibit the siting of wind turbines on this land, it was argued that the installation of wind turbines was not consistent with the features that led to this particular categorisation. The deployment of wind turbines and evacuation infrastructures could imply cutting down trees or getting rid of vegetation, which could be harmful to its forestry qualities.

The land on which the developer intended to locate the two remaining wind turbines was categorised as rural land unsuitable for development by Municipality 1's urban planning regulations. Article 112 of these regulations states that the land in question is not of special agricultural interest, is barren, or has historically been cultivated as rain-fed land and is not intended to be converted to irrigated land. The installation of renewable energy generation projects is not mentioned, either as an appropriate use or as a prohibited use for such land.

Project B's five wind turbines were planned to be located in Municipality 1. All five were to be located on land unsuitable for development. According to the municipality's urban planning regulations, three of these wind turbines were to be sited on agricultural land unsuitable for development. With regard to the compatibility of these wind turbines with urban planning regulations, the post-public consultation report referred to what had already been stated for Project A.

Lastly, Project C consisted of nine wind turbines, three of which were to be located in Municipality 3 and six in Municipality 2. One of the wind turbines planned for Municipality 3 was to be located on land categorised by the municipality's urban planning regulations as forest land not suitable for development. The other two were to be located on land categorised as agricultural land not suitable for development.

According to Article 116 of Municipality 3's urban planning regulations, on land unsuitable for development, '[...] urbanisation processes are not permitted, nor are uses contrary to the natural use of the land.' The land was categorised as suitable for forestry (Article 131) and agriculture (Article 130), respectively. In both these categories, structures of public interest or social utility were permitted. The suitability of renewable energy installations was not mentioned, however.

Of the six wind turbines to be located in Municipality 2, three were to be located on rural land unsuitable for development and the other three on forest land unsuitable for development. As mentioned earlier, Municipality 2's urban planning regulations neither expressly permit nor prohibit the installation of renewable energy projects on rural land unsuitable for development (Article 112).

On forest land unsuitable for development, Article 114 of these regulations states that the permitted use of such land is for forestry purposes. Despite the fact that Municipality 2's regulations on urban planning also do not expressly prohibit the wind turbines from being located on such land, it was argued that the installation of wind turbines was not consistent with the features that led to this particular categorisation.

Planned Roads

The installation of the wind turbines and associated infrastructure required both the modification of existing rural roads (already present in all four municipalities) and the opening of new roads. It was envisaged that the roads would be between 4.5 metres and 6 metres wide, to allow heavy vehicles to use them.

Around a third of all the planned roads, including access from the main road, ran through Municipality 1. Article 66 of Municipality 1's urban planning regulations stated that 'No new road, rural road, forest track, or any other type of road may be opened unless provided for in this Municipal Urban Land-Use Plan and/or in the special plans that develop it. The longitudinal and cross-sectional profiles of roads may not be altered without the appropriate municipal licence. Under no circumstances may any alteration involve damage to the natural environment.'

The creation of new roads was not permitted unless these were provided for in urban land-use plans, and when the Municipal Urban Land-Use Plan and other planning instruments were examined, no such provision was found.

With regard to alterations to road layout, it was established that the documentation submitted by the developer did not cover the changes implied by alterations to existing roads and opening new roads. For these reasons, it was argued that plans for new roads and alterations to existing roads in Municipality 1 were inappropriate from an urban planning perspective.

Impacts on Biodiversity, with Special Reference to Birds

The report analysed several potential environmental impacts of wind farm projects A, B, and C. This section describes the most significant potential impacts on biodiversity. The report also stated that according to the Catalonia Biodiversity Data Bank (*Banc de dades de la Biodiversitat de Catalunya*, n.d.), the areas where the wind farms were planned were home to a variety of plant communities. Some notable plant communities included *Asplenio-Quercetum ilicis*; *Cardamino hirsutae-Erophiletum praecocis*; *Dictamno-Arctostaphyleto-Quercetum valentinae*; *Diplotaxietum erucoidis*; *Erico-Thymelaeetum tinctoriae subassociatio Typicum; Moricandio-Carrichteretum annuae*; *Viburno-Quercetum ilicis.* It also mentioned plant species that were considered to be protected or threatened to some degree.

According to the National Biodiversity Information Node (*Nodo Nacional de Información sobre la Biodiversidad*), a total of 1071 plant species are located within a radius of 10 kilometres from a central point between the three projects (GBIF.ES, 2022).

For these reasons, it was argued that the developer should demonstrate how the construction and operation of these projects would avoid affecting protected, endangered, or rare plant species.

The Catalonia Biodiversity Data Bank states that 139 vertebrate species are located in the area surrounding the three projects. According to the

National Biodiversity Information Node, there are a total of 400 vertebrate species, including 28 mammal species and 222 bird species, within a radius of 5 kilometres from a central point between the three projects, and three sightings of Bonelli's eagles (*Aquila fasciata*) have been recorded in the triangle formed by the three wind farms.

The Bonelli's eagle is a protected species under Directive 2009/147/EC of the European Parliament and of the Council of 30 November 2009 on the conservation of wild birds, Annex I: species to be subject to special conservation measures concerning their habitats in order to ensure their survival and reproduction in their area of distribution. Taking into consideration the impact on birdlife, it was argued that the location chosen for these projects was not appropriate because of the potential impacts of wind turbines and power lines.

Project Fragmentation

From an analysis of the documentation (preliminary draft project documents, study of alternatives, and territorial and environmental diagnosis) submitted by the developer for each of the projects, it was observed that Projects A, B, and C were part of a much larger project that the developer was attempting to fragment.

As mentioned previously, the administrative procedure provided for in Decree-Law 16/2019 only allows for the processing of renewable energy projects with a capacity of up to 50 MW. Projects with a higher capacity have to follow a national-level administrative procedure which, at the time of writing, was known to be less streamlined.

Projects A, B, and C were promoted by limited companies belonging to the same business group, were planned in neighbouring municipalities, and the energy generated was directed to the same transformer substation. They totalled 126.5 MW: 49.5 MW for Project A, 27.5 MW for Project B, and 49.5 MW for Project C. In the post-public consultation report submitted, it was argued that given that there was, in reality, only one project, rather than three, the 50 MW limit for the procedure provided for in Decree-Law 16/2019 had been exceeded.

The documentation for Project A also included a 220-kW high-voltage overhead line to be processed along with the wind farm and its infrastructure. The function of this high-voltage overhead line was to transport the energy fed into the grid at the substation in the province of Lleida, mentioned previously, to another substation in the province of Barcelona, some 80 kilometres away. Although the developer repeatedly referred to it as feed-in infrastructure, we cannot agree with this description. The function of the 220-kW overhead line was not to feed energy into the electricity network but to transport it from one place to another so it could be distributed and used. It should be recalled that the scope of application of the procedures set out in Decree-Law 16/2019, Article 6, does provide for the processing of associated feed-in infrastructure but not high-voltage lines to transport electricity.

In relation to this, Article 4.2 of Royal Decree 1955/2000, of 1 December, Regulating the Transmission, Distribution, Marketing, and Supply of Electricity and the Authorisation Procedures for Electricity Facilities (*Real Decreto 1955/2000, de 1 de diciembre, por el que se regulan las actividades de transporte, distribución, comercialización, suministro y procedimientos de autorización de instalaciones de energía eléctrica*) states:

> Transmission will be carried out by transmission companies, which are those commercial companies whose purpose is to transport electricity, as well as to build, operate, and maintain transmission facilities in accordance with the provisions of Article 9 of Law 54/1997, of 27 November, on the Electricity Sector (*Ley 54/1997, de 27 de noviembre, del Sector Eléctrico*). In any event, 'Red Eléctrica de España, Sociedad Anónima' may carry out electricity transmission.

The fragmentation of projects and the consequent fragmentation of the corresponding environmental impact assessment procedures and environmental impact statements are strongly criticised in Spain and can even lead to the nullification of the project (e.g. Supreme Court Judgment No. 1933/2017, of 5 May [Administrative Division, Section 3], Fundamental Points of Law sections 5 and 6). Since the entry into force of Decree-Law 16/2019, the fragmentation of projects has been a common practice. The fragmentation of the environmental impact assessment could imply an incorrect evaluation of the cumulative impacts of renewable energy projects. These could lead to unexpected environmental impacts in the deployment of the installations. Another negative impact of the subsequent nullification of the project for this reason, and its dismantling, is a deceleration in the energy transition.

From my point of view, the fragmentation of projects to avoid national administrative procedures, and the resulting fragmentation of the corresponding environmental impact assessment procedures and environmental impact statements, constitutes a fraud against the law. Article 6.4 of the Spanish Civil Code published by the Royal Decree of 24th July 1889 (*Real Decreto de 24 de julio de 1889 por el que se publica el Código Civil*) states that 'Actions carried out under the text of a rule seeking a result prohibited by the legal system or contrary to it shall be deemed fraudulent and shall not prevent the appropriate application of the rule that was intended to be avoided.'

This point has been alleged on several occasions by different entities to the Renewable Energies Committee in diverse post-public consultation reports. There is no record of the Committee having expressed an opinion, either in favour or against, on whether the fragmentation of energy projects constitutes a fraud against the law. The fact that the Renewable Energy Committee has not taken a position on this issue has meant that while it has had advisory functions on the viability of project locations, many fragmented macro-projects have been submitted.

Resolutions of the Renewable Energies Committee

As required by Decree-Law 16/2019, the Renewable Energies Committee examined the sectorial reports from different administrations, in addition to the post-public consultation reports submitted by the interested parties. On the basis of the documentation submitted by the developer and the reports received, it issued a resolution on the viability of the sites selected for the wind energy project's infrastructure.

In February 2021, the Renewable Energies Committee resolved that the site of Project A was viable, as it was considered that there were no insurmountable obstacles. However, the route of the transmission power line (which the Committee understood to be the high-voltage line referred to earlier) had to be modified, as it affected two vital areas for golden eagles (*Aquila chrysaetos*) and Bonelli's eagles (*Aquila fasciata*). The Committee also resolved that the sites of Projects B and C were non-viable because there were impacts on endangered fauna that were considered insurmountable from the outset. It was specifically the potential impact on birdlife, especially Bonelli's eagles that led to the sites being declared non-viable. These impacts were not adequately assessed in the documentation submitted by the developer.

In the resolutions adopted by the Committee, reference was made to the representations submitted by the supra-municipal administration, but no express response was given to the issues it raised. The Committee did not give its opinion on the possible incompatibilities with urban planning regulations, the impossibility of laying out a significant part of the planned roads, the fragmentation of the project, or the failure to consider the high-voltage line to be a transmission structure. From a reading of the resolutions, it can be concluded that the Committee considered the high-voltage line included in Project A to be a feed-in infrastructure for the three wind farms. In my opinion, however, it cannot be understood as a feed-in infrastructure, for the reasons previously stated. If the Committee considered it to be a feed-in infrastructure, it should have ruled on the fragmentation of one large project into three smaller ones.

Conclusions

The installation of renewable energy facilities in Catalonia has been a source of social conflict in recent years. The concentration of renewable energy projects in particular areas, and the fact that the generated energy is consumed in places far from the affected area, is regarded unfavourably by some sectors of the community due to environmental impacts.

The acceleration of the administrative process has meant that numerous renewable energy projects have been submitted for approval in Catalonia. It should be borne in mind that the proposed sites for a significant proportion of the submitted projects analysed were declared non-viable (Observatori de les Energies Renovables, 2022).

In light of what is described in this chapter, it can be affirmed that the implementation of renewable energy projects must be more equitable and more evenly distributed across Catalonia, if it is to enjoy greater public acceptance. The case study shows the importance of an appropriate evaluation of impacts on biodiversity by developers. This case also shows how some aspects of crucial importance like the project's compatibility with urban planning regulations or fragmentation of massive projects sometimes are not analysed properly by the competent administrations. The resolutions of the Renewable Energy Committee linked to the case study were a missed opportunity to take a stand on issues of urban compatibility (especially regarding rural roads) and the fragmentation of large energy projects. A clear position on these issues could have had the ability to convey a message to project promoters about what aspects should be taken into account and what practices are preferable to avoid in the deployment of renewable energies. Likewise, a statement on these points would also have been able to provide legal certainty to promoting companies and entities that are interested parts in these procedures.

Taking into account what has been exposed in this chapter, planning and regulatory tools must be developed to ensure a balanced distribution of renewable energy projects across the Catalan territory. A balance needs to be struck between the efficiency of the site chosen for energy generation and genuine consideration of social and environmental impacts.

It is also essential for developers to take these factors into account before starting the administrative process for implementing renewable energy projects. Public administrations should also be acquainted with the regulatory framework (energy laws, environmental laws and urban planning regulations) and guarantee public participation to ensure that the energy transition in Catalonia is shared by all.

These conclusions, largely based on the case study and the advising in similar cases, can well be extrapolated to a global perspective (at least with regard to countries in the process of energy transition). The deployment of renewable energies is usually conditioned by regulations from different areas. The case study shows the need for both promoters and competent administrations to be knowledgeable not only about regulations related to energy but also about biodiversity regulations, urban planning compatibility, and court rulings on renewable energy projects. A lack of this knowledge can lead to the implementation of projects whose compatibility with legality is questionable and susceptible to being contested in the competent courts by affected citizens, environmental NGOs, or even other public administrations. Likewise, the development of regulatory frameworks must be favourable to real public participation.

The deployment of renewable energies is a complex challenge that requires a balance between economic, social, and environmental interests. Failure to do so may result in a significant investment of time, money, and effort in projects that may not be implemented or may be subject to appeals before the competent courts. Such consequences will slow down the energy transition, make it more difficult to achieve the renewable energy targets set for the next few years, and perpetuate society's dependence on fossil fuels.

References

Agència Catalana de Notícies. (2022). *"No passaran": la lluita de Bellprat contra tres projectes d'energies renovables.* 6 July. Retrieved February 1, 2023, from https://www.ccma.cat/324/no-passaran-la-lluita-de-bellprat-contra-tres-projectes-denergies-renovables/noticia/3174051/

Banc de Dades de la Biodiversitat de Catalunya. (n.d.). Retrieved March 3, 2023, from http://biodiver.bio.ub.es/biocat/index.jsp

Berbís Morelló, S. (2021). *La fotovoltaica estén la pressió de les renovables de les carenes a la plana*. Nació, 25 February. Retrieved November 11, 2022, from https://www.naciodigital.cat/lleida/noticia/42368/fotovoltaica-esten-pressio-renovables-carenes-plana

De la Varga Pastor, A. (2022). Las medidas autonómicas para mitigar y adaptar el cambio climático. Análisis comparativo de las leyes autonómicas de Catalunya, Andalucía e Illes Balears. In *Anuario de la Facultad de Derecho de la Universidad Autónoma de Madrid* (pp. 251–268). Retrieved February 1, 2023, from https://www.boe.es/biblioteca_juridica/anuarios_derecho/abrir_pdf.php?id=ANU-A-2022-10025100268

El Periódico de la Energía. (2022). *La eólica pide un empujón a la Generalitat de Catalunya*. 20 October. Retrieved January 15, 2023, from https://elperiodicodelaenergia.com/la-eolica-pide-un-empujon-a-la-generalitat-de-catalunya/

Environmental Pollution Panel, President's Science Advisory Committee. (1965). *Restoring the Quality of Our Environment* (pp. 126–127). The White House. Retrieved September 12, 2022, from https://legacy-assets.eenews.net/open_files/assets/2019/01/11/document_cw_01.pdf

Escolà, N. (2022). *Bellprat barra l'accés al poble per impedir macroprojectes energètics: "No és energia verda, és energia merdosa"*. Públic, 1 July 2022. Retrieved February 1, 2023, from https://www.publico.es/public/bellprat-barra-l-acces-poble-per-impedir-macroprojectes-energetics-no-energia-verda-energia-merdosa.html

Faneca, C. (2021). *El territori diu prou: Energies renovables sí, però així no*. Xarxanet.org, 28 May. Retrieved March 3, 2023, from https://xarxanet.org/ambiental/noticies/el-territori-diu-prou-energies-renovables-si-pero-aixi-no

Foster, S., & Elzinga, D. (2015). *The Role of Fossil Fuels in a Sustainable Energy System*. United Nations, December. Retrieved March 8, 2023, from https://www.un.org/en/chronicle/article/role-fossil-fuels-sustainable-energy-system

GBIF.ES Nodo Español de Información en Biodiversidad. (2022). Retrieved February 1, 2023, from https://www.gbif.es/

Generalitat de Catalunya. (2022). *Visor ambiental de les energies renovables*. Retrieved March 15, 2023, from https://agportal.dtes.gencat.cat/portal/apps/webappviewer/index.html?id=db72e7e7f81e4a9aa1ef367511fc1651&locale=ca

Generalitat de Catalunya, Medi ambient i sostenibilitat. (2022). *Visor Ambiental i dades d'energies renovables*. Retrieved January 12, 2023, from https://mediambient.gencat.cat/ca/05_ambits_dactuacio/avaluacio_ambiental/energies_renovables/visor/

General Secretariat of the Council. (2014). *European Council (23 and 24 October 2014)—Conclusions*. European Council, 24 October. Retrieved February 1, 2023, from https://data.consilium.europa.eu/doc/document/ST-169-2014-INIT/en/pdf

Gil Casión, M. I. (2022). Derecho y políticas ambientales en Cataluña (Primer Semestre 2022). *Revista Catalana de Dret Ambiental, 13*(1), 1–30. Retrieved April 4, 2023, from https://doi.org/10.17345/rcda3351

Gobierno de España. (2020). *Acuerdo de Consejo de Ministros por el que se aprueba la declaración del Gobierno ante la emergencia climática y ambiental*. 21 January. Retrieved February 1, 2023, from https://www.miteco.gob.es/es/prensa/declaracionemergenciaclimatica_tcm30-506551.pdf

Hidalgo, E., & Aymí, O. (2021). *No som el solar de Catalunya*. Cadena SER, 11 February. Retrieved February 1, 2023, from https://cadenaser.com/emisora/2021/02/11/ser_tarragona/1613026286_317128.html

Institut Català de l'Energia. (2022). *Prospectiva Energètica de Catalunya 2050—PROENCAT 2050*. 4 February. Retrieved March 8, 2023, from https://govern.cat/govern/docs/2022/02/04/13/14/3be61c8e-6048-4495-86be-c7b0b126158c.pdf

Jackson. P. (2007). *From Stockholm to Kyoto: A Brief History of Climate Change*. United Nations, June. Retrieved October 10, 2022, from https://www.un.org/en/chronicle/article/stockholm-kyoto-brief-history-climate-change

La Moncloa. (2020). *Declarada la emergencia climática*. 21 January. Retrieved October 10, 2022, from https://www.lamoncloa.gob.es/consejodeministros/Paginas/enlaces/210120-enlace-clima.aspx

Ministerio para la Transición Ecológica y el Reto Demográfico. (2020). *Plan Nacional Integrado de Energía y Clima 2021–2030*. 20 January. Retrieved February 1, 2023, from https://www.miteco.gob.es/images/es/pnieccompleto_tcm30-508410.pdf

Observatori de les Energies Renovables. (2022). *Progrés de la Implantació d'energies renovables a Catalunya—Objectius 2030 | 2050, Informe de situació 2021—Resum executiu—Juliol 2022*. Retrieved February 1, 2023, from https://observatorirenovables.cat/wp-content/uploads/2022/09/Informe-OBERcat-1-2021_FINAL.pdf

Rocaselva Soler, A. (2022). *El debat de les renovables Terres de l'Ebre: "Sí, però així no"*. ElPeriódico.cat, 3 July. Retrieved October 22, 2022, from https://www.elperiodico.cat/ca/medi-ambient/20210703/catalunya-ebre-energia-eolica-parcs-terres-de-lebre-11872282

Sala, A. (2022). *Catalunya sense vent ni sol. Regió7*, 4 October. Retrieved February 1, 2023, from https://www.regio7.cat/economia/2022/10/04/catalunya-energia-eolica-energia-solar-energies-renovables-moratoria-76775019.html

Saladié Gil, S. (2021). El model d'implantació de les energies renovables. Floc, *Lo*, (237), 20–25. Retrieved April 4, 2023, from https://raco.cat/index.php/LoFloc/article/view/405046

Legislation

Decree-Law 16/2019, of 26 November, on Urgent Measures for the Climate Emergency and to Encourage Renewable Energies (*Decreto-Ley 16/2019, de 26 de noviembre, de medidas urgentes para la emergencia climática y el impulso a las energías renovables*). Boletín Oficial del Estado No. 12, of 14 January 2020, pp. 3237–3259 (23 pages). Retrieved October 15, 2022, from https://www.boe.es/buscar/doc.php?id=BOE-A-2020-445

Decree-Law 24/2021, on Accelerating the Deployment of Distributed and Shared Renewable Energies (*Decreto-Ley 24/2021, de aceleración del despliegue de energías renovables distribuidas y participadas*). Boletín Oficial del Estado No. 305, of 22 December 2021, pp. 158676–158689 (14 pages). Retrieved October 14, 2022, from https://www.boe.es/buscar/doc.php?id=BOE-A-2021-21550

Directive 2009/147/EC of the European Parliament and of the Council of 30 November 2009 on the conservation of wild birds. Official Journal of the European Union No. 20, of 26 January 2010, pp. 7–25 (19 pages). Retrieved October 22, 2022, from https://eur-lex.europa.eu/legal-content/EN/TXT/?uri=CELEX per cent3A32009L0147

Directive (EU) 2018/2001 of the European Parliament and of the Council of 11 December 2018 on the promotion of the use of energy from renewable sources. Official Journal of the European Union No. 328, of 21 December 2018, pp. 82–209 (128 pages). Retrieved November 10, 2022, from https://eur-lex.europa.eu/legal-content/ES/TXT/HTML/?uri=OJ:L:2018:328:FULL&from=LV

Law 16/2017, of 1 August, on Climate Change (*Ley 16/2017, de 1 de agosto, del Cambio Climático*). Boletín Oficial del Estado No. 234, of 28 September 2017, pp. 94480–94518 (39 pages). Retrieved February 1, 2023, from https://www.boe.es/eli/es-ct/l/2017/08/01/16/con

Law 2/2021, of 29 December, on Fiscal, Financial, and Administrative Measures for the Public Sector (*Ley 2/2021, de 29 de diciembre, de medidas fiscales, financieras, administrativas y del sector público*). Boletín Oficial del Estado No. 18, of 21 January 2022, pp. 6489–6622 (134 pages). Retrieved November 20, 2022, from https://www.boe.es/buscar/doc.php?id=BOE-A-2022-953

Legislative Decree 1/2010, of 3 August, Approving the Recast Text of the Urban Planning Law (*Decreto Legislativo 1/2010, de 3 de agosto, por el que se aprueba el Texto Refundido de la Ley de Urbanismo*) Boletín Oficial del Estado No. 218, of 8 September 2010, pp. 77011–77125 (115 pages). Retrieved February 1, 2023, from https://www.boe.es/buscar/act.php?id=BOE-A-2010-13883

Organic Law 6/2006, of July 19, on the reform of the Statute of Autonomy of Catalonia. *(Ley Orgánica 6/2006, de 19 de julio, de reforma del Estatuto de Autonomía de Catalunya)* Boletín Oficial del Estado No. 172, of 20 July 2006, pp. 27269–27310 (42 pages). Retrieved October 30, 2022, from https://www.boe.es/eli/es/lo/2006/07/19/6/con

Regulation (EU) 2018/1999 of the European Parliament and of the Council of 11 December 2018 on the Governance of the Energy Union and Climate Action, amending Regulations (EC) No 663/2009 and (EC) No 715/2009 of the European Parliament and of the Council, Directives 94/22/EC, 98/70/EC, 2009/31/EC, 2009/73/EC, 2010/31/EU, 2012/27/EU and 2013/30/EU of the European Parliament and of the Council, Council Directives 2009/119/EC and (EU) 2015/652 and repealing Regulation (EU) No 525/2013 of the European Parliament and of the Council (Text with EEA elevance.) Official Journal of the European Union No. 328, of 21 December 2018, pp. 1–77 (77 pages). Retrieved November 1, 2022, from https://eur-lex.europa.eu/eli/reg/2018/1999/oj

Royal Decree of July 24, 1889, which publishes the Civil Code (*Real Decreto de 24 de julio de 1889 por el que se publica el Código Civil*). Gaceta de Madrid No. 206, of 25 July 1889, pp. 249–259 (11 pages). Retrieved November 1, 2022, from https://www.boe.es/buscar/act.php?id=BOE-A-1889-4763

Royal Decree 1955/2000, of 1 December, Regulating the Transmission, Distribution, Marketing, and Supply of Electricity and the Authorisation Procedures for Electricity Facilities (*Real Decreto 1955/2000, de 1 de diciembre, por el que se regulan las actividades de transporte, distribución, comercialización, suministro y procedimientos de autorización de instalaciones de energía eléctrica*). Boletín Oficial del Estado No. 310, of 27 December 2000, pp. 45988–46040 (53 pages). Retrieved October 15, 2022, from https://www.boe.es/buscar/act.php?id=BOE-A-2000-24019

Spanish Constitution (Constitución española). Boletín Oficial del Estado No. 311, of 29 December 1978, pp. 29313–29424 (112 pages). Retrieved November 12, 2022., from https://www.boe.es/buscar/act.php?id=BOE-A-1978-31229

Case Law

Spanish Supreme Court Judgment 1933/2017, of 5 May (Administrative Division, Section 3). Retrieved November 15, 2022., from https://www.poderjudicial.es/search/AN/openDocument/df7e521880dfc769/20170526

CHAPTER 11

Critical Mineral Supply in the European Energy Transition: Towards Just Institutional Approaches

Asmaa Khadim, Ron Janjua, and Chengjian Xu

Introduction

The European Green Deal (EGD) is an ambitious plan to transform the European Union (EU) economy to achieve climate and sustainability objectives (European Commission, 2019d). Yet, the EGD has met with fierce criticism, including concern over whether these environmental and economic objectives can be met, allegations that it prioritises private investment over the public interest, and increases regional inequality (Harvey & Rankin, 2020; Pérez, 2021a, 2021b; Khadim & van Eijken, 2022). Moreover, the EGD's 'deeply transformative policies' (European Commission, 2019d), especially regarding the clean energy transition

A. Khadim (✉) • C. Xu
Leiden University, Leiden, The Netherlands
e-mail: a.n.khadim@law.leidenuniv.nl

R. Janjua
University of Groningen, Groningen, The Netherlands

D. Iglesias Márquez et al. (eds.), *Legal Challenges at the End of the Fossil Fuel Era*, Global Issues,
https://doi.org/10.1007/978-3-031-61766-9_11

(CET), will have widespread impacts on communities and industries globally. The EU does not currently have a secure domestic critical mineral supply. Thus, the CET will necessitate the extraction of increasing quantities of critical minerals from other jurisdictions, thus potentially outsourcing environmental and human rights harms and contributing to socioenvironmental conflicts over resource extraction that originate in European decision-making but have impacts that are felt beyond its borders.

The EGD envisions careful attention to potential trade-offs between economic, environmental and social objectives, while emphasising a 'just transition' that ensures 'no one is left behind' (European Commission, 2019b). However, important questions arise about how policymakers will strike the appropriate balance (Khadim & van Eijken, 2022) and whether the EGD's conception of 'just transition' will include impacts upon individuals and communities beyond EU borders. Will these externalised impacts be factored into decision-making? If yes, how can EU decision-makers incorporate the perspectives and views of people in source jurisdictions? What are the imperatives of a just institutional approach? This chapter explores these questions from an interdisciplinary perspective, by including insights from the social sciences, law and environmental sciences.

The chapter begins, in Part 1, by discussing the necessary expansion of the global critical mineral supply chain and the implications for communities beyond the EU. The drivers of ever-increasing demand and the quantification of future mineral inputs for key battery materials are considered. Ongoing efforts at quantification have already demonstrated that the shift to electric vehicles (EVs) will require a drastic expansion in the global supply chain for lithium, cobalt and nickel (Xu et al., 2020). To contextualise the implications of this increasing demand, the authors discuss socioenvironmental conflicts in Latin America's 'Lithium Triangle,' an area of Argentina, Bolivia and Chile containing much of the world's lithium supply (Treadgold, 2021). This region already experiences high levels of socioenvironmental conflict over mining impacts, including pollution, access to water, climate change, human rights and Indigenous rights. The CET's mineral requirements risk exacerbating these impacts. For example, life-cycle water scarcity footprint tools have been used to evaluate anthropogenic contributions to regional water scarcity along global supply chains, showing that mining operations seem to have the highest impact (Schomberg et al., 2012). It is imperative that such assessments be factored into European energy policy to avoid simply outsourcing (and exponentially increasing) environmental and human rights harms.

Part 2 of this chapter highlights the key elements of a just transition reflected in the literature thus far. This provides a basis for discussing the EGD's conception of a just transition and what this means for communities affected beyond Europe, where critical raw materials may be sourced (i.e. 'source communities'). The current narrative around just transitions within the EGD does not appear to consider impacts beyond EU borders. The authors argue that for the European CET to be just, it must address considerations of distributive, recognition-based, restorative, procedural and interactional justice, and be participatory, deliberative, inclusive, legitimate, democratic, sustainable, socially just, and spatially and temporally just. Moreover, the EU's conceptualisation of a just transition must be broadened and operationalised to include communities along the entire supply chain. Otherwise, socioenvironmental conflict in source communities will not only hinder the ability to meet CET targets but will also call into question the values underpinning the EU's conceptualisation of a 'just transition.' Possible tools to allow European institutions to mitigate or prevent such conflicts are briefly considered, including environmental, social and human rights impact assessments and the mapping of existing socio-energy relationships. Applying these tools within the context of the EGD may allow policymakers to make informed decisions about whether critical minerals are being obtained in a sustainable and just manner, well *before* extracting and funnelling those resources to EU markets.

Global Supply Chain Expansion and Implications for Local Communities beyond the EU

Drivers of Increasing Demand for Critical Minerals: A Crisis on the Horizon?

Although fossil fuels have been central to human ways of living (Thompson, 2022), a global geopolitical and economic shift away from their extraction and utilisation is now in progress. Fossil fuel reliance is increasingly viewed as posing an intolerable risk to the environment (CCES, 2021; Denchak, 2022). The EU gas crisis has added to the multifactorial urgency driving the transition away from fossil fuels. The recent natural gas shortages, financial shockwaves and market volatility due to the Russia-Ukraine conflict have highlighted non-climate risks, including concerns over domestic energy security and strain on democratic institutions (Ali, 2022). European natural gas prices have skyrocketed some 700%, and global commodity prices have increased rapidly since early 2021 (Freitas et al., 2022). This market volatility has amplified what geographers call 'the social perception of risk' related to these commodities (Demski et al., 2014). As society grapples with increased costs and reduced availability, some countries (e.g. France, Germany) are contemplating gas sharing, rationing, and rolling blackouts (Alderman, 2022). Despite troubled histories, others are turning to gas extraction using hydraulic fracturing methods (e.g. the United Kingdom, The Netherlands, Germany, Hungary) (Peigné, 2022).

There are also firm political commitments in place to achieve 'net zero' carbon within certain time frames. The EGD, for example, proposes to cut GHG emissions by at least 55% of 1990 levels by 2030 (European Commission, 2020a). The United States (US) has re-joined the Paris Agreement and committed to slashing GHG emissions by 50% from 2005 levels by 2030 (Debusmann & Stallard, 2022). Emerging economies with large infrastructure needs (e.g. India, China) are expanding their renewable energy infrastructure (Unwin & Farmer, 2022). The financial sector is also undergoing investment shifts towards green energy technology, including changes to fossil fuel finance. Divestment by banks, pension plans, endowments and investors has shifted nearly USD14.5 trillion in investments away from carbon-based commodities and companies towards 'green' portfolios and environmentally conscious funds (Carlin, 2021). Many mining companies, including coal mining giants like BHP, are also pivoting to a 'greener model' and focusing on procuring CET minerals (BHP, 2022). These dynamics have spurred a new resource race with a massive increase in mining activity (Sonter et al., 2020).

The minerals required for a CET, termed critical minerals, are set to replace fossil fuels as the key to energy security (Thompson, 2022). Critical minerals include cobalt, lithium, graphite, copper, gold, bauxite and rare earth minerals,[1] which possess high-level magnetic, catalytic and optical properties that are integral to technological advancement (European Commission, 2017), including green technologies such as wind turbines, solar panels, battery storage and EVs. The dictates of forward policy

[1] Rare earth minerals include niobium, vanadium, platinoids, tungsten, germanium, antimony, fluorine, rhenium, tantalum, antimony and beryllium (Pitron 2020).

planning have committed governments to securing critical minerals via extraction projects that will affect societies and environments for decades to come (Sprecher & Kleijn, 2021).

While a reconceptualisation of energy production and consumption is an urgent matter, the rapid acceleration towards renewable energy creates a conundrum for source communities. Once green technologies are in finished form, they produce net-zero energy. However, the manufacture of renewable energy technology and infrastructure (including wind turbines, solar panels and geothermal power) is far more material intensive than existing technologies that rely upon non-renewables, such as petroleum and gas (Gloaguen et al., 2022). The mining of critical minerals drastically impacts landscapes. Based on current technology, mining tiny quantities of rare earth minerals is heavily polluting and requires displacing tonnes of material (Glenny, 2022). Further, the extraction of minerals is, itself, linked to climate change (Ali et al., 2017). Existing methodologies used to analyse the environmental impacts of global supply chains are currently insufficient in capturing the full suite of environmental, social and governance concerns (Lee et al., 2020). Although economists argue that material shortages will be short-lived, key factors governing critical mineral supply may not be physical in nature but rather socioenvironmental and economic (Mudd & Jowitt, 2018; Lee et al., 2020; Tilton et al., 2018; Klinger, 2023).

Quantifying Future Demand for Key Battery Materials

Batteries are expected to play an important role in achieving EGD targets (European Commission, 2019a). Batteries are largely deployed in the transportation and stationary energy storage sectors, including EVs which emit much lower GHG emissions, especially when charged with renewable-based electricity (Ma et al., 2012; Hoekstra, 2019). Stationary battery energy storage provides flexibility services for power systems, such as peak shaving, thus increasing the stability of the electricity system and lowering the demand for high-GHG power plants such as natural gas-based electricity generation (Peters et al., 2021). The EU plans a rapid deployment of alternative fuels infrastructure, including by boosting the uptake of zero and low emissions vehicles, renewable and low-carbon fuels, and related infrastructure for all modes of transport (European Commission, 2021b). In 2023, the European Parliament approved legislation banning the sale of internal combustion engine cars by 2035 (European Parliament, 2023).

This will require expansions in the battery supply chain ecosystem amongst member states, from extraction and processing of raw materials to manufacturing of battery cells and battery packs, and their use, second use, recycling and disposal in a circular economy context (European Commission, 2021c). The rapid influx of electric vehicles will also require massive investment in the electrical grid, with an estimated 400% increase in electrical charging capacity by 2025 (European Commission, 2021b).

The raw materials needed to produce lithium nickel cobalt manganese oxide (NMC) and lithium iron phosphate (LFP) batteries include lithium, nickel, cobalt, phosphate, aluminium, copper and other key elements (Xu et al., 2020). Studies assessing the demand for battery raw materials under different geographical scopes, time horizons, and battery and material types have found that raw material demand for batteries used in EVs, stationary energy storage and other applications will increase by a considerable factor in the short term (Huisman et al., 2020; Xu et al., 2020; Simon et al., 2015). Xu et al. (2020) estimate that global material demand for electric car batteries alone is estimated to increase by factors of 18–20 for lithium, 17–19 for cobalt, 28–31 for nickel and 15–20 for most other materials from 2020 to 2050. These trends indicate that the EV battery sector alone will consume as much as the 2019 global raw materials supply for all sectors.

The raw material supply challenge varies from region to region, but the EU faces one of the biggest challenges (Leonard et al., 2021). The EU lacks sufficient domestic raw material required for the CET, depending upon imports from abroad (Mancini et al., 2018; European Commission, 2022c). To address this, the European Commission released a list of priority critical minerals and proposed critical raw materials legislation (Schrijvers et al., 2020; European Commission, 2020b; 2023b). While EU member states rely extensively upon international markets (e.g. China), recent geopolitical tensions have highlighted domestic energy security concerns and prompted towards utilising domestic supplies (Yergin, 2020; Glenny, 2022). Finland, France, Sweden, Slovakia, Austria, the Czech Republic, Germany, Ireland, Portugal, Spain, Greece and Poland are potential sources of raw battery material (European Commission, 2018). However, domestic sourcing is not as simplistic as granting an extraction licence and permitting companies to commence operations where large deposits have been proven, due partly to community conflict, resistance and NIMBYISM[2] in EU member states (e.g. Spain, Serbia).

Concerns about the sustainable supply of battery raw materials have been voiced, especially in relation to the mining process (Sharma & Manthiram, 2020). Understanding the impacts of future demand for key battery materials requires an assessment of the supply risk and the socio-environmental effects of critical mineral extraction. Using such information, policymakers and industry players can make reliable strategic decisions to ensure a sustainable supply of battery raw materials towards a just transition to renewable energy (Xu et al., 2020). Efforts to achieve this are reflected in proposals around the responsible sourcing of raw materials (Young, 2018) and the Battery Identity Global Passport (Bai et al., 2020). Circular economy approaches (Sverko Grdic et al., 2020) have also been proposed, especially reducing battery material demand (Dominish et al., 2021), extending battery lifespan (Babbitt, 2020), and increasing the recycling rates of valuable battery materials (Gu et al., 2018). However, lithium is among the most challenging metals when it comes to reducing or offsetting its demand because of its dominant role in battery technologies and (currently) limited recycling (Zografos & Robbins, 2020).

Socioenvironmental Conflicts in the Lithium Triangle

Approximately 60% of the world's identified lithium resources lie along a mineral-rich stretch of the Andes bordering Argentina, Bolivia and Chile, known as the 'Lithium Triangle' (Lopez-Calva, 2022). The local ecosystem consists of salt flats, wetlands, grazing pastures, flamingo lakes and glaciers (Taillant, 2019; Pearce, 2022). The reserves are dissolved in ancient underground water, which must be pumped to the surface and evaporated in the sun to concentrate the lithium carbonate (Pearce, 2022). Mining is particularly contentious in the region due to concerns regarding the environment, food insecurity, crop damage, health and socioeconomic impacts, global warming, aesthetic impacts, crime, noise pollution, waste overflow and the scale of mega-projects (EJAtlas, 2022a, 2022b, 2022c; Khadim, 2021). Specific impacts to glacial environments are also concerning. Rock glaciers in the area supply regional water and are critical to ecosystem survival (Picolotti, 2017). Glacial environments are

[2] NIMBYSIM stands for 'Not In My Back Yard,' a succinct way of pejoratively describing local opposition to unwanted land uses (Devine-Wright, 2010).

especially vulnerable to mining because it accelerates their climate-related deterioration (Taillant, 2022), which in turn contributes to rising sea levels, extreme weather events, and potential alterations in the global carbon balance.

Socioenvironmental conflicts since the mid-1990s have affected the viability and success of many projects (Reboratti, 2012). Over the past decades, conflict and resistance have 'increased in vehemence, content and geographic scope' (Dondo, 2014: 6; Smart, 2017; Andrews et al., 2017), to the extent that social movements against mining in Latin America have become very organised and powerful (Bornschier & Vogt, 2021). Carruthers (2008) notes that Latin American environmentalism has a strong social justice component and that environmental concerns are deeply woven into the fabric of popular mobilisation for social justice and equity, because people face environmental threats in the arenas that are most directly salient to their lives and livelihoods. For example, local residents, church groups, environmental NGOs, farmers and wine producers organised against gold mining projects in a movement that swept across Latin America under the slogan 'Water is Worth More than Gold' (Taillant, 2015). In Bolivia, 42 socioenvironmental conflicts have been reported, more than half relating to extraction (EJAtlas, 2022b). In Argentina, 70 such conflicts have been reported, and again, more than half relate to extraction (EJAtlas, 2022a). Extractive conflicts comprise a similar share in Chile, where 60 environmental conflicts have been documented (EJAtlas, 2022c). Specifically with regard to lithium extraction, EJAtlas[3] has documented eight conflicts in Latin America, all located in Argentina, with the exception of one in Bolivia and one in Peru (see Table 11.1). However, this is likely the proverbial 'tip of the iceberg' as environmental consequences intensify from increased mining (e.g. drying up of rivers, aquifers) (Livingstone, 2019; Greenfield, 2022).

Extractive companies are increasingly viewed with suspicion as profit-driven entities that often generate relatively little direct benefit to local communities (Slack, 2009). Whether or not this is truly the case, local communities are legitimately concerned about the socioenvironmental impacts of extraction (Maurin & Vivoda, 2016), particularly where companies have a poor environmental record. Substandard environmental, social and labour practices in the mining sector have largely been the norm

[3] EJAtlas is a visual mapping system that catalogues socioenvironmental conflict, through participatory cartography involving civil society and academics.

Table 11.1 Socioenvironmental conflicts in Latin American: lithium (Source: Data drawn from EJAtlas)

Location	*Year*	*Issues*	*Mobilisation strategies*	*Current stage*
Salar del Hombre Muerto (Argentina)	1997	Water scarcity Pollution Desertification and deforestation Food insecurity Soil contamination Biodiversity impacts Waste overflow Global warming Consumption of fossil waters Socioeconomic impacts Tourism impacts	Widespread, mass mobilization Violence, arrests Creation of alternative reports and knowledge Development of network, collective action Development of alternative proposals Involvement of national and international NGOs Lawsuits, court cases, judicial activism Media based activism, alternative media Objections to the EIA Official complaint letters, petitions Public campaigns, street protest, marches	Mobilisation for reparations once impacts have been felt
Falchani, Carabaya (Peru)	2008	Water scarcity Destruction of glaciers Biodiversity impacts Pollution Deforestation Soil contamination Food insecurity Socioeconomic impacts Exposure to radioactivity Destruction of cave paintings and archeological sites	Some local organising Authorities express concern	Preventative resistance (precautionary phase)

(*continued*)

Table 11.1 (continued)

Location	*Year*	*Issues*	*Mobilisation strategies*	*Current stage*
Salinas Grandes and Laguna de Guayatayoc (Argentina)	2008	Water scarcity Desertification Biodiversity impacts Pollution Soil contamination Food insecurity Waste overflow Consumption of fossil waters Socioeconomic impacts Exposure to unknown risks (e.g. radiation) Impacts upon Indigenous communities Territorial appropriation	Street protests Visible mobilization Blockades Creation of alternative reports and knowledge Development of network, collective action Involvement of national and international NGOs Lawsuits, court cases, judicial activism Media based activism, alternative media Official complaint letters, petitions Public campaigns Street protest, marches Arguments for rights of nature	Preventative resistance (precautionary phase)
Salar de Uyuni (Bolivia)	2009	Water scarcity Lack of information and participation Biodiversity impacts Soil erosion and contamination Food insecurity Socioeconomic impacts	Visible mobilisation Blockades Media based activism, alternative media Objections to the EIA Official complaint letters, petitions Street protest, marches Strikes Hunger strikes, self-immolation	Latent (no visible resistance)

(*continued*)

Table 11.1 (continued)

Location	*Year*	*Issues*	*Mobilisation strategies*	*Current stage*
Tres Quebradas (Argentina)	2016	Pollution Biodiversity impacts Desertification Global warming Soil contamination and erosion Water scarcity Exposure to unknown risks (e.g. radiation) Socioeconomic impacts	Street protests and public campaigns Visible mobilization Artistic and creative actions (e.g. guerilla theatre, murals) Blockades Community-based participative research (popular epidemiology studies, etc.) Creation of alternative reports and knowledge Development of network, collective action Involvement of national and international NGOs Lawsuits, court cases, judicial activism Media based activism, alternative media Objections to the EIA Official complaint letters, petitions	Preventative resistance (precautionary phase)
Tapias, Córdoba (Argentina)	2017	Pollution Biodiversity loss Desertification and deforestation Soil contamination and erosion Water scarcity Exposure to unknown risks (e.g. radiation) Socioeconomic impacts	Street protests, marches Visible mobilization Creation of alternative reports and knowledge Development of network, collective action Media based activism, alternative media Official complaint letters, petitions Public campaigns	Preventative resistance (precautionary phase)

(*continued*)

Table 11.1 (continued)

Location	*Year*	*Issues*	*Mobilisation strategies*	*Current stage*
Sierra de Ancasti (Argentina)	2017	Pollution Biodiversity loss Desertification and deforestation Soil contamination and erosion Water scarcity Exposure to unknown risks (e.g. radiation) Socioeconomic impacts	Some local organizing Artistic and creative actions (eg guerilla theatre, murals) Creation of alternative reports and knowledge Development of a network, collective action Development of alternative proposals Official complaint letters, petitions Public campaigns Street protest, marches	Preventative resistance (precautionary phase)
Salares de Ororoz and Olroz-Cauchari (Argentina)	2018	Pollution Water scarcity Desertification Biodiversity loss Food insecurity Soil contamination Socioeconomic impacts Division of Indigenous communities	Some local organizing Development of network, collective action Lawsuits, court cases, judicial activism Media based activism, alternative media Objections to the EIA Public campaigns Arguments for rights mother nature	In reaction to implementation (during construction/ operation)

in Latin American source countries (Sagebien et al., 2008), increasing the level of distrust, particularly in foreign companies (Kemp et al., 2011). In response, many communities have galvanised and become quite vociferous in their resistance, utilising a variety of mobilisation strategies, including litigation.

Recourse to the courts has been spurred by the constitutional entrenchment of environmental rights and environmental protection in some Latin American countries. Each country in the Lithium Triangle has enacted constitutional provisions that protect the right to a healthy environment (*Constitución de la Nación Argentina*, 1994; *Constitución Política del Estado* [Bolivia], 2009; *Constitución Política de la República de Chile*, 1980). The Bolivian Constitution also recognises rights to water and Indigenous rights to self-determination. These provisions have dramatically expanded access to the courts (Mantecon, 2019; Khadim, 2021), including through an *amparo*, an effective, inexpensive and timely remedy for the protection of individual rights that allows for harmful mining impacts to be addressed (Carballo, 2009). The judiciaries of Argentina and Chile have made some interesting rulings stressing the importance of environmental protection in mining cases (*Matanza Riachuelo*, 2016; *Barrick Exploraciones*, 2019), even going so far as to close down operations where the risk of environmental harm was too great (Pascua Lama, 2022) or where consultation with local Indigenous communities was inadequate (*Coyo*, 2022).

Moreover, the region has seen a rise in just transition litigation that questions the distribution of benefits and burdens of transition policies amongst affected stakeholders (Tigre et al., 2023).[4] These cases invoke human rights, including environmental rights, but may also rely on other national and international laws and principles (ibid.). Just transition cases are not necessarily pro- or anti-climate action but rather demand that climate action is undertaken in a just and inclusive manner with attention to those in vulnerable situations (ibid.). Tigre et al. (2023) identified a number of such cases in Latin America, including five in Chile. One of these led to the invalidation of lithium mining contracts awarded to two foreign companies (*Coyo*, 2022). Given the extent of global demand for critical minerals, more such cases will likely appear before the courts. The judiciary may be tasked with broadening understandings of environmental justice and human rights within the context of just transitions.

Towards Just Institutional Approaches

Elements of a 'Just Transition'

The term 'just transition' has been examined extensively from the perspective of various disciplines, including labour, environmental justice, climate and energy scholarship, organisational studies, political science, political economy, environmental law, legal geography and urban planning. Given the diversity of disciplinary perspectives, there is no universally agreed-upon definition of or framework for just transitions and the concept remains ambiguous, rendering communication and debate difficult (Wang & Lo, 2021). Furthermore, there are currently only a limited number of

[4] See also Chap. 7 in this edited book.

empirical studies applying these frameworks in real-world settings (ibid.). Some scholars have attempted to provide detailed historical and conceptual overviews of the state of the art in just transitions literature (Stevis et al., 2019; Wang & Lo, 2021; Abram et al., 2022). Rather than amalgamate these to provide yet another definition or analytical framework, the authors have chosen to simply highlight the key elements reflected in the literature thus far (see Table 11.2). This provides a basis for discussing the EGD's conception of a just transition and what it means for affected source communities outside Europe.

The literature demonstrates that the most prominent attributes of a just transition are its distributive, procedural, recognition-based and restorative justice aspects, judging by the extent of scholarly focus on these. The broadest developments appear in conceptions of distributive justice, which have expanded over time to include the distribution of burdens and benefits, applied not just to affected workers but also other stakeholders who may disproportionately bear the burden of environmental harms and risks; access to land, water or other resources; and socioenvironmental impacts (Whiteman, 2009; EPA, 2010; Kemp et al., 2011; Newell & Mulvaney, 2013; Heffron & McCauley, 2018; Eisenberg, 2019). Thus, distributive justice considers fairness in terms of economic as well as other societal outcomes, including both intra- and intergenerational equity perspectives (Sovacool et al., 2016; Heffron & McCauley, 2018; Jasanoff, 2018; Fitzgerald, 2022; Sareen & Shokrgozar, 2022).

Recognition-based justice acknowledges that not all members of society are equally valued and that climate change and transitional policies threaten to exacerbate existing inequalities along gender, class and ethnic/racial lines (Healy & Barry, 2017; Delina & Sovacool, 2018; Heffron & McCauley, 2018; Schwanen, 2021; Abram et al., 2022; Tigre et al., 2023). This recognition fosters restorative justice aims, i.e. to rectify injustice caused by a transition (throughout the energy life-cycle) (Heffron & McCauley, 2017; Healy & Barry, 2017; Abram et al., 2022; Burke, 2022). Restorative justice cannot be simplified to a trickle-down approach that assumes that an 'energy transition for all' will benefit everyone.

Other elements of a just transition are that it must be participatory, deliberative, inclusive, legitimate, democratic, sustainable, socially just, as well as spatially and temporally just (see Table 11.1). Procedural and interactional justice guide engagement and decision-making processes around energy transitions. Essentially, the notion that a CET must encompass multiple forms of justice is gaining ground (Abram et al., 2022).

Table 11.2 Elements of a 'just transition'

Distributive	Requires fair treatment, redistribution of burdens and benefits. No group should bear a disproportionate burden of environmental harms/risks. Existing environmental inequalities (e.g. exposure to ill health, localised degradation) are not reproduced or exacerbated	EPA (, 2010); Newell and Mulvaney (2013); Eisenberg (2019)
	Includes mechanisms to ensure fair transition for workers impacted socioeconomically. Guarantees good jobs for people laid off, access to retraining, strong welfare state	Newell and Mulvaney (2013); Niemiec (2015); Abraham (2017); Eisenberg (2019)
	Includes energy justice (i.e. availability, access to energy for everyone)	Newell and Mulvaney (2013); Sovacool et al. (2016)
	Considers fairness of ends achieved (e.g. perceived fairness of concession rights, access to land/water, distribution of social/gendered/environmental impacts, compensation monies, distribution of taxes/royalties/benefits)	Whiteman (2009); Kemp et al. (2011); Heffron and McCauley (2018); Jasanoff (2018); McCauley and Heffron (2018); Schwanen (2021); Abram et al. (2022); Tigre et al. (2023)
	Direct, indirect and cumulative impact of mining benefits the most vulnerable	Hamann and Kapelus (2004); Kemp et al. (2011)
	Includes economic, societal (e.g. health, community, culture, identity, land rights), ecological, Indigenous spiritual outcomes	Whiteman and Mamen (2002)
	Considers equity/fairness of outcome of decision-making process in eyes of affected individuals/groups	Greenberg (1982); Whiteman and Mamen (2002); Whiteman (2009); COP16 (2010)
	Recognises and protects community ways of life, local knowledge, cultural difference	Bullard (2005); Evans and Phelan (2016)
	Includes intergenerational equity	Sovacool et al. (2016)
	Reduces inequality (including worldwide)	Heffron and McCauley (2018); Jasanoff (2018); Fitzgerald (2022); Sareen and Shokrgozar (2022)

(*continued*)

Table 11.2 (continued)

Procedural	Formalised decision-making processes ensure accountability and transparency	Whiteman (2009); Zandvliet and Anderson (2009); Kemp et al. (2011); Sovacool et al. (2016); Heffron and McCauley (2018); McCauley and Heffron (2018); Schwanen (2021); Abram et al. (2022); Tigre et al. (2023)
	Includes informational justice	Whiteman (2009)
	Includes people's perceptions of the fairness of decision-making procedures	Whiteman (2009)
Recognition-based	Recognises that not all members of society are equally valued in current socio-cultural, economic, and political arrangements, and that climate and transitional policies may exacerbate existing inequalities	Delina and Sovacool (2018); Healy and Barry (2017); Heffron and McCauley (2018); Schwanen (2021); Abram et al. (2022); Tigre et al. (2023)
Restorative	Considers how injustice caused by transition can be rectified. Considers energy life-cycle. Redresses past harm through compensation or measures to reduce likelihood of future harm	Heffron and McCauley (2018); Healy and Barry (2017); McCauley and Heffron (2018); Abram et al. (2022); Burke (2022); Tigre et al. (2023)
Interactional	Concerns relationship between conflicting parties (stakeholders treated with respect, honesty, consideration). Considers social conduct that affects human dignity	Whiteman and Mamen (2002); Kemp et al. (2011); Delina and Sovacool (2018)

Participatory and deliberative	Local participation in development/control over resources	Kemp et al. (2011); Scandrett et al. (2022)
	Solicits inclusive, authentic, and influential public participation/engagement on deliberation of energy issues	Delina and Sovacool (2018); Jasanoff (2018); Kumar et al. (2021)
	Requires meaningful involvement: (1) potentially affected communities have appropriate opportunity to participate in decisions about proposed activities; (2) public's contribution can influence regulatory decisions; (3) concerns of all participants are considered in decision-making process; and (4) decision-makers seek/facilitate the involvement of those potentially affected	EPA (, 2010)
	Provides conceptual space for dialogue to develop mutual understanding about interests/values (a measure of whether procedures capture the essence of understanding required to respond to the substance or heart of issues)	Kemp et al. (2011); Niemiec (2015); Abraham (2017); Harrahill and Douglas (2019); Scandrett et al. (2022)
Inclusive	Allows for participation of all affected, including those speaking for nature	Heffron and McCauley (2018); Jasanoff (2018); Stevis et al. (2019); Stevis and Felli (2020); Burke (2022); Fitzgerald (2022); Galgóczi (2022)
Legitimate	Transition must be legitimate in the eyes of citizens	COP16 (2010); Newell and Mulvaney (2013); Gürtler et al. (2021); Burke (2022)
	Considers perceptions of fairness of: decision-making procedures; distribution of outcomes; and interpersonal and informational aspects of encounters between parties	Whiteman (2009)

(*continued*)

Table 11.2 (continued)

Democratic	Strengthens role of democracy in sustainability transitions	Dietz et al. (2001); Evans and Phelan (2016); Healy and Barry (2017); Goddard and Farrelly (2018); Thombs (2019); Swilling (2019); Scandrett et al. (2022); Tschersich and Kok (2022)
Sustainable	Requires justice as unjust transition is not sustainable	Williams and Doyon (2019); Abram et al. (2022); Wang and Lo (2021)
Socially just	Considers human rights	Newell and Mulvaney (2013); Sovacool et al. (2016); Burke (2022)
	Considers Indigenous cultural dimensions (including traditional law and social/environmental/spiritual dimensions)	Whiteman and Mamen (2002); Whiteman (2009)
Spatially and temporally just	Considers spatial and temporal aspects of transition (e.g. where jobs may be lost versus gained, transition timelines)	Heffron and McCauley (2018); McCauley and Heffron (2018); Gürtler et al. (2021); Morena et al. (2020); Stevis and Felli (2020); Sharma and Banerjee (2021); Abram et al. (2022); Fitzgerald (2022); Sareen (2022); Pai (2021)

Just transitions literature highlights the importance of understanding patterns of exploitation, existing inequalities and power relations (Kemp et al., 2011; Newell & Mulvaney, 2013; Miller & Richter, 2014; Healy & Barry, 2017; Sovacool et al., 2016; Newell & Simms, 2021; Abram et al., 2022). Some, like Abram et al. (2022), recognise that the transition will be neither sustainable nor credible if it creates or worsens social inequalities, likely generating a backlash if it is not perceived to be just. However, just as there are clashing versions of what exactly constitutes a just transition (White, 2020), there are differences in understandings of the causes of injustice and who ought to be considered vulnerable (Wang & Lo, 2021). Whether to prioritise the struggles of displaced fossil fuel workers, interstate, interregional or intersectoral inequalities, intergenerational justice, colonised people against settler colonial states or the North-South ecological debt is far from settled (White, 2020).

Nevertheless, a just transition is not exclusive to just one group, nor one set of drivers or sources of transition (Burke, 2022). Here, it may be helpful to return to an earlier definition of 'environmental justice,' as set out by the US Environmental Protection Agency. Environmental justice entails 'the fair treatment and meaningful involvement of *all people* regardless of race, color, national origin, or income with respect to the development, implementation, and enforcement of environmental laws, regulations, and policies,' with particular attention to be paid to historically vulnerable populations (italics added) (EPA, 2010). Salter et al. (2018) view energy justice as necessitating an energy path that minimises and reverses the cumulative impacts of energy systems at local, regional and global levels, including adverse impacts on planetary ecosystems and on communities disproportionately burdened by the impacts of energy production, distribution and consumption.

A growing body of energy justice literature addresses energy supply chains and calls for 'whole energy system' approaches to prevent decarbonisation strategies from 'locking in' patterns of exploitation and dispossession that characterise the current global political economy (Heffron & McCauley, 2014; Jenkins et al., 2014; Healy & Barry, 2017). Forward planning can prevent or mitigate the creation of energy 'sacrifice zones' (Hernández, 2015; Healy & Barry, 2017). Otherwise, projects in the Global North may risk using the Global South as resource and sink (Ajl, 2018; Boyd, 2013; White, 2020), reproducing a form of climate colonialism in the name of CET (Zografos & Robbins, 2020; Sultana, 2022). Shifting attention upstream to supply-side policy analysis reorients attention and responsibility to a new set of actors and relations that may be responsible for energy injustices (Lazarus et al., 2015; Healy & Barry, 2017).

One thing is clear—job creation is a poor proxy for a just transition (Healy & Barry, 2017). Energy transitions must be 'humanised' (Jenkins, 2018). A just transition cannot be guaranteed if the interests and aspirations of communities are poorly negotiated between the parties involved (Evans & Phelan, 2016). White sees the 'just transition' as an analytic and policy discourse that alerts policymakers to the fact that unjust transitions are far more likely than just transitions, all things remaining equal (White, 2020). This is particularly the case where urgency is invoked to bypass regular democratic scrutiny (Newell & Simms, 2021; Kumar et al., 2021). Crisis management tends to frame responses in terms of top-down interventions by those with the power, resources and control over finance, production, infrastructure or the ability to call upon the coercive powers of the state (Newell & Simms, 2021). This limits opportunities for 'more plural, bottom-up, inclusive and deliberative pathways to sustainability' (ibid: 911). A focus on justice necessitates comprehensive, holistic and reflexive interventions that recognise heterogeneous and disparate processes and actors, and are framed around inclusiveness and strong public engagement (Delina & Sovacool, 2018). Given the interconnected nature of energy supply, this cannot be achieved in an energy transition that excludes large segments of (global) society in its conceptualisation and operationalisation of justice.

Current Approaches within the European Green Deal

The current narrative around just transitions within the EGD does not appear to account for impacts beyond European borders. The EGD is an internally focused plan that promises environmentally and socially sustainable objectives, while promoting a 'fair and prosperous' society that protects the health and well-being of *its* people from risk of environmental harm (italics added) (European Commission, 2019c). To facilitate these objectives, the EU has launched the Just Transition Mechanism (JTM), a set of programmes aimed at ensuring that the CET is implemented in a fair and inclusive manner that will 'leave no one behind' (European Commission, 2019b). The JTM provides targeted support within the EU based on a territorial approach (Strambo, 2020). Its focus is to act as a financial buffer to alleviate the socioeconomic impact of the transition on regions of the EU that experience adverse effects (Moesker & Pesch, 2022; European Commission, 2019b, 2022a; Manta et al., 2020).

Aside from funding, community and business engagement measures, such as the European Investment Bank's Just Transition Platform (JTP) and the European Climate Pact (ECP), bring together citizens, businesses, civil society organisations and local authorities to take action and support the CET. The JTP helps to identify and finance investment projects that contribute to the creation of new jobs and growth during the transition. Conferences under the JTP have brought together EU decision-makers, carbon-intensive businesses and regional governments that rely on these businesses, to discuss (among other things) stakeholder involvement and energy poverty (European Commission, 2023a). The ECP is an initiative aimed at fostering support and collaboration among different stakeholders, including governments, businesses and civil society, to develop innovative solutions and technologies that can help reduce GHG emissions and mitigate the impacts of climate change (European Commission, 2021a, 2022b).

Since EGD programmes are in the early stages, it remains to be seen what effects these dramatic restructuring efforts will have on society. Encouragingly, the above-noted measures seem to consider some elements of what a just transition entails, including efforts to address distributive, procedural, recognition-based and restorative justice, and provide for participatory and deliberative mechanisms. However, non-governmental organisations and scholars have already questioned the justness of the JTM from an internal EU perspective (Moesker & Pesch, 2022), and EGD policies appear to be absolutely silent on notions of justice for communities beyond EU borders. A review of policy documents illustrates that while EGD measures are motivated by a need to address climate change, i.e. a global concern that will require global cooperation, the European Commission's efforts focus on EU internalities such as local environment, economy, energy production, the workforce and citizens. There is a notable absence of discussion regarding how the EU will temper the impacts of its policies on external nations and their economies, citizens, and environments.

Abram et al. (2022) observe that there is a clear risk that the opportunity to enhance justice could be wasted if operationalisation of a just transition focuses only on labour impacts. In fact, studies indicate that renewable energy, the so-called green jobs creators may hardly benefit marginalised groups (Wang & Lo, 2021). Just transitions, in their dominant framing and operationalisation through policies such as the EGD, may serve to prolong the unfolding global crisis of climate degradation and social inequality (Bouzarovski, 2022). In addition, the EGD runs a real risk of placing severe pressure on lands held by historically vulnerable and marginalised communities (Zografos & Robbins, 2020), given that more than half of the critical mineral resource base is located on or near the lands of Indigenous and peasant peoples (Owen et al., 2023). More than a third of these lands face a co-occurrence of water risk and suboptimal food security conditions, with lithium being among the most exposed commodities to water risk (ibid.).

Clearly, the large supply of critical minerals in the Lithium Triangle coupled with high demand in the EU risks contributing to pre-existing North-South disparities, raising equity issues (Tigre et al., 2023; Zografos & Robbins, 2020). A just transition should not be viewed solely as an internal concern, but rather should be implemented along the entire supply chain. Otherwise, a 'justice hierarchy' may result in two classes of citizens, those living in Europe and those beyond its borders who shoulder the burden of the CET. Different classes of citizens who are entitled to varying levels of justice based on their regional location or socioeconomic status can hardly be the aim of a just energy transition.

Incorporating Global Dimensions into the EGD's Just Transition

Many regions that supply critical minerals have supplied resources since the colonial era, and EGD policymakers must consciously depart from mapping onto historical colonial relationships, to ensure a truly just and legitimate transition. Ensuring a just transition for communities around the globe may appear to be a daunting task. Yet, there are multiple facets of law, policy and economics that can be combined to create sophisticated tools for addressing current and evolving challenges around the CET (Harrington, 2022).

One possibility may be to establish an independent body within the European Commission that is responsible for operationalising a just transition throughout the critical mineral supply chain. Such a body may allocate funding and work with experts and stakeholders (e.g. national governments, source communities) to set standards, assess and review the impacts of sourcing critical minerals from a particular region. Environmental, social and human rights impact assessments can be instrumental in such assessments (Esteves et al., 2012; Vanclay, 2017; Götzmann,

2019; Khadim & van Eijken, 2022). These may be supplemented with data obtained by mapping existing socio-energy relationships, as demonstrated by Trutnevyte's work with a methodology called Exploration of Patterns in Near-optimal Energy ScEnarios (EXPANSE) (Trutnevyte, 2013, 2014; Miller & Richter, 2014). Trutnevyte examined diverse possible future energy mixes and their implications for achieving diverse community values based on deliberations and social surveys (Trutnevyte, 2013). Whole-systems approaches, as advocated by Abram et al. (2022), offer a conceptual framework that highlights the power of contextual learning and consideration of a plurality of perspectives. Thus, tools have been developed across disciplines that can be adapted to address this problem.

Applying these tools within the context of the EGD may allow policymakers to make informed decisions about whether critical minerals are being obtained in a sustainable and just manner, well *before* extracting and funnelling those resources to EU markets. These strategies can complement efforts to implement battery passport regulation. Though it may be possible, it is unclear how the battery passport can implement such broad conceptions of justice into its methodology. The authors are of the view that the better approach would be to leave questions of justice within the EGD's just transitions framework to be assessed independently and from a broader perspective. This would provide opportunities for timely and meaningful debate and deliberation about the potential trade-offs.

A far-reaching approach on the part of the European Commission may be criticised on a number of grounds, including national sovereignty concerns and the continuation of paternalistic North-South relations. The authors wish to clarify that what is proposed is not a high-handed or top-down approach but rather a deliberative and collaborative process that carefully considers the far-reaching *consequences* of decisions made within the EU, in consultation with affected communities (both internal and external), tailoring policy accordingly. The above are just a few suggestions, and there might be other, better ways to address the issues. However, a detailed discussion of policy alternatives is beyond the scope of this work.

Concluding Remarks

The accelerated pace of the European CET may very well precipitate a crisis in the coming years. Mining for critical minerals can create havoc in many arenas that touch directly upon the lives and livelihoods of individuals and communities in source countries, like those in the Lithium Triangle. Given that the personal stakes can be so high, widespread socio-environmental conflict over the extraction of critical minerals may be imminent in these regions as the CET unfolds. Not only will this affect the ability of the EU to meet its CET targets, but it will also call into question the EU's conceptualisation of the meaning of the term 'just' transition.

While the economic aspects of energy transitions are important, failing to take into account societal impacts limits the potential power of just transitions to affect long-term change (Harrington, 2022). Yet, there is potential for just transitions to be understood as 'imperfect, messy, and embodied projects, while incorporating diverse possibilities for disruptive politics and socioenvironmental justice' (Bouzarovski, 2022). While here is no clear path forward, genuine attention paid to justice issues amidst the CET may assist in forestalling at least some of the socioenvironmental conflict that might otherwise ensue.

Implementation of a broader conception of justice ought to begin with the battery supply chain, given that this is a priority area within the EGD and because of the serious implications for source communities. The authors propose using tools and methodologies such as environmental, social and human rights impact assessments, mapping of socio-energy relationships, and whole-systems approaches that account for conceptions of justice along the entire supply chain. These approaches can provide opportunities for learning and continual improvement that may be refined in the rollout of subsequent phases in the transition. Broad conceptions of justice that include source communities will help to overcome barriers and ensure sustainability within the supply chain. If policymakers are sincere in their desire to change the global environment for the better, they must secure the supply chain in a manner that upholds the principles of a just transition comprehensively.

References

Abraham, J. (2017). Just Transitions for the Miners: Labor Environmentalism in the Ruhr and Appalachian Coalfields. *New Political Science, 39*(2), 218–240.

Abram, S., Atkins, E., Dietzel, A., Jenkins, K., Kiamba, L., Kirshner, J., Kreienkamp, J., Parkhill, K., Pegram, T., & Santos Ayllón, L. M. (2022). Just Transition: A Whole-Systems Approach to Decarbonisation. *Climate Policy, 22*(8), 1033–1049.

Ajl, M. (2018). Beyond the Green New Deal. *The Brooklyn Rail*, November. Retrieved May 31, 2023, from https://brooklynrail.org/2018/11/field-notes/beyond-the-green-new-deal

Alderman, L. (2022). As Russia Chokes Europe's Gas, France Enters Era of Energy 'Sobriety'. *The New York Times*, 5 September. Retrieved May 31, 2023, from https://www.nytimes.com/2022/09/05/business/russia-gas-europe-france.html

Ali, S. H. (2022). The G7's Energy Awakening. *Project Syndicate*, 7 September. Retrieved May 31, 2023, from https://www.project-syndicate.org/commentary/g7-needs-science-based-energy-policies-by-saleem-h-ali-2022-09

Ali, S. H., Giurco, D., Arndt, N., Nickless, E., Brown, G., Demetriades, A., Durrheim, R., Enriquez, M. A., Kinnaird, J., Littleboy, A., Meinert, L. D., Oberhänsli, R., Salem, J., Schodde, R., Schneider, G., Vidal, O., & Yakovleva, N. (2017). Mineral Supply for Sustainable Development Requires Resource Governance. *Nature, 543*(7645), 367–372.

Andrews, T., Elizalde, B., Le Billon, P., Oh, C. H., Reyes, D., & Thomson, I. (2017). *The Rise in Conflict Associated with Mining Operations: What Lies Beneath?* Canadian International Resources and Development Institute.

Babbitt, C. W. (2020). Sustainability Perspectives on Lithium-Ion Batteries. *Clean Technologies and Environmental Policy, 22*(6), 1213–1214.

Bai, Y., Muralidharan, N., Sun, Y., Passerini, S., Whittingham, S., & Belharouak, I. (2020). Energy and Environmental Aspects in Recycling Lithium-Ion Batteries: Concept of Battery Identity Global Passport. *Materials Today, 41*, 304–315.

Barrick Exploraciones Argentina SA y otro c/Estado Nacional. (2019). Corte Suprema de Justicia de la Nación (Argentina), 4 June ('*Barrick Exploraciones*').

BHP. (2022). *Our Future.* Retrieved May 31, 2023, from https://www.bhp.com/about/our-future

Bornschier, S., & Vogt, M. (2021). *Resistance Against Mining in Latin America: Between Institutions and Protest.* Paper presented at the Annual Conference of the Swiss Political Science Association (SVPW/ASSP), 4–5 February, Berne, Switzerland.

Bouzarovski, S. (2022). Just Transitions: A Political Ecology Critique. *Antipode, 54*(4), 1003.

Boyd, D. R. (2013). *Constitutional Recognition of the Right to a Healthy Environment: Making a Difference in Canada.* David Suzuki Foundation.

Bullard, R. D. (2005). *The Quest for Environmental Justice: Human Rights and the Politics of Pollution.* Sierra Club Books.

Burke, M. J. (2022). Post-growth Policies for the Future of Just Transitions in an Era of Uncertainty. *Futures, 136*, 102900.

Carballo, J. (2009). Chapter 8: Argentina. In L. J. Kotze & A. R. Paterson (Eds.), *The Role of the Judiciary in Environmental Governance: Comparative Perspectives*. Kluwer Law International BV.

Carlin, D. (2021). The Case for Fossil Fuel Divestment. *Forbes*, 2 March. Retrieved May 31, 2023, from https://www.forbes.com/sites/davidcarlin/2021/03/02/the-case-for-fossil-fuel-engagement/

Carruthers, D. V. (2008). Introduction Popular Environmentalism and Social Justice in Latin America. In D. V. Carruthers (Ed.), *Environmental Justice in Latin America: Problems, Promise, and Practice*. The MIT Press.

Center for Climate and Energy Solutions (CCES). (2021). *Extreme Weather and Climate Change*. Retrieved May 31, 2023, from https://www.c2es.org/content/extreme-weather-and-climate-change/

Comunidad Atacameña de Coyo c/Ministerio de Minería y otros. (2022). La Tercera Sala de la Corte Suprema (Chile), Rol No 8.507-2022, 1 June ('*Coyo*').

Constitución de la Nación Argentina. (1994). [tr Jonathan Miller and Fang-Lian Liao]. Retrieved May 31, 2023, from https://www.constituteproject.org/constitution/Argentina_1994.pdf?lang=en

Constitución Política de la República de Chile. (1980). [tr Rodrigo Delaveau Swett]. Retrieved May 31, 2023, from https://www.constituteproject.org/constitution/Chile_2021?lang=en

Constitución Política del Estado (Bolivia). (2009). [tr Max Planck Institute]. Retrieved May 31, 2023, from https://www.constituteproject.org/constitution/Bolivia_2009?lang=en

COP16. (2010). Sixteenth session of the Conference of the Parties. United Nations Climate Change. 29 November 2023.

Debusmann, B., & Stallard, E. (2022). Biden unveils $2.3bn plan to fight climate change. *BBC*, 21 July. Retrieved May 31, 2023, from https://www.bbc.com/news/world-us-canada-62241954#:~:text=It%20includes%20%24385m%20to,protect%20workers%20from%20extreme%20weather

Delina, L. L., & Sovacool, B. K. (2018). Of Temporality and Plurality: An Epistemic and Governance Agenda for Accelerating Just Transitions for Energy Access and Sustainable Development. *Current Opinion in Environmental Sustainability, 34*, 1–6.

Demski, C., Poortinga, W., & Pidgeon, N. (2014). Exploring Public Perceptions of Energy Security Risks in the UK. *Energy Policy, 66*, 369–378.

Denchak, M. (2022). Fossil Fuels: The Dirty Facts. *NRDC*, 1 June. Retrieved May 31, 2023, from https://www.nrdc.org/stories/fossil-fuels-dirty-facts#sec-disadvantages

Devine-Wright, P. (2010). Public engagement with large-scale renewable energy technologies: breaking the cycle of NIMBYism. *WIREs Climate Change*, 2(1), 19–26.

Dietz, T., York, R., & Rosa, E. A. (2001). *Ecological Democracy and Sustainable Development*. Paper presented at the 2001 Open Meeting of the Human Dimensions of Global Environmental Change Research Community, 8 October, Rio de Janiero, Brazil.

Dominish, E., Florin, N., & Wakefield-Rann, R. (2021). *Reducing New Mining for Electric Vehicle Battery Metals: Responsible Sourcing Through Demand Reduction Strategies and Recycling*. Institute for Sustainable Futures, University of Technology Sydney.

Dondo, S. J. (2014). *Financial Assurance for Mine Closure: A Regulatory Perspective from the Argentine Context*. The University of Queensland. Occasional Paper Series.

Eisenberg, A. (2019). Just Transitions. *Southern California Law Review, 92*(101), 273.

Environmental Justice Atlas (EJAtlas). (2022a). *Environmental Conflicts in Argentina*. Retrieved May 31, 2023, from https://ejatlas.org/country/argentina

Environmental Justice Atlas (EJAtlas). (2022b). *Environmental Conflicts in Bolivia*. Retrieved May 31, 2023, from https://ejatlas.org/country/bolivia

Environmental Justice Atlas (EJAtlas). (2022c). *Environmental Conflicts in Chile*. Retrieved May 31, 2023, from https://ejatlas.org/country/chile

Environmental Protection Agency (EPA). (2010). *EPA's Action Development Process: Interim Guidance on Considering Environmental Justice During the Development of an Action*. United States Environmental Protection Agency. OPEI Regulatory Development Series.

Esteves, A. M., Franks, D., & Vanclay, F. (2012). Social Impact Assessment: The State of the Art. *Impact Assessment and Project Appraisal, 30*(1), 34–42.

European Commission. (2017). Communication from the Commission to the European Parliament, the European Council, the Council, the European Economic and Social Committee and the Committee of the Regions on the 2017 list of Critical Raw Materials for the EU. 13 September 2017, COM(2017) 490 final.

European Commission. (2018). Communication from the Commission to the European Parliament, the European Council, the Council, the European Economic and Social Committee and the Committee of the Regions: Annex 2 Strategic Action Plan on Batteries. 17 May 2018, COM(2018) 293 final.

European Commission. (2019a). *Delivering the European Green Deal*. Retrieved May 31, 2023, from https://commission.europa.eu/strategy-and-policy/priorities-2019-2024/european-green-deal/delivering-european-green-deal_en

European Commission. (2019b). *The Just Transition Mechanism: Making Sure No One is Left Behind*. Retrieved May 31, 2023, from https://commission.europa.

eu/strategy-and-policy/priorities-2019-2024/european-green-deal/finance-and-green-deal/just-transition-mechanism_en

European Commission. (2019c). *Protecting the Environment and Oceans with the Green Deal*. Retrieved May 31, 2023, from https://commission.europa.eu/strategy-and-policy/priorities-2019-2024/european-green-deal/protecting-environment-and-oceans-green-deal_en

European Commission. (2019d). Communication from the Commission to the European Parliament, the European Council, the Council, the European Economic and Social Committee and the Committee of the Regions: The European Green Deal. 11 December 2019, COM(2019) 640 final.

European Commission. (2020a). Communication from the Commission to the European Parliament, the European Council, the Council, the European Economic and Social Committee and the Committee of the Regions: Stepping up Europe's 2030 climate ambition, Investing in a climate-neutral future for the benefit of our people. 17 September 2020, COM(2020) 562 final.

European Commission. (2020b). Communication from the Commission to the European Parliament, the European Council, the Council, the European Economic and Social Committee and the Committee of the Regions: Critical Raw Materials Resilience: Charting a Path towards greater Security and Sustainability. 3 September 2020, COM(2020) 474 final.

European Commission. (2021a). *About the Climate Pact Community*. Retrieved May 31, 2023, from https://climate-pact.network.europa.eu/book-page/about-climate-pact-community

European Commission. (2021b). Communication from the Commission to the European Parliament, the European Council, the Council, the European Economic and Social Committee and the Committee of the Regions: A strategic rollout plan to outline a set of supplementary actions to support the rapid deployment of alternative fuels infrastructure. 14 July 2021, COM(2021) 560 final.

European Commission. (2021c). *European Battery Alliance*. Retrieved May 31, 2023, from https://single-market-economy.ec.europa.eu/industry/strategy/industrial-alliances/european-battery-alliance_en

European Commission. (2022a). *Climate Action: Free Allocation for the Modernisation of the Energy Sector*. Retrieved May 31, 2023, from https://climate.ec.europa.eu/eu-action/eu-emissions-trading-system-eu-ets/free-allocation/free-allocation-modernisation-energy-sector_en

European Commission. (2022b). European Climate Pact: The European Commission Takes Green Action with a Climate Pact Pledge—But What Does it Mean? Retrieved May 31, 2023, from https://climate-pact.europa.eu/news/european-commission-takes-green-action-climate-pact-pledge-what-does-it-mean-2022-04-22_en

European Commission. (2022c). Report from the Commission to the European Parliament and the Council: Progress on Competitiveness of Clean Energy Technologies. 15 November 2022, COM(2022) 643 final.

European Commission. (2023a). Energy: Just Transition Platform Conference—October 2022. Retrieved May 31, 2023, from https://energy.ec.europa.eu/topics/oil-gas-and-coal/eu-coal-regions/events/just-transition-platform-conference-october-2022_en

European Commission. (2023b). Proposal for a regulation of the European Parliament and of the Council establishing a framework for ensuring a secure and sustainable supply of critical raw materials and amending Regulations (EU) 168/2013, (EU) 2018/858, 2018/1724 and (EU) 2019/1020. 16 March 2023, COM(2023) 160 final.

European Parliament. (2023). *Regulation (EU) 2023/851 amending Regulation (EU) 2019/631 as regards strengthening the CO2 emission performance standards for new passenger cars and new light commercial vehicles in line with the Union's increased climate ambition*. Official Journal of the European Union.

Evans, G., & Phelan, L. (2016). Transition to a Post-carbon Society: Linking Environmental Justice and Just Transition Discourses. *Energy Policy, 99*, 329–339.

Fitzgerald, L. M. (2022). Winning Coalitions for Just Transitions: Insights from the Environmental Justice Movement. *Energy Research & Social Science, 92*, 102780.

Freitas, G., Stapczynski, S., & Shiryaevskaya, A. (2022). Natural Gas Soars 700%, Becoming Driving Force in the New Cold War. *Bloomberg*, 5 July. Retrieved May 31, 2023, from https://www.bloomberg.com/news/articles/2022-07-05/the-global-energy-crisis-just-got-even-worse-here-s-why?sref=QycjPqYS&leadSource=uverify%20wall

Galgóczi, B. (2022). From a 'Just Transition for Us' to a 'Just Transition for All'. *Transfer: European Review of Labour and Research, 28*(3), 349–366.

Glenny, M. (2022). The Scramble for Rare Earths Carries Big Geopolitical Risks. *Financial Times*, 27 September. Retrieved May 31, 2023, from https://www.ft.com/content/a7bf7029-9374-4a77-88b2-b4c21c865cbd

Gloaguen, R., Ali, S. H., Herrington, R., Ajjabou, L., Downey, E., & Stewart, I. S. (2022). Mineral Revolution for the Wellbeing Economy. *Global Sustainability, 5*, E15.

Goddard, G., & Farrelly, M. A. (2018). Just Transition Management: Balancing Just Outcomes with Just Processes in Australian Renewable Energy Transitions. *Applied Energy, 225*, 110–123.

Götzmann, N. (2019). *Handbook on Human Rights Impact Assessment (Research Handbooks on Impact Assessment)*. Edward Elgar.

Greenberg, J. (1982). Chapter 11—Approaching Equity and Avoiding Inequity in Groups and Organizations. In J. Greenberg & R. L. Cohen (Eds.), *Equity and Justice in Social Behavior*. Academic Press.

Greenfield, N. (2022). Lithium Mining Is Leaving Chile's Indigenous Communities High and Dry (Literally). *NRDC*, 26 April. Retrieved May 31, 2023, from https://www.nrdc.org/stories/lithium-mining-leaving-chiles-indigenous-communities-high-and-dry-literally

Gu, X., Ieromonachou, P., Zhou, L., & Tseng, M. (2018). Developing Pricing Strategy to Optimise Total Profits in an Electric Vehicle Battery Closed Loop Supply Chain. *Journal of Cleaner Production, 203*, 376–385.

Gürtler, K., Löw Beer, D., & Herberg, J. (2021). Scaling Just Transitions: Legitimation Strategies in Coal Phase-Out Commissions in Canada and Germany. *Political Geography, 88*, 102406.

Hamann, R., & Kapelus, P. (2004). Corporate Social Responsibility in Mining in Southern Africa: Fair Accountability or Just Greenwash? *Development, 47*(3), 85.

Harrahill, K., & Douglas, O. (2019). Framework Development for 'Just Transition' in Coal Producing Jurisdictions. *Energy Policy, 134*, 110990.

Harrington, A. R. (2022). *Just Transitions and the Future of Law and Regulation (Palgrave Studies in Environmental Transformation, Transition and Accountability)*. Palgrave Macmillan.

Harvey, F., & Rankin, J. (2020). What is the European Green Deal and will it really cost € 1tn?. *The Guardian*, 9 March. Retrieved May 31, 2023, from https://www.theguardian.com/world/2020/mar/09/what-is-the-european-green-deal-and-will-it-really-cost-1tn#:~:text=The%20green%20deal%20will%20work,incentives%20to%20encourage%20private%20sector

Healy, N., & Barry, J. (2017). Politicizing Energy Justice and Energy System Transitions: Fossil Fuel Divestment and a "Just Transition". *Energy Policy, 108*, 451–459.

Heffron, R., & McCauley, D. (2014). Achieving Sustainable Supply Chains Through Energy Justice. *Applied Energy, 123*, 435–437.

Heffron, R., & McCauley, D. (2017). The Concept of Energy Justice across the Disciplines. *Energy Policy, 105*, 658–667.

Heffron, R. J., & McCauley, D. (2018). What is the 'Just Transition'? *Geoforum, 88*, 74–77.

Hernández, D. (2015). Sacrifice Along the Energy Continuum: A Call for Energy Justice. *Environmental Justice, 8*(4), 151–156.

Hoekstra, A. (2019). The Underestimated Potential of Battery Electric Vehicles to Reduce Emissions. *Joule, 3*(6), 1412–1414.

Huisman, J., Ciuta, T., Mathieux, F., Bobba, S., Georgitzikis, K., & Pennington, D. (2020). *RMIS—Raw Materials in the Battery Value Chain—Final Content for the Raw Materials Information System—Strategic Value Chains—Batteries Section*. Publications Office of the European Union.

Jasanoff, S. (2018). Just Transitions: A Humble Approach to Global Energy Futures. *Energy Research & Social Science, 35*, 11–14.

Jenkins, K. (2018). Setting Energy Justice Apart from the Crowd: Lessons from Environmental and Climate Justice. *Energy Research & Social Science, 39*, 117–121.

Jenkins, K., McCauley, D., Heffron, R., & Stephan, H. (2014). Energy Justice, a Whole Systems Approach. *Queen's Political Review, 2*(2).

Kemp, D., Owen, J. R., Gotzmann, N., & Bond, C. J. (2011). Just Relations and Company—Community Conflict in Mining. *Journal of Business Ethics, 101*(1), 93–109.

Khadim, A. (2021). *Environmental Rights and Extractive Industries in Canada and Argentina: A Comparative Constitutional Analysis.* PhD thesis, The University of Queensland, Brisbane.

Khadim, A., & van Eijken, H. (2022). A Citizen-Centric Approach to Evidence-Based Decision-Making under the European Green Deal. *European Journal of Law Reform, 24*(1), 28–46.

Klinger, J. M. (2023). Social and Environmental Impacts of Rare Earth Mining, Processing, and Proliferation: Scoping and Preliminary Assessment. In S. Kalantzakos (Ed.), *Critical Minerals, the Climate Crisis and the Tech Imperium*. Springer Nature Switzerland.

Kumar, A., Höffken, J., & Pols, A. (2021). *Dilemmas of Energy Transitions in the Global South: Balancing Urgency and Justice*. Routledge.

Lazarus, M., Erickson, P., & Tempest, K. (2015). *Supply-Side Climate Policy: The Road Less Taken*. Stockholm Environment Institute.

Lee, J., Bazilian, M., Sovacool, B., & Greene, S. (2020). Responsible or Reckless? A Critical Review of the Environmental and Climate Assessments of Mineral Supply Chains. *Environmental Research Letters, 15*(10).

Leonard, M., Pisani-Ferry, J., Shapiro, J., Tagliapietra, S., & Wolff, G. B. (2021). The Geopolitics of the European Green Deal. *Bruegel Policy Contribution, 4*(21).

Livingstone, G. (2019). The Farmers Who Worry About Our Phone Batteries. *BBC*, 15 August. Retrieved May 31, 2023, from https://www.bbc.com/news/business-49207042

Lopez-Calva, L. F. (2022). Lithium in Latin America: A New Quest for "El Dorado"? *UNDP*, 26 May. Retrieved May 31, 2023, from https://www.undp.org/latin-america/blog/graph-for-thought/lithium-latin-america-new-quest-el-dorado

Ma, H., Balthasar, F., Tait, N., Riera-Palou, X., & Harrison, A. (2012). A New Comparison Between the Life Cycle Greenhouse Gas Emissions of Battery Electric Vehicles and Internal Combustion Vehicles. *Energy Policy, 44*, 160–173.

Mancini, L., Benini, L., & Sala, S. (2018). Characterization of Raw Materials based on Supply Risk Indicators for Europe. *The International Journal of Life Cycle Assessment, 23*(3), 726–738.

Manta, O., Gouliamos, K., Kong, J., Li, Z., Ha, N. M., Mohanty, R. P., Yang, H., Pu, R., & Yue, X. (2020). The Architecture of Financial Networks and Models of Financial Instruments According to the "Just Transition Mechanism" at the European Level. *Journal of Risk and Financial Management, 13*(10), 235.

Mantecon, C. L. (2019). Effectiveness of the Amparo Action in Environmental Matter from its Inclusion in the National Argentine Constitution. *Boletín Mexicano de Derecho Comparado, 52*(155), 939.

Maurin, C., & Vivoda, V. (2016). Shale Gas and the Energy Policy 'Trilemma'. In T. Hunter (Ed.), *Handbook of Shale Gas Law and Policy*. Intersentia.

McCauley, D., & Heffron, R. (2018). Just Transition: Integrating Climate, Energy and Environmental Justice. *Energy Policy, 119*, 1–7.

Mendoza, Beatriz Silvia y otros c/Estado Nacional y otros s/daños y perjuicios (daños derivados de la contaminación ambiental del Río Matanza Riachuelo. (2016). Corte Suprema de Justicia de la Nacion (Argentina), Fallos, M 1569 XL, 27 December 2016 ('*Matanza Riachuelo*').

Miller, C. A., & Richter, J. (2014). Social Planning for Energy Transitions. *Current Sustainable/Renewable Energy Reports, 1*(3), 77–84.

Moesker, K., & Pesch, U. (2022). The Just Transition Fund—Did the European Union Learn from Europe's Past Transition Experiences? *Energy Research & Social Science, 91*, 102750.

Morena, E., Krause, D., & Stevis, D. (2020). *Just Transitions: Social Justice in the Shift Towards a Low-Carbon World*. Pluto Press.

Mudd, G. M., & Jowitt, S. M. (2018). Global Resource Assessments of Primary Metals: An Optimistic Reality Check. *Natural Resources Research, 27*, 229–240.

Newell, P., & Mulvaney, D. (2013). The Political Economy of the 'Just Transition'. *The Geographical Journal, 179*(2), 132–140.

Newell, P., & Simms, A. (2021). How Did We Do That? Histories and Political Economies of Rapid and Just Transitions. *New Political Economy, 26*(6), 907–922.

Niemiec, J. (2015). What is Just Transition, and Why Do We All Need to Get Behind it? *Speech to European Trade Union Confederation*. 20 April.

Owen, J. R., Kemp, D., Lechner, A. M., Harris, J., Zhang, R., & Lèbre, E. (2023). Energy Transition Minerals and Their Intersection with Land-Connected Peoples. *Nature Sustainability, 6*, 203–211.

Pai, S. (2021). *Fossil Fuel Phase Outs to Meet Global Climate Targets: Investigating the Spatial and Temporal Dimensions of Just Transitions*. PhD thesis, University of British Columbia, Vancouver.

Pascua Lama. (2022). Tercera Sala de la Corte Suprema de Chile, Rol No 127.275-2020, 12 July 2022.

Pearce, F. (2022). Why the Rush to Mine Lithium Could Dry Up the High Andes. *YaleEnvironment360*, 19 September. Retrieved May 31, 2023, from https://e360.yale.edu/features/lithium-mining-water-andes-argentina

Peigné, M. (2022). Europe's Energy Crisis is Reviving the Fracking Industry. *Investigate Europe*, 1 September. Retrieved May 31, 2023, from https://www.investigate-europe.eu/en/2022/europes-energy-crisis-is-reviving-the-fracking-industry/

Pérez, A. (2021a). *Green Deals in a Time of Pandemics: The Future will be Contested Now*. Libros en Acción.

Pérez, A. (2021b). A Green New Deal for Whom?. *openDemcracy*, 23 April. Retrieved May 31, 2023, from www.opendemocracy.net/en/oureconomy/green-new-deal-whom/

Peters, I., Breyer, C., Jaffer, S. A., Kurtz, S., Reindl, T., Sinton, R., & Vetter, M. (2021). The Role of Batteries in Meeting the PV Terawatt Challenge. *Joule, 5*(6), 1353–1370.

Picolotti, R. (2017). ¿Qué protege la ley de glaciares y qué se esconde detrás de su modificación? *La Voz*, 10 December. Retrieved May 31, 2023, from https://www.lavoz.com.ar/opinion/que-protege-la-ley-de-glaciares-y-que-se-esconde-detras-de-su-modificacion/#:~:text=La%20ley%20de%20glaciares%20incorpora,glaciares%20y%20el%20ambiente%20periglacial

Reboratti, C. (2012). Socio-environmental Conflict in Argentina. *Journal of Latin American Geography, 11*(2), 3.

Sagebien, J., Lindsay, N., Campbell, P., Cameron, R., & Smith, N. (2008). The Corporate Social Responsibility of Canadian Mining Companies in Latin America: A Systems Perspective. *Canadian Foreign Policy Journal, 14*(3), 103–128.

Salter, R., Gonzalez, C. G., & Kronk Warner, E. A. (Eds.). (2018). *Energy Justice US and International Perspectives*. Edward Elgar Publishing.

Sareen, S. (2022). Drivers of Scalar Biases: Environmental Justice and the Portuguese Solar Photovoltaic Rollout. *Environmental Justice, 15*(2), 98–107.

Sareen, S., & Shokrgozar, S. (2022). Desert Geographies: Solar Energy Governance for Just Transitions. *Globalizations*, 1–17.

Scandrett, E., Dey, S., & Graham, S. (2022). Just Transitions. *Community Development Journal, 57*(1), 167–178.

Schomberg, A. C., Bringezu, S., & Flörke, M. (2012). Extended Life Cycle Assessment Reveals the Spatially-Explicit Water Scarcity Footprint of a Lithium-Ion Battery Storage. *Communications Earth & Environment, 2*(1), 11.

Schrijvers, D., Hool, A., Blengini, G. A., Chen, W., Dewulf, J., Eggert, R., van Ellen, L., Gauss, R., Goddin, J., Habib, K., Hagelüken, C., Hirohata, A., Hofmann-Amtenbrink, M., Kosmol, J., Gleuher, M. L., Grohol, M., Ku, A., Lee, M., Liu, G., et al. (2020). A Review of Methods and Data to Determine Raw Material Criticality. *Resources, Conservation and Recycling, 155*, 104617.

Schwanen, T. (2021). Achieving Just Transitions to Low-Carbon Urban Mobility. *Nature Energy, 6*(7), 685–687.

Sharma, A., & Banerjee, R. (2021). Framework to Analyze the Spatial Distribution of the Labor Impacts of Clean Energy Transitions. *Energy Policy, 150*, 112158.

Sharma, S. S., & Manthiram, A. (2020). Towards More Environmentally and Socially Responsible Batteries. *Energy & Environmental Science, 13*(11), 4087–4097.

Simon, B., Ziemann, S., & Weil, M. (2015). Potential Metal Requirement of Active Materials in Lithium-Ion Battery Cells of Electric Vehicles and its Impact on Reserves: Focus on Europe. *Resources, Conservation and Recycling, 104*, 300–310.

Slack, K. (2009). Digging Out from Neoliberalism: Responses to Environmental (Mis)governance of the Mining Sector in Latin America. In J. Burdick, P. Oxhorn, & K. M. Roberts (Eds.), *Beyond Neoliberalism in Latin America? Societies and Politics at the Crossroads* (pp. 117–134). Palgrave Macmillan.

Smart, S. (2017). Resistance Against Mining Extractivism in Chile. *Critical Planning, 23*, 59–81.

Sonter, L. J., Dade, M. C., Watson, J. E. M., & Valenta, R. K. (2020). Renewable Energy Production will Exacerbate Mining Threats to Biodiversity. *Nature Communications, 11*(1), 1–6.

Sovacool, B. K., Heffron, R. J., McCauley, D., & Goldthau, A. (2016). Energy Decisions Reframed as Justice and Ethical Concerns. *Nature Energy, 1*(5), 16024.

Sprecher, B., & Kleijn, R. (2021). Tackling Material Constraints on the Exponential Growth of the Energy Transition. *One Earth, 4*(3), 335–338.

Stevis, D., & Felli, R. (2020). Planetary Just Transition? How Inclusive and How Just? *Earth System Governance, 6*, 100065.

Stevis, D., Morena, E., & Krause, D. (2019). Introduction: The Genealogy and Contemporary Politics of Just Transitions. In E. Morena, D. Krause, & D. Stevis (Eds.), *Just Transitions: Social Justice in the Shift Towards a Low-Carbon World*. Pluto Press.

Strambo, C. (2020). Just Transition and the Geopolitics of Decarbonization in the EU. *Mistra Geopolitics*. Stockholm, Sweden: Stockholm Environment Institute. Policy Brief, vol 50.

Sultana, F. (2022). Critical Climate Justice. *The Geographical Journal, 188*(1), 118–124.

Sverko Grdic, Z., Krstinic Nizic, M., & Rudan, E. (2020). Circular Economy Concept in the Context of Economic Development in EU Countries. *Sustainability, 12*(7), 3060.

Swilling, M. (2019). *The Age of Sustainability: Just Transitions in a Complex World*. Routledge.

Taillant, J. D. (2015). *Glaciers: The Politics of Ice*. Oxford University Press.

Taillant, J. D. (2019). *Argentine Supreme Court Upholds Glacier Law*. Center for Human Rights and Environment. Retrieved May 31, 2023, from https://center-hre.org/argentine-supreme-court-upholds-glacier-law/

Taillant, J. D. (2022). *Meltdown: The Earth Without Glaciers*. Oxford University Press.

Thombs, R. P. (2019). When Democracy Meets Energy Transitions: A Typology of Social Power and Energy System Scale. *Energy Research & Social Science, 52*, 159–168.

Thompson, H. (2022). The Geopolitics of Fossil Fuels and Renewables Reshape the World. *Nature, 603*(7901), 364.

Tigre, M. A., Zenteno, L., Hesselman, M., Urzola, N., Cisterna-Gaete, P., & Luporini, R. (2023). *Just Transition Litigation in Latin America: An Initial Categorization of Climate Litigation Cases Amid the Energy Transition*. Sabin Center for Climate Change Law, Columbia Law School.

Tilton, J. E., Crowson, P. C. F., DeYoung, J. H., Jr., Eggert, R. G., Ericsson, M., Guzmán, J. I., Humphreys, D., Lagos, G., Maxwell, P., Radetzki, M., Singer, D. A., & Wellmer, F. (2018). Public Policy and Future Mineral Supplies. *Resources Policy, 57*, 55–60.

Treadgold, T. (2021). How The Metal Dubbed 'White Petroleum' Could Be The New Black For Big Oil. *Forbes*, 2 February. Retrieved May 31, 2023, from https://www.forbes.com/sites/timtreadgold/2021/02/02/white-could-be-the-new-black-for-big-oil/?sh=c86d58e207bb

Trutnevyte, E. (2013). EXPANSE Methodology for Evaluating the Economic Potential of Renewable Energy from an Energy Mix Perspective. *Applied Energy, 111*(C), 593–601.

Trutnevyte, E. (2014). The Allure of Energy Visions: Are Some Visions Better than Others? *Energy Strategy Reviews, 2*, 211–219.

Tschersich, J., & Kok, K. P. W. (2022). Deepening Democracy for the Governance Toward Just Transitions in Agri-food Systems. *Environmental Innovation and Societal Transitions, 43*, 358–374.

Unwin, J., & Farmer, M. (2022). The Top 10 Countries with the Largest Wind Energy Capacity in 2021. *Power Technology*, 14 March. Retrieved May 31, 2023, from https://www.power-technology.com/features/wind-energy-by-country/

Vanclay, F. (2017). Principles to Gain a Social Licence to Operate for Green Initiatives and Biodiversity Projects. *Current Opinion in Environmental Sustainability, 29*, 48–56.

Wang, X., & Lo, K. (2021). Just Transition: A Conceptual Review. *Energy Research & Social Science, 82*, 102291.

White, D. (2020). Just Transitions/Design for Transitions: Preliminary Notes on a Design Politics for a Green New Deal. *Capitalism Nature Socialism, 31*(2), 20–39.

Whiteman, G. (2009). All My Relations: Understanding Perceptions of Justice and Conflict between Companies and Indigenous Peoples. *Organization Studies, 30*(1), 101–120.

Whiteman, G., & Mamen, K. (2002). Examining Justice and Conflict Between Mining Companies and Indigenous Peoples: Cerro Colorado and the Ngabe-Bugle. *Journal of Business and Management, 8*(3), 293.

Williams, S., & Doyon, A. (2019). Justice in Energy Transitions. *Environmental Innovation and Societal Transitions, 31*, 144–153.

Xu, C., Dai, Q., Gaines, L., Hu, M., Tukker, A., & Steubing, B. (2020). Future Material Demand for Automotive Lithium-Based Batteries. *Communications Materials, 1*(1), 1–10.

Yergin, D. (2020). *The New Map: Energy, Climate, and the Clash of Nations.* Penguin Press.

Young, S. B. (2018). Responsible Sourcing of Metals: Certification Approaches for Conflict Minerals and Conflict-Free Metals. *The International Journal of Life Cycle Assessment, 23*(7), 1429–1447.

Zandvliet, L., & Anderson, M. B. (2009). *Getting it Right: Making Corporate-Community Relations Work.* Greenleaf Publishing.

Zografos, C., & Robbins, P. (2020). Green Sacrifice Zones, or Why a Green New Deal Cannot Ignore the Cost Shifts of Just Transitions. *One Earth, 3*(5), 543–546.

CHAPTER 12

'Sustainable Energy' under Reformed International Investment Agreements

Monika Feigerlová

Introduction

The tensions between climate change and international investment law have been recognised. In 2022 the climate mitigation report of the Intergovernmental Panel on Climate Change (IPCC) found that investor-state dispute settlement (ISDS) mechanism contained in international investment agreements (IIAs) that are designed to protect the interests of foreign investors can impede national mitigation efforts by allowing fossil fuel companies to block national phase-out legislation (Patt et al., 2022: 1505–1506). Similarly, climate scientists expressed concerns over the Energy Charter Treaty (ECT) and the achievement of the European Union (EU) climate neutrality target on the basis of the proposed treaty's modernisation (Saheb et al., 2022). Climate change has now become a mainstream issue for investment treaty policymakers (Paine, 2023) as well

M. Feigerlová (✉)
Institute of State and Law of the Czech Academy of Sciences, Prague, Czech Republic
e-mail: monika.feigerlova@ilaw.cas.cz

D. Iglesias Márquez et al. (eds.), *Legal Challenges at the End of the Fossil Fuel Era*, Global Issues,
https://doi.org/10.1007/978-3-031-61766-9_12

as academic debates with divergent conclusions (see, e.g. Brauch, 2022, for negative views on the investment treaty regime; Tietje, 2022, for opposite assessments).

International investment needs to be a leverage for and not a threat to the transition towards a low-carbon and sustainable economy. The 2021 NDC Synthesis report, prepared under the United Nations Framework Convention on Climate Change (UNFCCC), has shown that the most frequent mitigation measures adopted by States in order to combat climate change are measures for renewable energy (RE) generation (UNFCCC Secretariat, 2021: para. 19). Many governments set targets for increasing renewables in the energy mix and the electrification of key sectors of their economies. Yet, national renewable energy (RE) policies or their amendments have already been the subject of proceedings before international investment tribunals, and in many cases, States have been found liable for breaches of international investment agreements in relation to reductions in feed-in-tariffs and bonuses (notably Spain, Italy and the Czech Republic). In the so-called solar energy saga decisions of arbitral tribunals displayed divergent interpretations on certain key issues and heightened the controversy around the ISDS. The Energy Charter Treaty (ECT), which is undergoing a modernisation process with an uncertain outcome at the moment, has provided the legal basis for more disputes relating to renewable energy investments than those relating to fossil fuel investments (i.e. approximately 60% of the 150 investment cases known as of 1 June 2022 related to renewable energy; see ECT Secretariat, 2022).

Over the past decade, several countries have reformed or started to reform their bilateral investment treaties, especially the traditional IIAs, to better balance the rights and obligations of investors and States in response to the evolving nature of economic relations and the unwelcome case law of arbitral tribunals over unsettled matters. Traditional IIAs, often referred to as old-generation IIAs, are still prevalent agreements, mostly concluded before 2010. These agreements typically lack provisions aimed at safeguarding the host State's regulatory flexibility for the protection of environment, climate action and sustainable development. United Nations Conference on Trade and Development (UNCTAD), following a similar differentiation approach, uses the launch of UNCTAD's Investment Policy Framework for Sustainable Development in 2012 as a milestone where old-generation agreements encompass those signed between 1959 and 2011, while new-generation IIAs refer to agreements signed between 2012 and 2022 (UNCTAD, 2015, 2023).

The original motivation behind various initiatives, such as the long-standing debates on reforming the procedural aspects of international investment law, particularly the ISDS under the auspices of the United Nations Commission on International Trade Law, and the accompanying Academic Forum on ISDS; the modernisation of the Energy Charter Treaty, formally launched in July 2020; the termination of treaties (e.g. the plurilateral agreement for the termination of intra-EU bilateral investment treaties signed in 2020 that effectively terminated more than 100 existing treaties and limited exposure to ISDS claims among EU Member States), and the recalibration of substantive standards of protection in new treaties in line with UNCTAD and OECD recommendations and guides (e.g. UNCTAD, 2020), did not initially include climate change considerations. The intersection of climate change and investment treaties is the subject to ongoing research and public debate in the broader context of the future of investment treaties (see, e.g. Tuerk et al., 2019; OECD, 2022).

Modern treaty making has focused on elimination of shortcomings of the investment law regime and preservation of State's regulatory space. However, a narrow approach limited to safeguarding public policy space is insufficient in light of the necessary climate action. Protecting the right to regulate is not the primary function of investment agreements; if a State is primarily concerned with its right to regulate, the question arises whether that State should consider entering into an investment agreement (Titi, 2019: 182). International investment law is currently at a critical juncture and requires immediate reform. This reform cannot be delayed until the expiration of sunset clauses, which guarantee the continuation of investment protection for a certain period of time after the treaty's termination (van Aaken, 2022: 176). Negotiations on the modernisation of the ECT have shown that it is difficult to find agreement at a plurilateral level on ending the protection of investments in fossil fuels. The focus has shifted towards aligning the investment treaty regime with international climate commitments, in particular the Paris Agreement (Paine, 2023). In other words, safeguarding public policy space must be complemented by an active role of the IIAs in promoting and facilitating climate-friendly investments. The encouragement of investment that supports sustainable development is a central goal of the EU's investment policy (European Commission, 2023). Therefore, new international investment agreements should include more robust provisions aimed at achieving this objective. The European Council confirmed the commitment of the European

Union and its Member States to implementing the 2030 Agenda for Sustainable Development (United Nations, 2015). This commitment encompasses the goal of ensuring access to clean and affordable energy in a comprehensive, integrated and efficient manner, which builds synergies with the implementation of the Paris Agreement and other international commitments (European Council., 2017, para. 12). The critical need for a transition towards low greenhouse gas emissions and climate-resilient development that is fast and fair is recognised in the Paris Agreement.

Responding to these twin challenges—securing sustainable energy and mitigating climate change—is one of the urgent needs that the future international investment system will have to address. It is estimated that transitioning to a climate-resilient economy by mid-century will require significant private investment. Ensuring that investments contribute to decarbonisation, climate neutral economies and sustainable development becomes increasingly crucial as countries grapple with a range of social, pandemic, security and environmental challenges.

In general, very few international investment agreements contain provisions specifically related to climate change or energy transition (Polanco, 2017: 635). None of the modern investment treaties have entered into force yet. Treaties containing detailed environmental provisions (e.g. the United States-Mexico-Canada Agreement [USMCA] replacing NAFTA), sustainable development provisions or references to international environmental agreements, including to the UNFCCC and the Paris Agreement, are mostly part of more comprehensive free trade agreements (FTAs) outside the investment chapters (e.g. the Comprehensive Economic Trade Agreement [CETA] between Canada and the EU, the EU-Japan FTA, the Australia-UK FTA).

Several pioneering agreements explicitly emphasise the importance of 'the use of clean energy and renewable energy sources'. For example, the EU-UK Trade and Cooperation Agreement includes references to parties' commitment to 'climate change' and recognises 'the benefits of sustainable energy, renewable energy, in particular offshore generation in the North Sea'. One of the three main objectives of the EU's proposal to modernise the ECT was to ensure that the modernised ECT better reflects climate change and clean energy transition goals.

How do these recognitions translate into specific undertakings? This chapter will analyse the new provisions promoting sustainable energy, the function of these provisions and the extent to which these undertakings enhance the protection of investments in renewable or sustainable energy.

First, the issue of sustainable energy will be outlined and a review of relevant *expressis verbis* references to sustainable energy and similar concepts in the text of international investment agreements and free trade agreements containing investment chapters will be made. This will illustrate that mostly provisions of a preambular, reaffirming and cooperative nature have appeared in treaty practice. The final part will finish with a conclusion that will reflect on the potential roles of the provisions analysed in contributing to achieving the sustainable development goal (SDG) on sustainable energy and addressing the predictability concerns and interests of investors operating in renewable and clean energy sectors. Possible explanations for the inclusion of these provisions in international agreements on sustainable development will be considered, despite the still unclear implications of such provisions for case law.

Sustainable Energy under New-Generation IIAs

Sustainable Development Goals

IIAs are a relatively recent development in international law, with the first treaty signed in 1959 between Pakistan and Germany (Dolzer & Schreuer, 2008: 18). At the time of the rapid proliferation of IIAs in the late twentieth century, climate change and sustainable development were not considered as urgent as they are now, especially given the unequivocal scientific consensus on the human influence on global warming (IPCC, 2022: para. D.1.1). The UNFCCC on climate change, which addressed the need to stabilise greenhouse gases in the atmosphere at the global level, was adopted in 1992 and entered into force in 1994. The third Conference of the Parties (COP) to the UNFCCC, which laid the basis for negotiations on the Kyoto Protocol, was held in 1997. The Kyoto Protocol, which established a biding carbon emission reduction mechanism for developed countries, entered into force only in 2005, and its second commitment period ended in 2020, superseded by the 2015 Paris Agreement.

The primary objective of international bilateral and multilateral investment agreements has traditionally centred on economic development; even the 1992 UNFCCC goal was linked to 'economic development' that would 'proceed in a sustainable manner' (Article 2). The protection of investors' property rights was at the heart of international investment agreements, as a means for host States to foster their development by attracting foreign investment, and as a means for home States to protect their nationals from expropriation and guarantee of capital transfer. Traditionally, the impact of foreign direct investment (FDI) has not been measured under the investment treaty regime, including, for example, an assessment of its compatibility with host State environmental standards. Even if the investment project in question does not lead to development of the host country, it is still subject to protection under the relevant IIA, provided that it meets the IIA's definition of investment, which is usually broad, i.e. any kind of assets possessing the characteristics of an investment, including a certain duration, commitment of capital or other resources, expectation of gain or profit, or the assumption of risk (Dolzer & Schreuer, 2008: 64).

In 2015, the UN General Assembly adopted a new global framework for sustainable development, the 2030 Agenda, which has 17 sustainable development goals and 169 targets (SDGs) at its core. The SDGs build on the previous eight Millennium Development Goals from 2000 and expand them to include new pressing issues such as climate action or clean energy (United Nations, 2000). Adopted shortly before the Paris Agreement, the 2030 Agenda links the economic and social development goals and upholds the status of sustainable development as a guiding principle of international governance. As such, the concept of sustainable development has both a normative and interpretative function (Schacherer, 2019: 207). Sustainable development is understood as development that puts inclusive and sustainable development goals on an equal footing with economic growth and social development and environmental protection; it seeks to balance social, economic and environmental concerns. The primary objective of the Paris Agreement, which aims to strengthen the global response to climate change by, inter alia, ensuring that finance flows are consistent with a path towards low greenhouse gas emissions and climate-resilient development, is also set within the framework of sustainable development (see the chapeau of Article (2)(1) and paragraph (c) of the same Article of the Paris Agreement).

The EU and its Member States are committed to development that meets the needs of the present without compromising the ability of future generations to meet their own needs (European Council, 2017: para. 7). All EU internal and global actions and policy initiatives must take into account the SDGs from the outset, including international investment agreements. After being granted exclusive competence on foreign direct investment by the 2009 Lisbon Treaty, the EU started negotiating comprehensive, reform-oriented investment rules, mostly in the form of FTAs with investment protection chapters, such as the CETA, the EU-Singapore FTA or the EU-Vietnam FTA. EU Member States are parties to more than a third of all international investment treaties and their approach to treaty making is thus of fundamental importance for other instruments. The principle of sustainable development is enshrined in the Treaty on European Union and in the Treaty on the Functioning of the European Union (see Article 11), which sets out the Union's tasks.

In the context of foreign investment, the idea of sustainable development manifests itself in essentially two main ways. First, international investment agreements should strengthen the political and regulatory stability of the host State to encourage foreign investment without undermining the State's right to regulate on social and environmental matters (see, e.g. Schacherer & Hoffmann, 2019: 570–580). The sustainable development prerogative functions here as an element of balance between economic, social and environmental considerations. Second, the investment law regime can contribute to the goal of sustainable development by providing positive signals to invest in economic activities that are sustainable in the long term and to shape investment flows.

While the incentive function is primarily implemented at the domestic policy level, international investment agreements can play a role in promoting sustainable investment and helping to ensure effective implementation of domestic investment legislation. As research shows, not all aspects of foreign direct investment (FDI) have positive impacts on economic growth and development, on technology transfer or on the environment (see, e.g. Johnson, 2019: 127–128 and references cited therein on voluminous literature concerning FDI's effects on the host country industry). The SDGs are increasingly used to analyse foreign investment and to assess whether international investment rules advance or undermine sustainable development. For example, the protection afforded by international agreements to sustainable development-friendly investments can indicate the host State's objectives and policy priorities and increase predictability for investors. In particular, developing countries that need to attract foreign investment can develop sustainable development criteria to focus on desirable infrastructure and projects that provide relevant benefits to the host country.

Sustainable Energy and Foreign Direct Investment

Among the 17 sustainable development goals, one is dedicated exclusively to energy, which calls for 'affordable, reliable, sustainable and modern energy for all' to be achieved by 2030 (SDG 7). As the UN explains, ensuring access to 'clean' and 'affordable' energy is key to economic development, as the energy system supports all sectors from medicine and education to agriculture, infrastructure, communications and high technology (United Nations, undated). Lack of access to energy hinders economic and human development, 2017. The 17 SDGs are interconnected, and SDG 7 is closely linked to SDG 13, which addresses climate change.

Sustainable energy can be defined as energy derived from natural resources that are capable of being replenished, such as water, wind or solar power, and therefore can be sustained in the long term (definition by Oxford University Press, 2023). The consumption of traditional energy sources such as oil and gas leads to their depletion and excludability and is a major cause of global warming (Leal-Arcas, 2016: 823). Sustainable energy thus must reorient historic development patterns that relied on fossil fuels to a more sustainable trajectory. Decarbonised energy is essential to achieving both the 2030 Agenda for Sustainable Development and the Paris Agreement on climate change (UN Secretary-General, 2021). According to Leal-Arcas, renewable energy can provide clean, secure and affordable energy, and meet all three attributes of the energy sustainability trilemma, i.e. achieving energy security, energy equity and environmental sustainability (Leal-Arcas, 2016: 823–824).

The SDG 7 requires a substantial increase in the share of renewable energy in the global energy mix and improvement in energy efficiency by 2030. This will also lead to the reduction of CO_2 emissions and contribute to SDG 13 on climate change. Adequate investment in energy infrastructure and clean energy technologies will therefore be needed on an unprecedented scale to address the twin energy challenges of ensuring access to clean and affordable energy services for all by 2030 and accelerating the energy transition towards net-zero emissions by 2050 (UN Secretary-General, 2021). Renewable energy deployment still lags behind, especially in transport, industry, heating and cooling (UN Secretary-General, 2021), and FDI can be an important tool to promote renewable energy generation (Schacherer & Hoffmann, 2019: 582–583).

In the context of international investments, the analysis of the SDGs focused primarily on identifying relevant infrastructure needs and how to channel FDI into development of sustainable infrastructure (Johnson, 2019: 135). However, Johnson points out the inaccuracy of the widely held narrative that protection through the IIAs is needed for infrastructure investments where the investor is allegedly most at risk of government actions after having committed significant fixed assets in the given country (Johnson, 2019: 135–136). For example, the EU has ensured stability of financial support for the development of renewable energy (RE) generation in legislative instruments at the national level. The EU Directive on Renewable Energy Sources prohibits EU Member States from changing their RE support schemes in an unpredictable way that negatively affects the rights derived from them, while emphasising that 'policy unpredictability and instability have a direct impact on overall costs of deploying the RE in the EU' (recital 29 and article 6). EU investors should benefit from the stability of feed-in-tariffs, if they are used in the future, under national legislation before national courts.

So-called 'energy transition cases', which we are already seeing in the investment space, may arise from government measures to reduce or phase out fossil fuel use and from changing conditions of markets, including the renewable energy generation market built by regulation (Viñuales, 2022). These disputes are classified in the literature also as sustainable development cases (Schacherer & Hoffmann, 2019: 582–583). They include the numerous claims by foreign investors in the renewable energy sector. A number of States that have revoked or changed their renewable energy commitments and modified the existing legal regime have faced monetary disputes before international investment tribunals. The alignment of RE support schemes with international investment protection norms raises a number of complex issues. This is because the extent of public support directly affects anticipated profits, and investors are therefore sensitive to modifications of the incentive regime.

The approximately 60 disputes that have arisen, mostly based on the Energy Charter Treaty, involve maily three countries, Spain, Italy and the Czech Republic (see ECT and UNCTAD databases for the most recent numbers). In these cases, the climate commitments of the countries were not explicitly at issue, but the core of the proceedings was the measures taken by the governments to meet their clean energy targets. The decisions of arbitral tribunals displayed divergent interpretations of some key issues (e.g. the source of the investor's legitimate expectations and a duty of due diligence) and assessments of factual patterns (e.g. whether the relevant domestic law on renewable energy contained a stability commitment designed to induce investors to invest by protecting them from legislative or regulatory changes). The Czech Republic is a country that has prevailed in all but one of the investment cases, however, with decisions adopted by divided tribunals on the impact of the 'solar levy' (see, e.g. *Antaris*; *WA Investments-Europa Nova*; *I.C.W. Europe Investments*; *JSW Solar*; *Photovoltaik Knopf*; *Voltaic Network*; and *Natland Investment*).

Despite the large number of renewable energy cases, the current practice of investment tribunals does not provide a simple answer as to the question of what protection existing treaties provide for renewables. Some arbitrators have been of the view that the countries that have been most affected by investment claims have adopted inflexible regulatory structures and have been able to largely avoid the urgent and problematic measures they have subsequently sought to use to stabilise the sector (see, e.g. *Stadtwerke München*, para. 260).

Heightened security concerns, particularly following the Russian invasion of Ukraine, are also playing a significant role in energy investment. The relationship between economics and security is likely to be reframed if States increasingly resort to national security considerations as a means to avoid obligations under investment standards (van Aaken, 2022: 173). Van Aaken points to the proliferation of foreign investment review rules on screening investments in recent years and estimates that security exceptions will increase both in terms of their occurrence in the IIAs and their use by the States, which in turn will lead to an increase in adjudication.

Sustainable Energy in the Contractual Design of the New-Generation IIAs

Modern treaty drafting focuses on reforming the IIAs to ensure that investments are consistent with the host State's environmental, climate change and sustainable development measures, and to minimise the risk of these measures being challenged by investors through the ISDS.

Old-generation IIAs rarely included sustainable development in their preambles, except in some North and South American treaties (Schacherer & Hoffmann, 2019: 569). Currently, over 200 international investment treaties contain the term 'sustainable development' in the text of the IIAs collected by UNCTAD (the search result comprises treaties that were signed, are in force and/or were terminated), while as of 2023 a total of

3267 IIAs are in existence (i.e. 2832 bilateral treaties and 435 treaties with investment provisions). The term IIA as used in this contribution includes bilateral treaties, multilateral treaties or investment chapters as a part of larger trade and investment partnerships (FTAs; e.g. Trans Pacific Partnership, CETA). Unlike most bilateral investment agreements, the FTAs commonly contain separate chapters addressing non-economic interests such as environmental protection, labour or sustainable development.

In 2020, UNCTAD formulated recommendations on the eight IIA provisions most in need of reform in line with the sustainable development goals. Modern treaties now contain separate provisions on climate action and sustainable development; sustainable development chapters are now part of EU treaty practice that incorporates investment protection provisions in preferential trade agreements (Gehring & Tokas, 2022).

References to energy transition or sustainable energy are emerging. The EU-UK Trade and Cooperation Agreement incorporates a preambular declaration on 'sustainable energy'. In several treaties, references to 'the use of renewable energy sources' or 'clean energy' appear in preambles or as an area of cooperation, while two treaties mention 'off-shore energy' as a specific point of interest for the contracting parties. The main examples of such references in the preambles, the scope of treaties' coverage and the provisions on subsidies will be analysed below. Other provisions of the IIAs, such as those dealing with sustainable development in general, investor responsibilities or investment protection standards, without mentioning sustainable energy, may also create a positive or negative environment for the energy transition to a low-carbon economy and will not be addressed in this contribution.

Preambular Prerogatives and Object of Treaties

The preamble of any international treaty reveals the intentions, ideas and goals of its drafters. Investor protection has historically been the most important function of the IIAs. The treaty's primary purpose has usually been expressed in the preambles of most IIAs as being to yield economic benefits for the concluding States. Nowadays, new and broader objectives of investment treaties, such as the goal of sustainable development, are gaining importance (Makane & Negm, 2019: 241).

In the UNCTAD database, only four treaties contain references to 'sustainable energy'; one reference appears in the preamble and the others in the treaties' provisions dealing with cooperation. All treaties involve the EU as a contracting party. None of the treaties is a traditional stand-alone investment treaty that would contain substantive standards of investment protection and dispute settlement mechanisms between the respondent State and a claimant investor. Furthermore, a number of FTAs containing investment provisions, notably those concluded by the UK, also emphasise the importance of 'renewable energy' in their preambles.

First, the EU-UK Trade and Cooperation Agreement refers to the parties' commitment to 'climate change' and recognises 'the benefits of sustainable energy, renewable energy, in particular offshore generation in the North Sea' (Preamble 17). The agreement also recognises 'the benefits of trade and investment in energy and raw materials' and 'the importance of supporting the delivery of cost efficient, clean and secure energy supplies' to the EU and the UK (Preamble 14). A separate chapter is devoted to safe and sustainable energy in which the parties reaffirm relevant targets (EU) and ambitions (UK) for the share of gross final energy consumption from renewable energy sources in 2030 (Article 319). In addition to the agreed support for electricity from renewable sources, specific cooperation is envisaged in relation to offshore grid development and the large renewable energy potential of the North Seas region, including cooperation on standards (Articles 321 and 323). Renewable energy is mentioned as particularly important for climate change mitigation and adaptation (Article 401(2)(c)). The EU-UK Trade and Cooperation Agreement is specific in nature due to the UK's previous membership of the EU as it abandons most investment protection standards normally found in IIAs and focuses on investment liberalisation schemes of the trade in services that include international investment (Gehring & Tokas, 2022: 780). Its provisions declaring the fight against climate change as one of the essential elements of the partnership or on environmental protection can be inspiration for the particularly strong language used (see, e.g. Articles 393 and 401).

Similarly, the development of 'sustainable energy technologies' is included as an area of cooperation in the 2017 Australia-EU Framework Agreement (Article 48). The IIA does not provide any specific rights for investments and investors in renewable energy (see also Article 16). The preamble of the treaty refers to sustainable development in economic, social and environmental dimensions (preamble).

The EU Framework Agreements on Comprehensive Partnership with Vietnam and Korea express a commitment to promote sustainable development objectives, including environmental protection and cooperation in combating climate change (preambles), and address cooperation in the energy sector, including fostering the transfer of technologies aimed at 'sustainable energy production and use' (Article 42(1) (c) of the EU-Viet Nam Framework Agreement; Article 17 (1)(d) of the EU-Korea Framework Agreement). However, these treaties do not contain any provisions on investment protection for investors.

Specific references to 'renewable energy' are made in the preambles of a number of UK treaties. For example, a 'commitment to promote the use of renewable energy sources' is included in the preamble of the 2020 Strategic Partnership between Moldova and the UK (not yet in force, in provisional application). Nonetheless, the treaty does not provide for investment protection provisions to investors; it focuses on trade and cooperation in the energy sector, including the development and promotion of renewable energies in an economic and environmentally sound manner, and the reduction of greenhouse gas emissions through renewable energy projects (Article 72 (f) and (g), Chapter 17). A simple reference to cooperation or promotion of the use of renewable energy sources can be found in the preambles of the 2029 Georgia-UK Partnership and Cooperation Agreement or the 2020 Ukraine-UK Political, Free Trade and Strategic Partnership Agreement, where again, no investment protections provisions addressed to investors are present.

From a legal perspective, the impact of preambles is limited because they do not create separate treaty obligations. They do, however, articulate the considerations and goals of the contracting States in entering into the treaty and, thus, provide the context for treaty interpretation within the meaning of Article 31(1) of the Vienna Convention on the Law of Treaties. The 'ordinary meaning' of the terms of the treaty can be ascertained by reference to their 'context' in the treaty (which includes the text of the treaty, its preamble, and annexes) and the 'object and purpose' of the treaty (Article 31(2)). An investment tribunal should therefore take into account the IIA's preamble and object provisions to assist it resolving any question of interpretation.

In the past, investment arbitral tribunals have often resorted to the preamble of the treaty when interpreting the provisions of an investment treaty (see, e.g. *SGS Société Générale*, para. 116; or *Occidental Exploration*, para. 183). For example, in order to reconcile investment protection and environmental policy, arbitral tribunals have considered the two values to be mutually supportive, referring to the wording of the preamble, which mentions both economic and sustainable development (see *Bilcon*, para.

596). Arbitral case law shows that care and caution should be exercised in the formulation of preambles to ensure a balanced interpretation of treaty obligations (Titi, 2019: 168).

Treaties that endorse non-investment values seek to deprioritise investment protection as a principle objective and suggest other values, such as sustainable development or environmental concerns, to be used by arbitral tribunals in interpreting the relevant provisions. The preamble serves as an interpretative tool in that it favours an interpretation that is consistent with the preamble. Preamble references to the importance of renewable energy, sustainable development, climate change mitigation or the clean energy transition are intended to inform the arbitral tribunal's interpretation of the investment treaty standards and to encourage the interpreter of the treaty to arrive at a meaning of treaty's provisions that enable States' efforts to address decarbonisation, the energy transition and climate change. It will depend on the specific language used. For example, preambular statements that simply recognise the importance of the use of renewable energy sources (e.g. the Georgia-UK Partnership and Cooperation Agreement) may provide less interpretative guidance than, for example, text that states that the treaty's investment protection objectives must be achieved in a manner consistent with climate change mitigation or 'in a manner conductive to sustainable development' (e.g. the EU-UK TCA, Article 355).

Although 'objective of the treaty' clauses are not typical of investment treaties, they are not excluded and may be included in the main body of the treaty. Stating the objective of the treaty in one succinct provision can give the objective as an interpretative guide weight beyond that usually given to the preamble (Makane & Negm, 2019: 242). For example, a number of tribunals have invoked the stated purpose of the ECT captured in its Article 2 and in interpreting the FET standard, pursuing the objectives of legal stability and transparency. Given that the stated purpose of the ECT highlights the role of the treaty in providing a legal framework that promotes long-term cooperation in the energy field, tribunals have suggested that the ECT is conceived as enhancing the stability required for such cooperation (e.g. *Antin*, paras. 518–23; *Eiser*, para. 382; or *Stadtwerke München*, para. 378). The ECT's preamble that makes a reference to the UNFCCC has played a much smaller role in renewable energy disputes.

Other proposals for provisions supporting the interpretation of investment treaties in a manner compatible with climate change initiatives and clean energy transition might include rules that give primacy to agreements related to climate change. For example, an IIA could provide that actions in pursuit of the UNFCCC or the Paris Agreement do not violate the IIA (VanDuzer, 2016: 454).

A new trend in treaty making concerns so-called joint interpretation clauses, which give contracting States the power to issue joint interpretations that are explicitly recognised as binding on adjudicative bodies (Marotti, 2020). These provisions may provide that interpretative determinations adopted jointly by treaty parties will be binding on tribunals constituted under the agreement or may even provide for specific bodies that will be competent to issue such interpretation statements. For example, an additional interpretation of CETA's substantive standards of protection is at the time of writing under negotiation between the EU and Canada to clarify certain terms of CETA's provisions on fair and equitable treatment and indirect expropriation with respect to climate change (Draft Decision of the CETA Joint Committee). The Draft Decision requires a CETA's adjudicative body deciding an investment dispute to interpret and apply CETA's investment provisions with 'a due consideration of the commitments under the Paris Agreement and respective climate neutrality objectives of the relevant countries and in a way that allows the States to pursue their respective climate change mitigation and adaptation policies'.

In summary, the preambles of some recent FTAs contain references to general commitments to combat climate change as well as specific references to the importance of sustainable energy and the use of renewable energy sources. Such provisions do not necessarily enhance protection for investors in renewable energy but rather provide governments with flexibility to implement climate change programmes. However, they generally do not promote specific transition policies.

Admission and Protection of Sustainable Energy Investments

While the concept of distinguishing between investments and prioritising certain sectors is beginning to appear in the preambles of recently agreed IIAs, to the best of my knowledge, no treaty has explicitly differentiated between low-carbon and high-carbon investments in a manner that restricts the treaty's scope of protection, as suggested by some authors (e.g. Bernasconi-Osterwalder & Brauch, 2019; Cima, 2021). Limiting the definition of investment to sustainable investments can be one of the avenues of how future investment treaties could potentially support governments to facilitate decarbonisation and climate-resilient investment. The only example of sectoral carve-out for climate purposes is the modernised ECT, which introduces a flexibility mechanism to voluntarily phase out existing investments in coal, oil and fossil gas and exclude the protection of future fossil fuel-related investments (Agreement in Principle, 2022). The design of the proposed mechanism follows two years of negotiations on the ECT reform, during which only the UK and the EU expressed an interest in participating in the phasing out of protection over fossil fuel-related investments listed in Annex NI to the Agreement in Principle. However, due to inconsistent and uncertain assessments of the modernised treaty amongst EU Member States and within the EU institutions, the prospect of a vote on the ECT's amendments by the Energy Charter Conference is no longer realistic. This shift is attributed both to the withdrawal of several EU Member States from the ECT and the European Commission's recent proposal for a coordinated withdrawal from the ECT (European Commission, 2023).

The concept of the carve-out from treaty coverage is seen as an effective way of controlling damage and providing certainty to States, as opposed to policy based carve-outs which aim to limit the protection standards and leave the determination to the relevant adjudicative body (Paine & Sheargold, 2023: 12). Although Paine and Sheargold advocate for a purpose-based carve-out, a sectoral carve-out that removes treaty protection for fossil fuel investments would also bring investment treaties into compliance with Article 2(1)(c) of the Paris Agreement by not encouraging finance flows that are inconsistent with a low-carbon future (Paine & Sheargold, 2023: 13).

Even the progressive FTA between New Zealand and the UK, which recognises 'the need to reduce the use of fossil fuels and to support the global transition to clean energy in order to further the implementation of the SDGs of the 2030 Agenda and the objectives of the UNFCC and the Paris Agreement' (Article 22.8), does not distinguish between low-carbon and high-carbon investments in its investment chapter (Chapter 14). It does, however, contain a commitment by each Party to end unabated coal-fired electricity generation on its territory as part of the clean energy transition and in the context of the parties' membership in the Powering Past Coal Alliance (Article 22.8 (2) (b)).

Another question concerning the definition of investment is whether the concept of sustainable development, which would include sustainable energy, could be part of such a definition. In the past, some arbitral tribunals have attempted to transpose the criterion of the significance of the relevant investment for the host country's economic development into the definition of investment in order for the project to qualify as a protected investment. This idea, which first appeared in the *Fedax* decision (para 43), was restated in the *Salini* decision in 2001 and applied in subsequent decisions, has proved to be highly controversial (Schacherer & Hoffmann, 2019: 571). According to the well-known *Salini* test, four criteria must be established in order to determine the existence of an investment, i.e. a contribution of the investor, a certain duration of the project, the existence of an operational risk, and '[i]n reading the Convention's preamble, one may add the contribution to the economic development of the host State of the investment as an additional condition' (*Salini*, para 52). As there have been different views on what actually constitutes economic development, many subsequent tribunals have questioned the *Salini* test, its feasibility and benefits of the economic development requirement (e.g. *Phoenix*, para 82).

Subsequent treaty practice confirms that there is no consensus in this regard. For example, India 2015 Model Bilateral Investment Treaty (BIT) requires a significance of the foreign investment for the development of the host country, or the 2012 Southern African Development Community Model BIT aims to protect only investments that foster sustainable development. On the other hand, no such requirement is contained in model treaties of European countries, such as the Italy 2022 Model BIT or the Czech Republic 2016 Model BIT. Rare treaties limit the treaty's scope to investments that contribute to the sustainable development of the host State (see article 1(3) of the Reciprocal Investment Promotion and Protection Agreement between Morocco and Nigeria of 2016; not in force).

The suggestion that international investment treaties should protect those energy investments that support the transition towards a climate-neutral energy system has emerged in the context of the ECT's modernisation process referred to above. Providing protection for investments in sustainable energy, while ensuring that investments that undermine sustainable development or exacerbate climate change problems are not similarly covered, could indicate the government's priorities in promoting desirable investments. However, the recognition of renewable energy projects as low-carbon investments does not in itself afford rights to renewable energy investors or lead them to invest more. In this context, Tropper and Wagner express doubts about the usefulness of the modernised ECT for clean energy investors, as the proposed amendments, including, for example, the fair and equitable treatment, are too restrictive for climate-friendly investors and reduce the level of investment protection for renewable energy investments (Tropper & Wagner, 2022: 847). The implications of IIAs for encouraging investments that might not have otherwise occurred depend on the nature of the protection offered.

A smaller number of IIAs regulate the rights of entry for foreign investment (the pre-establishment phase, which addresses whether and how the treaty applies to the acquisition and establishment of investment). With the shift from self-standing agreements on investment to more integrated treaties with investment provisions (TIPs) observed in the last decade, the legal framework governing protection of foreign investment tends to cover both the pre-establishment and post-establishment phases of investment (Tuerk et al., 2019: 154–155). Economic partnerships or free trade agreements combine trade and investment and contain provisions on admission of investment and liberalisation obligations, unlike traditional BITs, especially those of European countries. The United States, Canada and new EU treaties grant, albeit to a limited extent, the right of admission, usually on the basis of national treatment clauses. Typically, certain sectors are excluded through exception clauses, using either a positive list of sectors open to foreign investors or a negative list of sectors that are closed.

Regulating market access through IIAs highlights the importance and potential in defining the appropriate investments to be attracted by the host State, i.e. investments that contribute to decarbonisation, the achievement of sustainable development goals or a climate-neutral economy. In the context of treaties addressing the treatment of investors at the stage of establishment and acquisition of their investments, the focus on sustainable energy investment becomes more important (Gehring & Tokas, 2022: 808).

Most renewable energy sources are considered sustainable based on their rate of natural replenishment and environmental impact. Renewable energy projects, in principle, should meet this requirement. For instance, the EU and Japan, in their economic partnership agreement, have specifically emphasised support for renewable energy resources, which are sustainable (article 16.5). Investments in goods and services related to 'sustainable renewable energy' are considered particularly relevant to climate change mitigation and for achieving sustainable development goals. Japan and EU shall, inter alia, seek to facilitate trade and investment in goods and services of particular relevance to climate change mitigation. Similarly, sustainable renewable energy goods and related services are considered climate-friendly investments in the 2018 EU-Singapore FTA (Article 12.11.2.). Singapore and the EU further stipulate that they 'share the objective of increasing the generation of energy from renewable and sustainable non-fossil sources', in particular through facilitating trade and investment (Article 7.1.). The promotion of 'foreign direct investment in environmental goods and services, including environmental technologies, sustainable renewable energy' is further included in the free trade agreement between the European Union and the Republic of Korea (art 13.6.2.) or the Australia-UK Free Trade Agreement of 2021 (Articles 13.18 and 22.6.2).

Subsidies

Investment disputes concerning renewables have predominantly concerned claims over reduced levels of incentive schemes. In response to this experience, the EU has begun to incorporate provisions in new treaties which provide that a State's decision not to issue, renew or maintain a subsidy in the absence of any specific commitment under law or contract to issue, renew or maintain that subsidy does not constitute a breach of treaty (e.g. Article 8.9.3. of CETA, Article 2.3(3) of EU-Singapore Investment Protection Agreement, Article 17bis of the Modernised ECT). These provisions will help States avoiding similar claims in the future once the government adopts *ex post* regulatory changes over incentive schemes.

Some IIAs exclude subsidies from the national treatment obligation (Article 2.1(2) of the 2018 EU-Singapore Investment Protection Agreement). Such exceptions have rarely been tested in investment arbitration practice but could be useful for governments to avoid challenges to certain climate change measures, such as financial support for local businesses in renewable energy sectors, which might otherwise be problematic on the grounds of discrimination (VanDuzer, 2016: 452).

Given the urgency of the climate crisis and the inadequacy of current measures limited to promoting renewables, countries will increasingly have to resort to direct reductions in fossil fuels. The global phase-out of coal power generation and fossil fuel subsidies will need to be accelerated (Glasgow Climate Pact, 2021).

Some FTAs already limit the use of fossil fuel subsidies with the aim to promote sustainable development. For example, in the EU-Singapore

FTA, countries recognise the objective of progressively reducing fossil fuel subsidies and each party shall ensure the need to reduce greenhouse gas emissions is duly taken into account in the design of public support programmes for fossil fuels (Article 12.11(3)). New Zealand and the UK in their newly signed free trade agreement emphasise the importance of reforming fossil fuel subsidies, with each party committing to take steps to eliminate harmful fossil fuel subsidies with limited exceptions (Article 22.8 (2)(a)). For example, the commitment to phase out environmentally harmful fossil fuel subsidies is also contained in the negotiated partnership between the EU and African, Caribbean and Pacific countries which would encompass over 100 countries (ACP-EU Partnership Agreement (2021), Article 24).

Conclusion

Issues of energy supply and climate change are a matter of common concern given the value of the rights at stake (Leal-Arcas, 2016: 822). Energy is key to achieving both the objectives of the 2030 Agenda for Sustainable Development and the Paris Agreement. Without rapid and deep decarbonisation, the 1.5 °C temperature goal will become unachievable. Massive and rapid deployment of renewable energy, i.e. tripling global renewable power capacity by 2030, is one of the key actions outlined in the Global Roadmap for Accelerated SDG 7 Action.

Modern treaty making has mostly focused on preserving the State's regulatory space in the context of sustainable development, to minimise the risk that climate mitigation or adaptation measures will be challenged by investors through investor-State arbitration. Innovative provisions safeguard governments' flexibility to implement and refine climate change policies and laws to achieve climate goals (e.g. including the right to regulate, reformulation of indirect expropriation or FET clauses, specific provisions clarifying that a State's decision not to issue or maintain a subsidy in the absence of any specific commitment under law or contract does not constitute a breach of the treaty). These provisions are largely a response by States to earlier arbitral decisions in which States were exposed to liability for regulatory measures taken in support of public interests but which compromised the interests of foreign investors. These provisions do not necessarily enhance the protection of renewable energy sources but rather provide governments with flexibility to implement climate change programmes.

Provisions in new IIAs recognising explicitly 'the benefits of sustainable energy' or 'importance of the use of renewable energy sources' are emerging in small numbers, mostly in the form of (weak) preambular commitments and in FTA's with or without investment chapters or investment protection. In general, the focus, in particular in the EU treaties, is on regulating and facilitating investment rather than protecting investment. References to sustainable development or the importance of sustainable and clean energy in preambles have implications for the interpretation of substantive investment norms and help determine the overall objective of the treaty. Some recently concluded post-2020 IIAs refer to commitments to implement the Paris Agreement. However, provisions reiterating the Paris Agreement commitments in IIAs without advancing specific mitigation or transitioning policies have more symbolic value. In addition to the preambular clauses, IIAs also integrate provisions on sustainable development and cooperation in the main text of the treaties, which aim to promote and facilitate more sustainable trade and investment in general, including in renewable energy and new technologies. Again, such provisions appear in FTAs rather than in typical stand-alone investment treaties. Several IIAs address changes in subsidies from an investment perspective, with rare examples of FTA's commitments to reduce fossil fuel subsidies.

These provisions illustrate a shift in the perception of investment regime reform from a purely defensive strategy to a positive mechanism that supports the transition to a low-carbon economy and facilitates investments that are consistent with the Paris Agreement, such as renewable energy. Certainly more than piecemeal changes are needed to achieve a deeper paradigm shift in the international investment system. At the very least, this means stopping providing protection for energy investment in fossil fuels. The concept of fossil fuel carve-out has made its way into the text of the modernised ECT, but this has not been sufficient for several countries given the length of the phase-out periods and the voluntary nature of the phase-out mechanism.

Climate change and the energy transition following the narrative of a just transition are important issues for the future of international investment treaties. It can be observed that investment treaty specialists and environmental and climate experts have started to come to the same table to discuss whether to abandon international investment treaties or how to redesign the purpose and form of international investment treaties to ensure that the investment treaty regime is aligned with climate change priorities (e.g. recent public consultations and OECD surveys).

Foreign direct investment can have varying impacts on economic, social and governance conditions in the host State and does not necessarily lead to or support sustainable development by itself (Johnson, 2019: 129). Much more needs to be done in IIAs' treaty design in terms of facilitation of sustainable investment, including transfers of new technologies relating to sustainable energy, and development of local capacities and infrastructures. At the time of writing, the EU and Angola have just concluded the negotiations for a Sustainable Investment Facilitation Agreement that proposes a new design of investment facilitation treaty without direct rights of investors, seeking to promote investments contributing to sustainable development, including provisions on technical assistance and capacity building for the implementation of the agreement and establishment of a Committee on Investment Facilitation (see Articles 5.6, 7.1 and 7.2 of Draft EU-Angola SIFA). Energy transition pathways will vary from country to country. Despite EU's ambitious climate and sustainable development policies, the effective decarbonisation of the economy cannot occur if the EU Member States act in isolation. The EU, being the world's largest destination and source of inbound and outbound international investments, is in a strong position to influence the reform of the investment treaty regime towards more sustainable (energy) investments (European Parliament, 2022).

Acknowledgments The research for this chapter has been supported/subsidised within the *Lumina quaeruntur* award of the Czech Academy of Sciences for the project 'Climate law' conducted at the Institute of State and Law.

References

Bernasconi-Osterwalder, N., & Brauch, M. D. (2019). Redesigning the Energy Charter Treaty to Advance the Low-Carbon Transition. *Transnational Dispute Management, 1*, 11–12. Retrieved March 1, 2023, from https://www.transnational-dispute-management.com/article.asp?key=2632

Brauch, M. D. (2022, October 17). 'The Agreement in Principle on ECT "Modernization": A Botched Reform Attempt that Undermines Climate Action' Kluwer arbitration blog. Retrieved March 1, 2023, from https://arbitrationblog.kluwerarbitration.com/2022/10/17/the-agreement-in-principle-on-ect-modernization-a-botched-reform-attempt-that-undermines-climate-action/

Cima, E. (2021). Retooling the Energy Charter Treaty for Climate Change Mitigation: Lessons from Investment Law and Arbitration. *Journal of World Energy Law and Business, 14*, 80–81.

Dolzer, R., & Schreuer, C. (2008). *Principles of International Investment law* (1st ed.). Oxford University Press.

ECT Database. *List of Cases.* Retrieved March 1, 2023., from https://www.energychartertreaty.org/cases/list-of-cases/

ECT Secretariat. (2022, June 1). *Statistics of ECT Cases.* Retrieved March 1, 2023, from https://www.energychartertreaty.org/fileadmin/user_upload/All_statistics_-_1_June_2022.pdf

European Council. (2017, June 20). *A Sustainable European future: The EU Response to the 2030 Agenda for Sustainable Development—Council Conclusions.* 10370/17. Retrieved March 1, 2023, from https://www.consilium.europa.eu/media/23989/st10370-en17.pdf

IPCC. (2022). Summary for Policymakers. In P. R. Shukla, J. Skea, R. Slade, A. Al Khourdajie, R. van Diemen, D. McCollum, M. Pathak, S. Some, P. Vyas, R. Fradera, M. Belkacemi, A. Hasija, G. Lisboa, S. Luz, & J. Malley (Eds.), *Climate Change 2022: Mitigation of Climate Change. Contribution of Working Group III to the Sixth Assessment Report of the Intergovernmental Panel on Climate Change.* Cambridge University Press. https://doi.org/10.1017/9781009157926.001

Gehring, M., & Tokas, M. (2022). Synergies and Approaches to Climate Change in International Investment Agreements: Comparative Analysis of Investment Liberalization and Investment Protection Provisions in European Union Agreements. *The Journal of World Investment & Trade, 23*(5–6), 778–812.

Johnson, L. (2019). FDI, International Investment Agreements and the Sustainable Development Goals. In M. Krejewski & R. T. Hoffmann (Eds.), *Research Handbook on Foreign Direct Investment* (pp. 126–148). Edward Elgar Publishing.

Leal-Arcas, R. (2016). Sustainability, Common Concern, and Public Goods. *George Washington International Law Review, 49*, 801–877.

Makane, M., & Negm, M. H. (2019). An African View on the CETA Investment Chapter. In M. Makane & S. Schacherer (Eds.), *Foreign Investment Under the Comprehensive Economic and Trade Agreement (CETA)* (pp. 239–269). Springer.

Marotti, L. (2020). The Proliferation of Joint Interpretation Clauses in New International Investment Agreements: A Mixed Blessing? *ICSID Review, 35*(1–2), 63–81. https://doi.org/10.1093/icsidreview/siaa001

OECD. (2021, November 16). *FDI Qualities Policy Toolkit. Policies for Improving the Sustainable Development Impacts of Investment.* Retrieved March 1, 2023, from https://www.oecd.org/investment/investment-policy/FDI-Qualities-Policy-Toolkit-Consultation-Paper-2021.pdf

OECD. (2022). *Investment Treaties and Climate Change. OECD Public Consultation (January–March 2022)*. Retrieved March 1, 2023, from https://www.oecd.org/investment/investment-policy/OECD-investment-treaties-climate-change-consultation-responses.pdf

Oxford University Press. (2022). *Oxford English Dictionary Online*. Retrieved March 1, 2023, from https://www.oed.com/view/Entry/195210?redirectedFrom=sustainable#eid

Paine, J. (2023, January 27). *2022 in Review: Climate Change as a Mainstream Issue for Investment Treaty Policymakers*. Kluwer Arbitration Blog. Retrieved March 1, 2023, from https://arbitrationblog.kluwerarbitration.com/2023/01/27/2022-in-review-climate-change-as-a-mainstream-issue-for-investment-treaty-policymakers/

Paine, J., & Sheargold, E. (2023). A Climate Change Carve-Out for Investment Treaties. *Journal of International Economic Law, 00*, 1–20. https://doi.org/10.1093/jiel/jgad011

Patt, A., Rajamani, L., Bhandari, P., Ivanova Boncheva, A., Caparrós, A., Djemouai, K., Kubota, I., Peel, J., Sari, A. P., Sprinz, D. F., & Wettestad, J. (2022). 2022: International Cooperation. In P. R. Shukla, J. Skea, R. Slade, A. Al Khourdajie, R. van Diemen, D. McCollum, M. Pathak, S. Some, P. Vyas, R. Fradera, M. Belkacemi, A. Hasija, G. Lisboa, S. Luz, & J. Malley (Eds.), *IPCC, 2022: Climate Change 2022: Mitigation of Climate Change. Contribution of Working Group III to the Sixth Assessment Report of the Intergovernmental Panel on Climate Change*. Cambridge University Press. https://doi.org/10.1017/9781009157926.016

Polanco, R. (2017). Integrating Trade, Investment and Climate Change. In T. Cottier & K. Scheffer (Eds.), *Elgar Encyclopedia of International Economic Law* (pp. 239–269). Edward Elgar Publishing.

Saheb, Y., Sallee, J.-B., Masson-Delmotte, V., et al. (2022). *Letter from climate scientists to EU Leaders on the IPCC Climate Mitigation Report Findings Related to the Energy Charter Treaty*. Retrieved March 1, 2023, from https://endfossilprotection.org/sites/default/files/2022-06/2022-06-21%20Letter%20from%20climate%20scientists%20to%20EU%20leaders.pdf

Schacherer, S., & Hoffmann, R. T. (2019). International Investment Law and Sustainable Development. In M. Krejewski & R. T. Hoffmann (Eds.), *Research Handbook on Foreign Direct Investment* (pp. 563–595). Edward Elgar Publishing.

Schacherer, S. (2019). The CETA Investment Chapter and Sustainable Development: Interpretative Issues. In M. Makane & S. Schacherer (Eds.), *Foreign Investment Under the Comprehensive Economic and Trade Agreement (CETA)* (pp. 207–238). Springer.

Tietje, C. (2022, November 15). 'Der Energiecharta-Vertrag im Kreuzfeuer der Kritik. Ist der Rücktritt vom Energiecharta-Vertrag wirklich ein großer Wurf der Ampelkoalition?' Verfassungsblog. Retrieved March 1, 2023, from https://verfassungsblog.de/der-energiecharta-vertrag-im-kreuzfeuer-der-kritik/

Tropper, J., & Wagner, K. (2022). The European Union Proposal for the Modernisation of the Energy Charter Treaty—A Model for Climate-Friendly Investment Treaties? *The Journal of World Investment & Trade, 23*(5–6), 813–848.

Tuerk, E., Baumgartner, J., & Atansaova, D. (2019). Trends and Reform Debates. In M. Krejewski & R. T. Hoffmann (Eds.), *Research Handbook on Foreign Direct Investment* (pp. 150–172). Edward Elgar Publishing.

Titi, C. (2019). The Right to Regulate. In M. Makane & S. Schacherer (Eds.), *Foreign Investment Under the Comprehensive Economic and Trade Agreement (CETA)* (pp. 159–183). Springer.

UN Secretary-General. (2021). *Global Roadmap for Accelerated SDG7 Action in Support of the 2030 Agenda for Sustainable Development and the Paris Agreement on Climate Change*. Retrieved March 1, 2023, from https://www.un.org/sites/un2.un.org/files/2021/11/hlde_outcome_-_sdg7_global_roadmap.pdf

UNCTAD Database. *Investment Dispute Settlement Navigator*. Retrieved March 1, 2023., from https://investmentpolicy.unctad.org/investment-dispute-settlement/advanced-search

United Nations. (undated). *SDG 7 on Affordable and Clean Energy*. Retrieved March 1, 2023, from https://sdgs.un.org/topics/energy

United Nations. (2000). *Millennium Declaration. UN Document A/RES/55/2. General Assembly resolution 55/2*. Retrieved March 1, 2023, from https://www.un.org/en/development/desa/population/migration/generalassembly/docs/globalcompact/A_RES_55_2.pdf

United Nations. (2015). *Transforming Our World: The 2030 Agenda for Sustainable Development. UN Document A/RES/70/1*. Retrieved March 1, 2023, from https://documents-dds-ny.un.org/doc/UNDOC/GEN/N15/291/89/PDF/N1529189.pdf?OpenElement

UNCTAD. (2015). *UNCTAD's Investment Policy Framework for Sustainable Development*. United Nations. Retrieved March 1, 2023, from https://unctad.org/publication/investment-policy-framework-sustainable-development-2015-edition

UNCTAD. (2020). *International Investment Agreements Reform Accelerator*. United Nations. Retrieved March 1, 2023, from https://unctad.org/publication/international-investment-agreements-reform-accelerator

UNCTAD. (2023). *Trends in the Investment Treaty Regime and a Reform Toolbox for the Energy Transition. IIAs Issues Note, 2023*. United Nations. Retrieved March 1, 2023, from https://unctad.org/system/files/official-document/diaepcbinf2023d4_en.pdf

UNFCC Secretariat. (2021, September 17). *Nationally Determined Contributions under the Paris Agreement—Synthesis Report by the Secretariat. FCCC/PA/*

CMA/2021/8. Retrieved March 1, 2023, from https://unfccc.int/documents/306848

van Aaken, A. (2022). Investment Law in the Twenty-First Century: Things Will Have to Change in Order to Remain the Same. *Journal of International Economic Law, 26*(1), 166–176. https://doi.org/10.1093/jiel/jgac054

VanDuzer, J. A. (2016). The Complex Relationship Between International Investment Law and Climate Change Initiatives: Exploring the Tension. In P. Delimatsis (Ed.), *Research Handbook on Climate Change and Trade Law* (pp. 434–461). Edward Elgar Publishing.

Viñuales, J. E. (2022). *The International Law of Energy*. Cambridge University Press.

Treaties, Laws and Legal Documents

ACP-EU Partnership Agreement. (2021). Partnership Agreement between [the European Union/The European Union and its Member States], of the one part, and Members of African, Caribbean and Pacific States, of the other part. Negotiated Agreement text initialled by the EU and OACPS chief negotiators on 15th April 2021. Retrieved March 1, 2023, from https://investmentpolicy.unctad.org/international-investment-agreements/treaty-files/6457/download

Agreement between the EU and Japan for an Economic Partnership (signed 17 July 2017, entered in force 1 February 2019).

Agreement for the termination of Bilateral Investment Treaties between the Member States of the European Union (29.5.2020) OJ L 169, 1–41.

Agreement in Principle on the Modernisation of the Energy Charter Treaty (2022, June 24). Retrieved March 1, 2023, from https://www.bilaterals.org/IMG/pdf/reformed_ect_text.pdf

Australia-EU Framework Agreement. (2017). Framework Agreement between the European Union and its Member States, of the one part, and Australia, of the other part (signed 7 August 2017, entered in force 21 October 2022).

Australia-UK Free Trade Agreement. (2021). Free Trade Agreement between the United Kingdom of Great Britain and Northern Ireland and Australia (signed 16 December 2021, not yet in force).

CETA. (2016). The Comprehensive Economic and Trade Agreement (CETA) between Canada on the one hand and the European Union and its Member States on the other (signed 30 October 2016, certain provisions are provisionally applied as of 21 September 2017).

Directive (EU) 2018/2001 on the promotion of the use of energy from renewable sources (2018), OJ L 328, (the transposition into member state's laws by 30 June 2021).

Draft Decision of the CETA Joint Committee. (2022). Decision No X/2022 of the CETA Joint Committee on the interpretation of certain terms in Article

8.10, Annex 8-A and Article 8.39. Retrieved March 1, 2023, from https://www.gerechter-welthandel.org/wp-content/uploads/2022/09/20220905_Draft-Decision-and-Declarations_CETA.pdf

Draft EU-Angola SIFA. (2022). Draft EU-Angola Sustainable Investment Facilitation Agreement. Retrieved March 1, 2023, from https://ec.europa.eu/commission/presscorner/detail/en/ip_22_6136

ECT. (1994). Energy Charter Treaty and Energy Charter Protocol on Energy Efficiency and Related Environmental Aspects (signed 17 December 1994, entered into force 16 April 1998).

EU-Singapore FTA. (2018). Free trade Agreement between the European Union and the Republic of Singapore. OJ L 294, 14.11.2019, p. 3–755 (signed 19 October 2018, entered into force 21 November 2019).

EU—Singapore Investment Protection Agreement. (2018). Investment Protection Agreement between the European Union and its Member States, of the One Part, and the Republic of Singapore, of the Other Part (signed 15 October 2018, not yet in force).

EU-Korea Framework Agreement. (2010). Framework Agreement between the European Union and its Member States, on the one part, and the Republic of Korea, on the other part (signed 27 June 2012, not yet in force).

EU-Viet Nam Framework Agreement. (2017). Framework Agreement on Comprehensive Partnership and Cooperation between the European Union and its Member States, of the one Part, and the Socialist Republic of Viet Nam, of the other Part (signed 27 June 2012, not yet in force).

EU-UK Trade and Cooperation Agreement. (2021). Trade and Cooperation Agreement between the European Union and the European Atomic Energy Community, of the one part, and the United Kingdom of Great Britain and Northern Ireland, of the other part, OJ L 149, 30.4.2021, p. 10–2539.

European Commission. (2023). Proposal for a Council decision on the Union withdrawal from the Energy Charter Treaty COM(2023) 447 final.

European Parliament. (2022). Resolution of 23 June 2022 on the Future of EU International Investment Policy, 23 June 2022, 2021/2176(INI).

Free Trade Agreement between the European Union and the Republic of Korea. (2011). (signed 6 October 2010, entered in force 1 July 2011).

Free Trade Agreement between the United Kingdom of Great Britain and Northern Ireland and New Zealand. (2022). (signed 28 February 2022, not in force).

Georgia-UK Partnership and Cooperation Agreement. (2019). Strategic Partnership and Cooperation Agreement between Georgia and the United Kingdom of Great Britain and Northern Ireland (signed 21 October 2019, entered in force 1 January 2021).

Glasgow Climate Pact. (2021). UNFCCC, Decision -/CMA.3, advance unedited version. Retrieved March 1, 2023, from https://unfccc.int/documents/310497 [1/CMA.3].

Kyoto Protocol to the United Nations Framework Convention on Climate Change (adopted 10 December 1997, entered into force 16 February 2005) 2303 U.N.T.S. 162.

NAFTA. (1993). North American Free Trade Agreement (1993) 32 ILM 289, 605 (signed in 1992, entered into force 1 January 1994).

Paris Agreement under the United Nations Framework Convention on Climate Change (signed 12 December 2015, entered into force 4 November 2016).

Reciprocal Investment Promotion and Protection Agreement between the Government of the Kingdom of Morocco and the Government of the Federal Republic of Nigeria (signed 3 December 2016, not in force).

Strategic Partnership, Trade and Cooperation Agreement between the Republic of Moldova and the United Kingdom of Great Britain and Northern Ireland (signed 24 December 2020, not in force).

Treaty of Lisbon (signed 17 December 2007, entered into force on 1 December 2009), OJ C 306.

Ukraine-UK Political, Free Trade and Strategic Partnership Agreement. (2020). Political, Free trade and Strategic Partnership Agreement between Ukraine and the United Kingdom of Great Britain and Northern Ireland (signed 8 October 2020, entered into force on 1 January 2021).

UNFCC. (1992). United Nations Framework Convention on Climate Change (signed 9 May 1992, entered into force 21 March 1994).

Vienna Convention on the Law of Treaties (published 23 May 1969, entered into force on 27 January 1980).

Decisions

Antaris Solar GmbH and Dr. Michael Göde v Czech Republic (Award, 2 May 2018) UNCITRAL.

Infrastructure Services Luxembourg S.à.r.l. and Energia Termosolar B.V. (formerly Antin Infrastructure Services Luxembourg S.à.r.l. and Antin Energia Termosolar B.V.) v Kingdom of Spain, ICSID Case No. ARB/13/31(Award, 15 June 2018).

Bilcon of Delaware et al v Government of Canada, PCA Case No. 2009-04 (Award on Jurisdiction and Liability, 17 March 2017).

Eiser Infrastructure Limited and Energía Solar Luxembourg S.à r.l. v Kingdom of Spain, ICSID Case No. ARB/13/36 (Award, 4 May 2017).

Fedax N.V. v The Republic of Venezuela, ICSID Case No. ARB/96/3 (Decision of the Tribunal on Objections to Jurisdiction, 11 July 1997).

I.C.W. Europe Investments Limited v Czech Republic (Award, 15 May 2019) UNCITRAL.

Jürgen Wirtgen, Stefan Wirtgen, Gisela Wirtgen, JSW Solar (zwei) GmbH & Co. KG v Czech Republic (Award, 19 October 2017) UNCITRAL.
Natland Investment Group N.V., Natland Group Limited, G.I.H.G. Limited, Radiance Energy Holding S.à.r.l. v Czech Republic (Partial Award, 20 December 2017) UNCITRAL.
Occidental Exploration and Production Company v The Republic of Ecuador, LCIA Case No UN3467, Final Award (1 July 2004).
Phoenix Action, Ltd. v The Czech Republic, ICSID Case No. ARB/06/5 (Award, 15 April 2009).
Photovoltaik Knopf Betriebs v Czech Republic (Award, 15 May 2019) UNCITRAL.
Salini Costruttori S.p.A. and Italstrade S.p.A. v Kingdom of Morocco, ICSID Case No. ARB/00/4 (Decision on Jurisdiction, 23 July 2001).
SGS Société Générale de Surveillance S.A. v Republic of the Philippines, ICSID Case No ARB/02/6, Decision of the Tribunal on Objection to Jurisdiction (29 January 2004).
Stadtwerke München GmbH, RWE Innogy GmbH, and others v Kingdom of Spain, ICSID Case No ARB/15/1, Award (2 December 2019).
Voltaic Network GmbH v Czech Republic (Award, 15 May 2019) UNCITRAL.
WA Investments-Europa Nova Limited v Czech Republic (Award, 15 May 2019) UNCITRAL.

CHAPTER 13

Energy Charter Treaty and Investor-State Dispute Settlement: Legal Challenges in Energy Transition

Magdalena Bas-Vilizzio

Introduction

Investor-state dispute settlement has come under scrutiny over the past two decades. In 2003, the United Nations Conference on Trade and Development (UNCTAD) sounded the alarm in its World Investment Report, highlighting that international investment agreements "limit the sovereign autonomy of the parties" (UNCTAD, 2003: 146). Investor-state dispute settlement has become the quintessential arena for the tension between investment protection and the right to regulate, especially in areas such as human rights, environmental protection, climate change and energy transition. Four years later, Bolivia became the first country to withdraw from the International Centre for the Settlement of Investment Disputes Convention, followed by Ecuador (2009, although it later returned in 2021) and Venezuela (2011). In addition, both Bolivia and

M. Bas-Vilizzio (✉)
University of the Republic, Montevideo, Uruguay

D. Iglesias Márquez et al. (eds.), *Legal Challenges at the End of the Fossil Fuel Era*, Global Issues,
https://doi.org/10.1007/978-3-031-61766-9_13

Ecuador withdrew from most of their bilateral investment treaties (BITs). South Africa also followed the trend of withdrawing from BITs, while India underwent a complex process to regain access to international arbitration after an earlier U-turn. What is driving this trend? Primarily, it is rooted in their experiences with investment arbitration and new constitutional provisions.

The Bolivian case may be illustrative. Aguas del Tunari dispute was the prelude to Bolivia's withdrawal from the International Centre for Settlement of Investment Disputes (ICSID) in 2007. In 1999, a few months after the privatisation of Cochabamba's water and sanitation services by Aguas del Tunari—owned by the US company Bechtel—the water tariff rose by more than 50%, leading to significant popular protests dubbed the "water war" (Ministerio de Relaciones Exteriores y Cultos de Bolivia, 2008: 100).

Based on the Bolivia-Netherlands Bilateral Investment Treaty, Aguas del Tunari initiated arbitration proceedings at the ICSID in 2001 (Case No. ARB/02/3), but the case was dropped under public pressure. The company sold its shares to Bolivia for the symbolic sum of two bolivianos—less than one dollar. In 2007, Bolivia withdrew from the ICSID Convention, citing incompatibility with the Calvo doctrine enshrined in the 1967 Constitution, which was in force at the time; procedural issues—conflicts of interest between actors and arbitrators, double jeopardy, forum shopping, treaty shopping, and cherry picking; and the high cost of proceedings, among others (Ministerio de Relaciones Exteriores y Cultos de Bolivia, 2008). Similarly, the 2009 Bolivian Constitution includes a clause stating that all foreign investments are subject to the jurisdiction, laws, and authorities of Bolivia (Article 320, Paragraph II).

Although the arguments differ, this situation is not solely a concern of the Global South.[1] At present, this issue has gained global prominence, encompassing voices from both multilateral organisations and the Global North. The International Monetary Fund's Staff Climate Note 2022/007 suggests that the investor-state dispute settlement regime could impede energy transitions, leading to regulatory chill or regulatory paralysis

[1] Despite the fact that the Global South is not a monolithic notion, countries usually argue that investor-state dispute settlement clashes with their Constitutions—e.g. Brazil, Bolivia, Ecuador—or were plaintiffs in disputes that question human rights or environmental protection measures. Global North, specifically the European Union, claims for an extension of policy space in order to address the negative externalities of the regime (Ratton Sanchez Badin & Morosini, 2018).

(Prasad et al., 2022). Similar arguments emerged in October–November 2022, when seven European Union members announced their withdrawal from the Energy Charter Treaty, joining Italy, which had previously withdrawn in 2016. The message from European Union members is a significant development, especially as it involves countries from the US-Europe axis, where the regime was shaped by norm entrepreneurs in the 1950s and 1960s.[2]

Against this background, this chapter aims to address the legal challenges posed by foreign investment in the context of energy transition and to explore the role of regulatory sovereignty in this scenario. This chapter follows the concepts introduced by Martti Koskenniemi (1990) regarding international law, emphasising that it is a fusion of both legal and political elements, rather than being exclusively one or the other. Therefore, this work aims not only to establish itself on a strictly legal basis but also to approach the subject within the intersecting fields of International Law and International Relations. Moreover, Juliana Peixoto (2022) asserts that investment protection represents a "wide space of intersection and mutual fertilization" in the ongoing dialogue between international law and international relations.

The chapter is divided into three parts. The first section examines the Energy Charter Treaty and its costs in terms of fighting climate change and facilitating the energy transition. The second section examines the legal and political context of 'European withdrawal', including individual withdrawals and the path towards coordinated withdrawals. The third section focuses on the implications for regulatory sovereignty (Bas Vilizzio, 2022, 2023a), highlighting how the Energy Charter Treaty underscores the need to manage tensions between the international investment regime and other international frameworks. The chapter concludes with some final thoughts.

The Energy Charter Treaty: A 'Fossil Energy Club'?

The Energy Charter Treaty is an international agreement signed in Lisbon on 12 December 1994. It entered into force on 16 April 1998. The treaty strives to stimulate global collaboration within the energy industry, laying special emphasis on advancing and protecting investments across all forms

[2] The legal imagination of norm entrepreneurs can be studied in Perrone (2021), particularly in chapter 2 (pp. 51–80).

of energy, either fossil or renewable. Article 26 introduces investor-state arbitration mechanisms to resolve disputes between foreign investors—nationals of a Party—and host States that are Parties to the Treaty. This provision allows investors to choose to arbitrate through ICSID or its Additional Facility, a sole arbitrator or an ad hoc tribunal under the Rules of the United Nations Commission on International Trade Law, or to choose to arbitrate through the Arbitration Institute of the Stockholm Chamber of Commerce.

According to the UNCTAD Policy Investment Hub, a total of 157 disputes invoking this legal mechanism have been recorded as of 31 December 2012.[3] It is the most frequently invoked agreement in the history of investor-state dispute settlement (Bas Vilizzio & Stanley, 2021). This is particularly notable in cases concerning Spain (51 out of 56 disputes), Italy (13 out of 13 disputes), and Hungary (5 out of 17 disputes).

To provide a further example, Masdar Solar & Wind Cooperatief U.A. v. Kingdom of Spain dispute—ICSID Case No. ARB/14/1—should be examined. The main argument of the claim focused on the violation of the principle of fair and equitable treatment, in addition to the realistic expectations within the legal framework on which they had relied and invested, as promoted by Spain for the renewable energy sector. The implementation of reductions by Spain led to changes in this legal framework, followed by the cancellation of the renewable energy subsidy programme (Royal Legislative Decree 14/2010, Act 15/2012, Royal Legislative Decree 9/2013, Order IET/1045/2014). These actions were found to be inconsistent with the commitment to stability.[4] The award ultimately favoured the investors.

Masdar v. Spain dispute, as well as the wide majority of disputes against Spain,[5] does not focus on the energy transition. On the other hand, the

[3] Data available at https://investmentpolicy.unctad.org/investment-dispute-settlement, accessed 17 August 2023.

[4] Following Cotula (2009: 72), stabilisation clauses "aim to stabilise the terms or conditions of an investment project". They originally refrain the host State to nationalise the investment or to change the contract unilaterally, but nowadays, stabilisation clauses have a broader scope that includes modifications in the fiscal regime, tariff structure, or regulatory framework for investment projects.

[5] Other disputes against Spain are: Charanne B.V. and Construction Investments S.a.r.l. v. Spain—Stockholm Chamber of Commerce Case No. 062/2012; Eiser Infrastructure Limited and Energía Solar Luxembourg S.àr.l. v. Kingdom of Spain—ICSID Case No. ARB/13/36; Infrastructure Services Luxembourg S.à.r.l. and Energia Termosolar B.V. (formerly Antin Infrastructure Services Luxembourg S.à.r.l. and Antin Energia Termosolar B.V.) v. Kingdom of Spain—ICSID Case No. ARB/13/31; RWE Innogy GmbH and RWE Innogy

claims challenge the steps taken to diminish or remove the advantages of renewable energy implementation, enacted to alleviate the consequences of the economic crisis between 2008 and 2014. During such crises, States frequently assert a pressing requirement for policy autonomy (Sornarajah, 2015). Nonetheless, it is vital to point out that the Energy Charter Treaty belongs to a group of 'traditional' agreements within International Investment Law. It was negotiated during a phase of liberal transformation characterised by rigid investment protection (Sornarajah, 2015: 43–45). This period saw a lack of active commitment by countries towards achieving Sustainable Development Goals.

Moreover, the Working Group on the issue of human rights and transnational corporations and other business enterprises (Working Group) (2021: 6) points out that the international investment agreements celebrated before 2010 are characterised by the three 'i': 'imbalance, inconsistency and irresponsibility'. In consequence, these treaties "enabled arbitrators who believed in furthering investment protection to discard context and emphasise the strict interpretation of the treaty so as to promote investment protection and downplay competing interests such as public interests" (Sornarajah, 2021).

The Agreement in principle on the modernised Energy Charter Treaty (Agreement in Principle) was adopted on 24 June 2022, after 15 rounds of negotiations between July 2019 and June 2022. In 2021, Martin Dietrich Brauch, lead researcher at Columbia Center on Sustainable Investment, suggested that the optimal course for contracting States would be to terminate the treaty and commence afresh. Nonetheless, this new beginning should be anchored in fundamental questions: "How can international law help our countries to phase out fossil fuel investments, ramp up low-carbon energy investment, and promote a fair transition, in line with the Paris Agreement and Sustainable Development Goal 7.1 target to ensure universal access to affordable, reliable, and modern energy services by 2030?"

The Agreement in Principle was intended to be adopted at the 33rd session of the Energy Charter Conference on 22 November 2022. However, Germany, France, Spain, and the Netherlands staunchly opposed the modernised text, leading to the removal of the discussion from the

Aersa S.A.U. v. Kingdom of Spain—ICSID Case No. ARB/14/34; Novenergia II—Energy & Environment (SCA), SICAR v. Kingdom of Spain—Stockholm Chamber of Commerce Case No. 063/2015; among many others.

meeting agenda. Their main argument was that there is a misalignment between the Agreement in principle and Sustainable Development Goals 7 ("Ensure access to affordable, reliable, sustainable and modern energy for all") and 13 ("Take urgent action to combat climate change and its impacts"), according to Brauch (2021, 2022). It does not align either with the goals and commitments of the Paris Agreement, which objective is to "strengthen the global response to the threat of climate change, in the context of sustainable development and efforts to eradicate poverty". Article 2 paragraph 1 refers to specific measures that are closely linked to this issue: "holding the increase in the global average temperature to well below 2°C above pre-industrial levels and pursuing efforts to limit the temperature increase to 1.5°C above pre-industrial levels" (section a) and "increasing the ability to adapt to the adverse impacts of climate change and foster climate resilience and low greenhouse gas emissions development" (section b).

What are the legal challenges posed by foreign investment in energy transition? Brauch (2022) states that the Energy Charter Treaty Agreement in Principle is "a botched reform attempt that undermines climate action". This is due to the fact that it restricts subsidies for fossil fuels without providing explicit undertakings or instructions for implementation. The writer highlights the lack of guidance for a fair transition, which ought to incorporate methods of global collaboration for nations or regions that suffer the most from climate change.

Furthermore, the impacts of the Energy Charter Treaty can be seen in the disputes initiated due to the Dutch 'coal phase-out'. In 2019, the Dutch Parliament enacted the Prohibition of Coal in Electricity Production Act, which orders an end to coal-based energy consumption by 2030. Aligned with the objective of reducing greenhouse gas emissions by 49% compared to 1990 levels, this legislation promotes coal-fired power stations to transition into other sectors to sustain their operations. In this context, under Energy Charter Treaty, in 2021 two foreign investors

(RWE[6] and Uniper,[7] both companies based on an EU member: Germany[8]) sued the Netherlands. In both cases, the claimants maintain that their rights have been infringed due to insufficient time provided for adjusting to the energy transition. The resolution of both disputes is pending.[9]

Considering Sustainable Development Goal 7 and the objectives and obligations specified in the Paris Agreement, interaction with the investor-state dispute settlement mechanism is restricted. The majority of global investment agreements were concluded 20–60 years ago when environmental issues, including climate change and energy transition, were not of paramount importance on the international agenda as they are today (UNCTAD, 2020: 12).[10] These agreements often do not include exemptions related to these matters, even when the measures are implemented as part of a collective framework and constitute a legitimate exercise of regulatory power. It is noteworthy that modern equitable agreements are frequently interpreted by arbitrators using the reasoning of outdated unfair agreements (Alschner, 2022).

Eco Oro v. Colombia[11] is an illustrative example of the aforementioned. The controversy centres around the demarcation of the Santurbán Páramo

[6] RWE AG and RWE Eemshaven Holding II BV v. Kingdom of the Netherlands, International Centre for Settlement of Investment Disputes, Case No. ARB/21/4. Information and documents available at https://www.italaw.com/cases/9156, accessed 17 August 2023.

[7] Uniper SE, Uniper Benelux Holding B.V. and Uniper Benelux N.V. v. Kingdom of the Netherlands, ICSID Case No. ARB/21/22. Information and documents available at https://www.italaw.com/cases/9146, accessed 17 August 2023.

[8] This is not a minor point, taking into consideration the position of the European Commission and the Court of Justice of the European Union regarding arbitrations between national investors of one member and another member, following the Achmea and Micula cases. Court of Justice of the European Union, Case C-284/16. International Centre for Settlement of Investment Disputes, Case No. ARB/05/20, Ioan Micula, Viorel Micula, S.C. European Food S.A, S.C. Starmill S.R.L. and S.C. Multipack S.R.L. v. Romania. For further analysis refer to Bas Vilizzio (2019).

[9] Additionally, the Russian invasion of Ukraine—started on 24 February 2022—has slowed down the Netherlands measures, particularly because of Russian gas supply restrictions. Nevertheless, the commitment to phase out coal energy by 2030 remains resolute (Meijer & Deutsch, 2022).

[10] UNCTAD (2012: 102) affirms that there is a new generation of agreements that dialogues with sustainable development and investors' responsibility, but it is more recent.

[11] Eco Oro Minerals Corp. v. Republic of Colombia, International Centre for Settlement of Investment Disputes, Case No. ARB/16/41. Information and documents available at https://www.italaw.com/cases/6320, accessed 17 August 2023.

(Santander, Colombia), in line with Colombia's efforts to mitigate climate change.[12] The Canadian mining firm Eco Oro has stated that the ban on activities in the Angostura project site (Páramo de Santurbán) has left its investment at a disadvantage. Moreover, there are inconsistencies to note within the decision on jurisdiction, responsibility, and guidelines on quantum (1 September 2021). The tribunal found that the State was liable for breaching the minimum standard of treatment while also recognising the State's right to regulate. As a result, the importance of arbitral tribunals in shaping the enforcement of norms that uphold the right to regulate or in determining whether the State's actions fall within the limits of this framework is vital (Perrone, 2019: 16–17).[13]

The Energy Charter Treaty: The Perfect Storm for a European Exit

The increase in investment arbitration cases between the 1990s and the first decade of the twenty-first century correlates with the growing international emphasis on environmental protection (Sornarajah, 2015: 333), climate action, and energy transition. This has deepened in the last decade, both globally and regionally, with new additions to the international legal landscape such as the Escazú Agreement (in force since 2021) and the European Green Deal (2019). Almost a decade earlier, Italy announced its withdrawal from the Energy Charter Treaty, which entered into force on 1 January 2016. Six European Union members deposited the instrument of withdrawal from the Energy Charter Treaty: Poland (7 December 2022), France (7 December 2022), Germany (19 December 2022), Slovenia (13 October 2023), Luxembourg (16 June 2023), and Spain (16 April 2024). Although the Netherlands announced the withdrawal in 2022, it has not taken the legal step by 31 July 2024.

These countries presented similar positions. For example, on 18 October 2022, the Dutch Minister for Climate and Energy Policy, Rob Jetten, announced to the Parliament that "despite many of the

[12] According to the non-governmental organisation World Wildlife Fund Colombia (2018), 50% of the world's páramos are located in Colombia. Although they cover only 1.7% of Colombian territory, they provide 70% of the water consumed by the population and 70% of the water destined for hydroelectric energy production.

[13] For a further analysis regarding Eco Oro v. Colombia, refer to Sierra (2021) and Bas Vilizzio (2023b).

modernisations that are now in the negotiation outcome, we do not see how the ECT has been sufficiently aligned with the Paris Agreement" (Euroactiv, 2022). This idea is closely linked to the report of the High Council for Climate—an advisory body to the French government. It made its position clear: the Energy Charter Treaty and its modernisation are not compatible with the goals of decarbonisation and greenhouse gas emission reduction for 2030 (Haut Conseil pour le Climat, 2022: 4).

In addition, the investor-state dispute settlement included in the treaty "has generated an increased risk of loss of sovereignty for signatory States in the development or implementation of their energy and climate policies, and led to an increase in litigation and arbitration awards in conflict with the decisions of national courts and European tribunals"[14] (Haut Conseil pour le Climat, 2022: 1). In terms of compensation, it is estimated that climate action could generate investor-state claims of more than $340 billion (Tienhaara et al., 2022: 701). Therefore, the report of the Council on Climate Change concludes that a European coordinated exit would be necessary to restore member states' sovereignty in public energy and climate policy.

The European Parliament has also supported the withdrawal of the European Union and its Member States.[15] In particular, paragraph 18 of the European Parliament resolution on the outcome of the modernisation of the Energy Charter Treaty (2022/2934(RSP))—24 November 2022—emphasises that "the EU can only ratify the modernised ECT with the final consent of Parliament, and that Parliament will consider its previous positions and the shortcomings of the modernisation if requested to consent to it; takes the position that Parliament will support the EU's exit from the ECT when requested to consent to it".

On 8 February 2023, a leaked 'non-paper' prepared by the European Commission came to light. The paper claimed that the Energy Charter

[14] Original sentence: "Ce mécanisme a généré un risque accru de perte de souveraineté pour les États signataires dans l'élaboration ou la mise en œuvre de leurs politiques énergétiques et climatiques, et entraîné une multiplication des contentieux et des sentences arbitrales en conflit avec les décisions des juridictions nationales et européennes".

[15] International investment agreements usually include clauses that establish that substantial provisions of the treaty remain in force for a specific period after the termination—generally 5, 10, 15 or 20 years. Thus, sunset clauses give foreign investors the possibility to file a claim despite the fact that the agreement has already terminated.

Treaty remained 'unavoidable' despite its incompatibility with climate goals, particularly the goal of achieving climate neutrality by 2050, a cornerstone of the European Green Deal (Euroactiv, 2023). The treaty posed a threat to the reduction of carbon emissions from fossil fuels, an issue that had been a central concern for the European Union in recent decades. As a result, according to Gijs et al. (2023), the Commission underwent a significant shift, opting to follow the regional trend by recommending a coordinated withdrawal.

This shift materialised on 7 July 2023, when the Commission presented two proposals for Council decisions to authorise the withdrawal of both the European Atomic Energy Community—COM(2023) 446 final—and the European Union—COM(2023) 447 final—from the Energy Charter Treaty. The arguments put forward were based on two main premises. First, the investor-state dispute settlement provisions, which allow arbitration between an investor who is a national of one Member State and another Member State, are incompatible with the principle of autonomy in European Union law, as concluded in Opinion 1/17 of the European Union Court of Justice.[16] Furthermore, in case C-741/19 Republic of Moldova v. Komstroy—judgement of 2 September 2021—the Court concluded that the investor-state arbitration provision in the Energy Charter Treaty is incompatible with the principle of autonomy and therefore does not apply to disputes within the European Union—paragraph 66.[17]

Secondly, the agreement in principle still offers protection to fossil fuels and is therefore not in line with the commitments made in the European Green Deal, the REPowerEU, or the Paris Agreement. For example, the agreement in principle on the modernised Energy Charter Treaty could not only hamper Europe's ambition to become the world's first climate neutral continent by 2050 but also the 'Fit for 55' initiative, which aims to reduce net greenhouse gas emissions by at least 55% by 2030 compared to 1990 levels. It could also undermine the plan to reduce Europe's dependence on Russian fossil fuels in the wake of the Ukraine conflict.

It is important to note that under Article 47 of the Energy Charter Treaty, unilateral withdrawal takes effect one year after written notification. However, any withdrawal, whether unilateral or coordinated, would

[16] European Union position towards investor-state dispute settlement is dual, and for that reason, it can be weak: only intra-European Union arbitrations are incompatible—not the extra ones.

[17] For a specific study of the judgement in case C-741/19, see Odermatt (2021).

trigger Article 47.3, also known as the survival or sunset clause. This clause provides that the provisions of the Treaty will continue to apply to investments made for 20 years from the date of notification of withdrawal. Such clauses, which are common in international investment agreements, deviate from the conventional effects of treaty termination by unilateral withdrawal, as set out in Article 70 of the Vienna Convention on the Law of Treaties between States (1969), by preserving rights and obligations even after termination (Reinisch & Mansour Fallah, 2022: 104).

In this framework, foreign investors could potentially bring claims even after the defendant state has withdrawn from the ECT. In order to avoid this, the Contracting Parties could decide to terminate the sunset clause or to provide for a shorter 'survival' period. A similar solution was reached by the 23 Member States in 2020 for the Agreement on the Termination of Bilateral Investment Treaties between the Member States of the European Union.[18] As argued by Voon (2019), disregarding this clause becomes an unviable option when certain states have unilaterally withdrawn or are discussing a coordinated withdrawal. This inevitability, as pointed out by Eckes et al. (2023), underlines the need to address Article 47.3 of the Energy Charter Treaty.

The European Commission's proposal to the Council of the European Union marks a different approach to the issue. While the Commission emphasises that the ECT as a whole "has never, does not and will never apply [to intra-European Union investor-state dispute settlement], including its Article 47.3" (European Commission, 2023a: 3; 2023b: 3), it recognises the inherent risk. The Commission therefore recommends that an interpretative agreement or declaration be concluded between the European Union, Euratom, and the Member States on the interpretation of the Treaty or the application of its provisions in accordance with Article 31(3)(a) of the Vienna Convention on the Law of Treaties. While Skovgaard Poulsen and Gertz (2021) argue that interpretative statements are closely linked to the 'dual role' of states in investor-state arbitration, encompassing their role as respondent in disputes and as interpreter of treaties, this solution may "limit the influence of arbitrators to maintain the regime of the past, as in Eco Oro v. Colombia" (Bas Vilizzio, 2022: 13).

Finally, after the European Parliament gave the green light, on 27 June 2024, the Council of the European Union notified the depositary of the Energy Charter Treaty of the withdrawal of the European Union and Euratom from the treaty.

[18] For a further analysis, see Tropper and Reinisch (2022).

Regulatory Sovereignty and the Need to Readdress Investor-State Dispute Settlement

The coordinated withdrawal of the European Union from the Energy Charter Treaty acts as a protective barrier for the regional policy space, in particular for the regional and national right to regulate. Consequently, the role of regulatory sovereignty in this context becomes a key consideration. As articulated by Stephen Krasner (2009), sovereignty remains unparalleled in both international relations and international law. Nevertheless, the concept of sovereignty has changed over time.

Currently, there is a shift towards a new concept of sovereignty that is "is rooted in the defence of policy space, where the right to regulate is the heart and can therefore be called regulatory sovereignty" (Bas Vilizzio, 2022: 14). In line with Krasner's analytical categories in 'Sovereignty: Organised Hypocrisy' (Krasner, 1999), regulatory sovereignty has elements of authority and legitimacy but lacks elements of control.

While the boundaries of regulatory sovereignty are still being defined, it includes regulation of national security, the environment, climate change, public health, and, most importantly, the protection of human rights. These are the areas of regulation that states can choose to exclude from investor-state dispute settlement, thereby creating a framework to prevent regulatory chill or regulatory paralysis (Bas Vilizzio, 2023a: 97). The Morocco-Nigeria Bilateral Investment Treaty, with its explicit recognition of the right to regulate, and the tobacco curve-out in the Comprehensive and Progressive Agreement for Trans-Pacific Partnership are examples of agreements that tend to protect regulatory sovereignty. Although this new concept of sovereignty originated in the field of investor-state dispute settlement, it is gradually extending to other areas, especially in times of multidimensional crises such as the present, characterised by simultaneous economic, security, climate, geopolitical, and other crises.

The pursuit of more policy space to address the challenges of energy transition and the recognition of the incompatibility of the principle of autonomy within the Union of the modernised Energy Charter Treaty, in particular with regard to investor-state arbitration within the European Union, lay the groundwork for a coordinated exit. Neither Climate Law, the Sustainable Development Goals, the Paris Agreement nor the European Green Deal provides explicit guidance on how states should fulfil their climate commitments, leaving states with a wide margin of manoeuvre in their implementation. In this context, Iglesias Márquez (2021: 41–42) notes that the United Nations Guiding Principles on Business and Human Rights (2011) offer a potential means of bridging this gap.

In particular, Guiding Principle No. 9, under the section 'Ensuring policy coherence' of the first pillar: protect, states that "States should maintain adequate domestic policy space to fulfil their human rights obligations when pursuing business-related policy objectives with other States or business enterprises, for example through investment treaties or contracts". However, as highlighted by the Working Group (2021: 3), international investment agreements, if not properly drafted, have the potential to encroach on the right of States to regulate, further exacerbating the asymmetry between the rights and obligations of investors.

However, the history of international law shows that "decoupling rights and obligations is a central aspect of the legal imagination that dominates the international law on foreign investment" (Perrone, 2022: 376). It is therefore relevant to ask today: what is the current landscape? What references are there in international investment agreements to human rights, the right to regulate, climate change and related concepts?

As of 26 August 2023, out of the 3268 international investment agreements—comprising bilateral investment treaties and treaties with investment provisions—catalogued in the UNCTAD Investment Policy Hub database,[19] 172 incorporate the notion of 'human rights', 82 pertain to the 'right to regulate', and merely 7 encompass the concept of 'policy space'. In the context of more specialised topics connected to the Energy Charter Treaty, 600 agreements refer to the term 'environment', 51 include 'climate change', and only one explicitly references 'energy transition'.[20] The most recent example is the China-Ecuador Bilateral Investment Treaty, ratified on 10 May 2023, which recognises the concept

[19] See UNCTAD Investment Policy Hub database (https://investmentpolicy.unctad.org/international-investment-agreements [Accessed on 26 August 2023]).

[20] All the terms were searched in English, Spanish, Portuguese, and French. That is: (1) human rights, derechos humanos, direitos humanos, droits humains; (2) right to regulate, derecho de regular, direito de regular, droit à réglementer (droit de réglementer); (3) policy space, espacio para políticas (espacio de política), espaço político, espace politique; (4) environment, medioambiente (medio ambiente), ambiente, environnement; (5) climate change, cambio climático, mudança climática, changement climatique; (6) energy transition, transición energética, transição energética, transition énergétique.

of energy transition under Article 16.13 in the context of environmental cooperation.

Agreements addressing human rights, climate change or even the right to regulate remain a minority. After analysing the European Union's treaty-making practice since 2009, Gehring and Tokas (2022: 135) conclude that it is more valuable to include positive obligations (such as non-regression obligations) and investment facilitation provisions (such as regulatory cooperation, technical assistance, exchange of best practices, etc.) than prohibitions. However, this innovation in treaty practice does not prevent foreign investors from challenging regulatory measures.

Moreover, arbitrators often interpret modern BITs similarly to the older, unbalanced ones (Alschner, 2022). Cases where investors' claims juxtapose two different, self-contained frameworks highlight the need for a more fluid dialogue within a fragmented international law. Some disputes may be instructive, such as the case of Philip Morris v. Uruguay—ICSID Case No. ARB/10/7—concerning tobacco control measures—Uruguay to protect the human right to public health. Following the World Health Organisation's Framework Convention on Tobacco Control, in 2009 Uruguay implemented two measures related to cigarette packaging: the 80/80 rule[21]—health warnings on 80% of the front and back of the packaging—and the single presentation rule—a single presentation per brand in order to avoid misleading terms or descriptors. The tribunal concluded that Uruguay had the right to regulate in order to protect public health.

Urbaser v. Argentina—ICSID Case No. ARB/07/26—represents another interesting dispute in water and sewer services sector. The controversy has origin in the emergency measures taken by Argentina to alleviate the effects of the economic and social crisis of 2001, particularly those linked to updating public service rates. The dispute is relevant because for the first time an investment tribunal analyses the obligation of foreign investment companies in relation to the human right to water, based on Argentina's counterclaim. The tribunal recognised the human right to water, despite concluding that companies have a negative obligation in terms of human rights. Positive obligations must be expressly acknowledged in the treaty, a provision that the Argentina-Spain Bilateral Investment Treaty does not include.

[21] The prelude to plain packaging.

Finally, Odyssey v. Mexico dispute—ICSID Case No. UNCT/20/1—also demonstrates that international investment protection and environmental protection go in different ways. The environmental impact of Odyssey's seabed mining project to dredge phosphate sands from Gulf of Ulloa—Baja California Sur—involves potential damages to biodiversity—fish, marine mammals, caguama turtles. The company argued that Mexican denial of the environmental authorisation was illegal and infringed fair and equitable treatment provided in North American Free Trade Agreement. The dispute is pending.

In terms of the fragmentation of international law, (International Law Commission, 2006) this scenario creates a dilemma: states should choose between protecting foreign investment or human rights—including the environment or climate change. Given that the international investment protection regime focuses on the expectations of foreign investors (Cotula, 2023: 786), is there room for human rights protection or climate change mitigation? The Special Rapporteur on human rights and the environment, David R. Boyd, affirms that "more than 10,000 fossil fuel assets worldwide are covered by investor-state dispute settlement provisions, including three quarters of foreign-owned coal-fired power plants, raising the spectre that States may be reluctant to phase out coal in a timely manner" (Boyd, 2022: 15).

Therefore, with regard to the Energy Charter Treaty, the options are to maintain an unbalanced treaty that hinders the energy transition, or to withdraw from it in order to align with the commitments and obligations outlined in climate law, the Paris Agreement and the Sustainable Development Goals. The ongoing reform of the Energy Charter Treaty is not sufficient to achieve these goals, and if treaty makers were to design a new international investment regime at this time, it would put climate action at its heart rather than using the current form as a starting point (Cotula, 2023: 788–791).

The report on human rights-compatible international investment agreements emphasises that "including public policy exceptions explicitly in international investment agreements will also assist States in taking legal and policy measures to regulate investment to promote human rights, protect the environment or realise the Sustainable Development Goals" (Working Group, 2021: 19). Since investor-state dispute settlement is the area in which the dilemma is expressed, the reports recommend that it be replaced by a system that is "fairer for all" (Working Group, 2021: 21–22). The replacement should also include the elimination of survival clauses, a key part of the current regime.

Final Thoughts

Climate change is a 'common concern of mankind', as stated in the preamble to the United Nations Framework Convention on Climate Change. This chapter has examined the legal challenges of the energy transition, focusing on the Energy Charter Treaty, an international treaty established to regulate both fossil and renewable energy. Given the historical context in which it was concluded, it focused on the unwavering protection of foreign investment. Today, the momentum of energy transition and climate change mitigation has led the European Union to question its modernisation. Six member states—Poland, France, Germany, Slovenia, Luxembourg, and Spain—withdrew from the treaty, following a similar move by Italy in 2016. After the European Parliament gave the green light for a coordinated withdrawal, the European Union and the Euratom exited on 27 June 2024. All withdrawals take effect one year after the formal notification to the depositary of the treaty.

The modernisation of the Energy Charter Treaty and the inevitable coordinated withdrawal of the European Union and Euratom from the Energy Charter Treaty confirm that investor-state arbitration highlights the need to rethink the tension between climate change/energy transition and investment protection/investor-state dispute settlement, in particular its impact on regulatory sovereignty. The cases against Spain, Italy, and the Czech Republic, and more recently the controversies surrounding the Dutch coal phase-out, are important examples of this tension.

However, the institutions of the European Union do not have a direct impact beyond its borders; the energy transition is not only important for Europe but also for the entire global community. Can the European Union maintain its dual position on investor-state dispute settlement—within and outside the Union? The legal and political implications of a coordinated withdrawal from the Energy Charter Treaty, and even the withdrawal of seven member states, create a significant shift both regionally and globally. Critical voices continue to take steps towards the construction of a more equitable and balanced system for all actors: Global North and Global South states, corporations, and local communities.

References

Alschner, W. (2022). *Investment Arbitration and State-Driven Reform: New Treaties, Old Outcomes.* Oxford University Press.

Bas Vilizzio, M. (2019). Arbitrajes inversor-Estado intra-Unión Europea: Problemática actual y escenarios futuros. *Revista Diplomática. Segunda Época., 1*(2), 61–70. Retrieved August 17, 2023, from https://www.gub.uy/ministerio-relaciones-exteriores/sites/ministerio-relaciones-exteriores/files/documentos/publicaciones/Revista%20Diplom%C3%A1tica%202.1%20%281%29_0.pdf

Bas Vilizzio, M. (2022). Back to Sovereignty? Policy Space in Investor-State Dispute Settlement. *Revista Brasileira de Politica Internacional, 65*(2), 2022. Retrieved August 23, 2023, from https://doi.org/10.1590/0034-7329202200216

Bas Vilizzio, M. (2023a). *Hacia la soberanía regulatoria. Espacio de política pública y solución de controversias inversor-Estado.* Comisión Sectorial de Investigación Científica. Unidad de Comunicaciones. Universidad de la República.

Bas Vilizzio, M. (2023b). Eco Oro contra Colombia: protección ambiental bajo presión. In D. Iglesias Márzquez & A. Pigrau (Eds.), *Litigación en materia de empresas y derechos humanos: estudio de casos.* Tirant Lo Blanch.

Bas Vilizzio, M., & Stanley, L. (2021). Protección de inversiones y transición energética: nuevo dilema político. *La Diaria*, 17 February. Retrieved August 17, 2023, from https://ladiaria.com.uy/opinion/articulo/2021/2/proteccion-de-inversiones-y-transicion-energetica-nuevo-dilema-politico/

Boyd, D. R. (2022). *Paying Polluters: The Catastrophic Consequences of Investor-State Dispute Settlement for Climate and Environment Action and Human Rights.* Report of the Special Rapporteur on the issue of human rights obligations relating to the enjoyment of a safe, clean, healthy and sustainable environment, David R. Boyd. Document A/78/168. Retrieved October 27, 2023, from https://documents-dds-ny.un.org/doc/UNDOC/GEN/N23/205/29/PDF/N2320529.pdf

Brauch, M. D. (2021) Should the European Union Fix, Leave or Kill the Energy Charter Treaty?. *Columbia Center on Sustainable Investment.* Retrieved August 17, 2023, from https://doi.org/10.7916/d8-snt7-sa38

Brauch, M. D. (2022). The Agreement in Principle on ECT 'Modernization': A Botched Reform Attempt that Undermines Climate Action. *Kluber Arbitration Blog.* Retrieved August 17, 2023, from https://arbitrationblog.kluwerarbitration.com/2022/10/17/the-agreement-in-principle-on-ect-modernization-a-botched-reform-attempt-that-undermines-climate-action/

Cotula, L. (2009). Regulatory Takings, Stabilization Clauses and Sustainable Development. *OECD Investment Policy Perspectives, 2008*, 69–90. Retrieved October 31, 2023, from https://read.oecd-ilibrary.org/finance-and-investment/oecd-investment-policy-perspectives-2008/regulatory-takings-stabilisation-clauses-and-sustainable-development_ipp-2008-4-en

Cotula, L. (2023). International Investment Law and Climate Change: Reframing the ISDS Reform Agenda. *Journal of World Investment & Trade, 24*(4–5), 766–791. Retrieved October 27, 2023, from https://doi.org/10.1163/22119000-12340310

Eckes, C., Main-Klingst, L., & Schaugg, L. (2023). Why a Coordinated Withdrawal from the Energy Charter Treaty is Inevitable. *Euroactiv*, 25 January. Retrieved August 17, 2023, from https://www.euractiv.com/section/energy/opinion/why-a-coordinated-withdrawal-from-the-energy-charter-treaty-is-inevitable/

Euroactiv. (2022). Netherlands Follows Spain in Quitting Energy Charter Treaty. *Euroactiv*, 19 October. Retrieved October 31, 2023, from https://www.euractiv.com/section/energy/news/netherlands-follows-spain-in-quitting-energy-charter-treaty/

Euroactiv. (2023). LEAK: Exit from Energy Charter Treaty 'Unavoidable', EU Commission Says. *Euroactiv*, 8 February. Retrieved August 17, 2023, from https://www.euractiv.com/section/energy/news/exit-from-energy-charter-treaty-unavoidable-eu-commission-says/

European Commission. (2023a). *Proposal for a Council Decision on the Withdrawal of the Union from the Energy Charter Treaty, Document COM(2023) 447 Final.* Retrieved August 17, 2023, from https://eur-lex.europa.eu/legal-content/EN/TXT/PDF/?uri=CELEX:52023PC0447

European Commission. (2023b). *Proposal for a Council Decision on the Withdrawal of the European Atomic Energy Community from the Energy Charter Treaty, Document COM(2023) 446 Final.* Retrieved August 17, 2023, from https://eur-lex.europa.eu/legal-content/EN/TXT/PDF/?uri=CELEX:52023PC0446

Gehring, M., & Tokas, M. (2022). Synergies and Approaches to Climate Change in International Investment Agreements Comparative Analysis of Investment Liberalization and Investment Protection Provisions in European Union Agreements. *Journal of World Investment & Trade, 23*(5–6), 778–812. Retrieved October 27, 2023, from https://doi.org/10.1163/22119000-12340270

Gijs C., Di Sario, F., & Mathiesen, K. (2023). Brussels Backtracks: EU Prepares to Quit Dirty Energy Club. *Politico*, 7 February. Retrieved August 17, 2023, from https://www.politico.eu/article/european-union-quit-dirty-energy-treaty-commission-climate-goals/

Haut Conseil pour le Climat. (2022). *Avis sur la modernisation du Traité sur la Charte de l'énergie*. Un avis du Haut Conseil pour le Climat. Retrieved

October 31, 2023, from https://www.hautconseilclimat.fr/wp-content/uploads/2022/10/2022-10-19_Avis-TCE_HCC.pdf

Iglesias Márquez, D. (2021). La implementación de los Principios Rectores sobre empresas y derechos humanos en tiempos de emergencia climática: Perspectiva latinoamericana. In H. Cantú Rivera & D. A. Pamplona (Eds.), *A una década de los Principios Rectores sobre las Empresas y los Derechos Humanos*. Tirant Lo Blanch.

International Law Commission. (2006). *Fragmentation of International Law: Difficulties Arising from the Diversification and Expansion of International Law*. Document A/CN.4/L.702. Retrieved August 26, 2023, from https://legal.un.org/ilc/documentation/english/a_cn4_l702.pdf

Koskenniemi, M. (1990). The Politics of International Law. *European Journal of International Law, 1*(1), 4–32.

Krasner, S. D. (1999). *Sovereignty: Organized Hypocrisy*. Princeton University.

Krasner, S. D. (2009). *Power, the State, and Sovereignty: Essays on International Relations*. Routledge.

Meijer, B. H., & Deutsch, A. (2022). Update 2. Netherlands Activates Energy Crisis Plan, Removes Cap on Coal Plants. *Reuters*, 20 June. Retrieved August 17, 2023, from https://www.reuters.com/business/energy/netherlands-activates-energy-crisis-plan-removes-cap-coal-plants-2022-06-20/

Ministerio de Relaciones Exteriores y Cultos de Bolivia. (2008). Bolivia y el CIADI: crónica de un divorcio anunciado. In S. Valdomir & C. Santos (Eds.), *Soberanía de los pueblos o intereses empresariales* (pp. 103–128).

Odermatt, J. (2021). Is EU Law International? Case C-741/19 Republic of Moldova v Komstroy LLC and the Autonomy of the EU Legal Order. *European Papers, 6*(3), 1255–1268. Retrieved August 26, 2023, from https://doi.org/10.15166/2499-8249/522

Peixoto, J. (2022). Legal Perspectives in IR and the Role of Latin America. In *Oxford Research Encyclopedia of International Studies*. International Studies Association and Oxford University Press. Retrieved August 17, 2023, from https://doi.org/10.1093/acrefore/9780190846626.013.686

Perrone, N. M. (2019). The 'Invisible' Local Communities: Foreign Investor Obligations, Inclusiveness, and the International Investment Regime. *American Journal of International Law Unbound, 113*, 16–21. Retrieved August 17, 2023, from https://doi.org/10.1017/aju.2018.92

Perrone, N. M. (2021). *Investment Treaties and the Legal Imagination. How Foreign Investors Play by Their Own Rules*. Oxford University Press.

Perrone, N. M. (2022). Bridging the Gap between Foreign Investor Rights and Obligations: Towards Reimagining the International Law on Foreign Investment. *Business and Human Rights Journal, 7*, 375–396. Retrieved August 26, 2023, from https://doi.org/10.1017/bhj.2022.16

Prasad, A., et al. (2022). Mobilizing Private Climate Financing in Emerging Market and Developing Economies. *IMF Staff Climate Note 2022/007*, International Monetary Fund. Retrieved August 17, 2023, from https://www.imf.org/en/Publications/staff-climate-notes/Issues/2022/07/26/Mobilizing-Private-Climate-Financing-in-Emerging-Market-and-Developing-Economies-520585

Ratton Sanchez Badin, M., & Morosini, F. (2018). Reconceptualizing International Investment Law from the Global South An Introduction. In M. Ratton Sanchez Badin & F. Morosini (Eds.), *Reconceptualizing International Investment Law from the Global South*. Cambridge University Press.

Reinisch, A., & Mansour Fallah, S. (2022). Post-Termination Responsibility of States?—The Impact of Amendment/Modification, Suspension and Termination of Investment Treaties on (Vested) Rights of Investors. *ICSID Review—Foreign Investment Law Journal, 37*(1-2), 101–120. Retrieved August 17, 2023, from https://doi.org/10.1093/icsidreview/siab023

Sierra, J., (2021). Is the Arbitral Award in the Eco Oro v Colombia Dispute 'Bad Law'? *Afronomics Law*, 11 November. Retrieved August 17, 2023, from https://www.afronomicslaw.org/category/analysis/arbitral-award-eco-oro-v-colombia-dispute-bad-law

Skovgaard Poulsen, L., & Gertz, G. (2021). Reforming the Investment Treaty Regime: A 'Backward Looking' Approach. *Brief Paper, Global Economy and Finance Programme, March 2021*. Retrieved August 23, 2023, from https://www.chathamhouse.org/sites/default/files/2021-03/2021-03-10-reforming-investment-treaty-regime-poulsen-gertz.pdf

Sornarajah, M. (2015). *Resistance and Change in the International Law on Foreign Investment*. Cambridge University Press.

Sornarajah, M. (2021). *International Law on Foreign Investment* (5th ed.). Cambridge University Press.

Tienhaara, K., et al. (2022). Investor-State Disputes Threaten the Global Green Energy Transition. *Science, 376*(6594), 701–203. Retrieved October 31, 2023, from https://doi.org/10.1126/science.abo4637

Tropper, J., & Reinisch, A. (2022). The 2020 Termination Agreement of Intra-EU BITs and Its Effect on Investment Arbitration in the EU—A Public International Law Analysis of the Termination Agreement. *The Austrian Yearbook on International Arbitration, 2022*(16), 301–340. Retrieved August 23, 2023, from https://ssrn.com/abstract=3947835

UNCTAD. (2003). *World Investment Report 2003. FDI Policies for Development: National and International Perspectives*. UNCTAD. Retrieved August 17, 2023, from https://unctad.org/system/files/official-document/wir2003light_en.pdf

UNCTAD. (2012). *World Investment Report 2012. Towards a New Generation of Investment Policies.* UNCTAD. Retrieved August 17, 2023, from https://unctad.org/system/files/official-document/wir2012_embargoed_en.pdf

UNCTAD. (2020). Investment Policy Responses to the COVID-19 Pandemic. *Investment Policy Monitor.* UNCTAD. Retrieved August 17, 2023, from https://unctad.org/system/files/official-document/diaepcbinf2020d3_en.pdf

Voon, T. (2019). Modernizing the Energy Charter Treaty: What About Termination?. *Investment Treaty News, 10*(4). Retrieved August 17, 2023, from https://www.iisd.org/itn/en/2019/10/02/modernizing-the-energy-charter-treaty-what-about-termination-tania-voon/

Working Group on the Issue of Human Rights and Transnational Corporations and Other Business Enterprises. (2021). *Human Rights-Compatible International Investment Agreements.* Document A/76/238. 27 July. Retrieved August 26, 2023, from https://documents-dds-ny.un.org/doc/UNDOC/GEN/N21/208/09/PDF/N2120809.pdf

World Wildlife Fund Colombia. (2018). *Ley de páramos: comienza la implementación*, 1 August. de agosto. Retrieved August 17, 2023, from https://www.wwf.org.co/?332290/Ley-de-paramos-comienza-la-implementacion

Printed by Printforce, the Netherlands